The Space Shuttle Program

Challenger's final crew — STS 51-L. ***Left to right, rear:*** Ellison S. Onizuka, S. Christa McAuliffe, Gregory B. Jarvis, Judith Resnik. ***Front:*** Michael J. Smith, Francis R. Scobee, Ronald E. McNair (NASA).

Columbia's final crew — STS-107. ***Left to right:*** David M. Brown, Rick Husband, Laurel Clark, Kalpana Chawla, Michael P. Anderson, William "Willie" C. McCool, Ilan Ramon (NASA).

The Space Shuttle Program

How NASA Lost Its Way

R. MICHAEL GORDON

McFarland & Company, Inc., Publishers
Jefferson, North Carolina, and London

LIBRARY OF CONGRESS CATALOGUING-IN-PUBLICATION DATA

Gordon, R. Michael, 1952–
The Space Shuttle Program : how NASA lost its way /
R. Michael Gordon.
p.   cm.
Includes bibliographical references and index.

ISBN 978-0-7864-3434-3
softcover : 50# alkaline paper ∞

1. Space shuttles—Accidents—United States.   2. Space shuttles—United States—Reliability.   3. Space Shuttle Program (U.S.)—Evaluation.   4. United States. National Aeronautics and Space Administration—Decision making.   I. Title.
TL867.G65 2008   629.44'10973—dc22   2008020396

British Library cataloguing data are available

©2008 R. Michael Gordon. All rights reserved

No part of this book may be reproduced or transmitted in any form or by any means, electronic or mechanical, including photocopying or recording, or by any information storage and retrieval system, without permission in writing from the publisher.

On the cover: The launch of STS-121 *Discovery* and
the mission patch of STS-1 *Columbia* (NASA)

Manufactured in the United States of America

McFarland & Company, Inc., Publishers
Box 611, Jefferson, North Carolina 28640
www.mcfarlandpub.com

For the final crews of *Challenger* and *Columbia*
who in a real sense challenged the whole of space.
Thank you for your courage and dedication.

Contents

Introduction   1

SECTION I. PROBLEMS IN DEVELOPMENT
One. In the Beginning   5
Two. A History of Success? The Solids   39

SECTION II. COUNTDOWN TO DISASTER
Three. 51-C: Too Cold for the Military   55
Four. 1985 Was Not a Good Year   68
Five. Pre-Flight: Dangers Not Seen   84

SECTION III. *CHALLENGER*'S FINAL MISSION
Six. A Cold Morning Like No Other   93
Seven. The Final Ascent: The Flight of *Challenger*   106
Eight. Post-Flight Operations   129

SECTION IV. INVESTIGATIONS AND RECOVERY
Nine. The *Challenger* Investigation   153
Ten. Unanswered Questions: Beyond the Commission   171
Eleven. Recovery for Flight? STS-26   179

SECTION V. FROM *CHALLENGER* TO *COLUMBIA*
Twelve. Beyond the Recovery: Was It Really Fixed?   199
Thirteen. The Fall of Space Shuttle *Columbia*   210
Fourteen. On the Road Again   236

Section VI. The Future of Manned Spaceflight
Fifteen. A Look to the Future   247

Appendix I. Shuttle Flights Pre-*Challenger* Disaster   253
Appendix II. The Flights of *Columbia*   263
Appendix III. The Mission and Crew of *Challenger* 51-L   267
Appendix IV. The Crew of *Columbia* STS-107   278
Appendix V. Personal Observations on the Reliability of the Shuttle by Dr. Richard P. Feynman   280
Appendix VI. The *Columbia* Emails: Opportunities Missed   288
Appendix VII. Mission 51-L Press Kit   296
Abbreviations and Definitions   317
Chapter Notes   321
Bibliography   331
Index   341

Introduction

"It's burst into flames ... it's crashing, it's crashing terrible, oh my, get out of the way please. It's burning, bursting into flames... This is terrible, this is one of the worst catastrophes in the world."

With these dramatic words, an earlier generation of Americans, along with the rest of the world, was rocked from their comfortable reliance upon a new technology. It would be a technology that was never to recover from that day's events. The date was May 6, 1937, and the location was an airfield outside of Lakehurst, New Jersey.

It was at Lakehurst where a small group of interested individuals and one reporter, Herb Morrison, had gathered to welcome and record the arrival of the German Zeppelin *Hindenburg*. This arrival was like all other Zeppelin arrivals— majestic, but not of great interest to the press or general public. It was becoming commonplace to see great airships. The only live coverage was a general-interest story to be broadcast to a limited audience. Even the reporter assigned to the story was not present to record its passage. He was at a bar waiting for the new guy to come back from the job.

It became a landmark event, never to be forgotten by those who heard the broadcast live or saw the results of the inferno on newsreels. Those images of disaster and human frailty were forever burned into the minds of men and women everywhere.

For a younger generation of Americans, their catastrophic landmark event would occur on January 28, 1986. This event would forever be linked to the words, "*Challenger*, go at throttle up." This time the disaster would be shown to the world, live and in color. This live coverage would add to the almost unbelievable quality of the images unfolding before their eyes. At first it could not be accepted. Those images would show *Challenger* and her seven-member crew moving upward toward space in her final graceful and powerful moment. They would show *Challenger*'s dark passage. Seventeen years later, America was again shocked to witness a second Shuttle crew lose their lives on live TV as *Columbia* was destroyed by the heat of reentry only fifteen minutes from landing.

In its final chapters, this book looks at some of the unanswered questions from *Challenger*'s flight and reviews *Columbia*'s final mission, as well as looks to the future of the U.S. manned space program which includes a return to the moon and on to Mars.

This work relies heavily on live audio and video coverage of those events broadcast by the National Aeronautics and Space Administration (NASA). Details of the history of Shuttle flight problems and investigations were taken from the *Report of the Presidential Commission on the*

Space Shuttle Challenger *Accident*, and the Columbia *Accident Investigation Board*. In addition, daily newspaper and monthly magazine coverage from several sources, most notably *Countdown* magazine, were used to fill in details not readily available from NASA sources. All photos and technical drawings are from NASA unless otherwise indicated.

This book traces the development of the lifting-body and early Shuttle programs, and details the construction of Shuttle OV-099, which was to become *Challenger*, as well as the problems associated with the development of OV-102, *Columbia*. We then look at the flights made by *Challenger*, *Columbia* and other orbiters before the first accident. The top-secret Shuttle flight 51-C is reviewed, as we ask why that mission was not launched when *Challenger* was operating under almost identical conditions.

The reader is then taken through a careful look at the final ascent, the post-flight search-and-rescue operations, and the investigation of the *Challenger* and *Columbia* failures.

In addition, the reader will find references in the back of the book, including a copy of the NASA 51-L mission press kit and a chart showing all of the United States manned Shuttle flights up to the *Challenger* 51-L mission, as well as a listing of all flights made by *Columbia*. Also included are the personal and outspoken observations of Dr. Richard Feynman, a member of the *Challenger* Commission, which were included in the final *Challenger* Commission report.

We learned much after *Challenger*'s final flight — about ourselves and how we go about the business of discovery. *Columbia*'s final flight showed that we have not learned enough. We must remember that as far as space flight is concerned, we are still in our infancy.

Section I

Problems in Development

Chapter One

In the Beginning

"They had the hunger to explore the universe and discover its truths."
— President Ronald Reagan, January 28, 1986

11:38 A.M. EST, January 28, 1986

"Liftoff. Liftoff of the 25th Space Shuttle mission, and it has cleared the tower. Houston is now controlling, roll program initiated. *Challenger* now heading down range. Engines beginning throttling down now at 94 percent. Normal throttle for most of the flight, 104 percent. Will throttle down to 65 percent shortly. Engines at 65 percent. Three engines running normally. Three good fuel cells. Three good APUs. Velocity 2,257 feet per second, altitude 4.3 nautical miles, downrange distance 3 nautical miles. Engines throttling up. Three engines now at 104 percent."

"*Challenger*, go at throttle up."

A Germ of an Idea

When President Eisenhower ordered the National Aeronautics and Space Administration to proceed with a program of human space flight on August 18, 1958, he never envisioned what adventures were to come, and in the beginning of space flight neither could anyone else. What was known was that risks would be everywhere, but that the rewards would be unbelievable.

Before there could be a fleet of Space Shuttles, designers and test pilots would need to find answers to basic questions. The most basic of these was: could a craft such as this actually fly? Some of these answers were to be found in a new type of experimental aircraft called a lifting body. A lifting body is an aircraft with no wings that acquires its aerodynamic lift from the shape of its fuselage.

In the early days of the space race with the now defunct Soviet Union, it was decided that the fastest and easiest way to orbit and bring men back from space was to use a ballistic capsule. This was fine for a one-shot expendable program, but it was not a long-term solution for a continued presence in space. A better, less expensive and more controllable method would have to be found. It would also have to be more reliable and able to be used much more often by more than just highly trained professional astronauts. Indeed, NASA, through its X-plane series, was moving higher and faster, and would eventually have entered orbit with a small Shuttle-type craft if the agency had not been interrupted by the race to the moon.

What the space agency would need in the future was a space-plane that would have the

M2-F1, the world's first lifting-body test aircraft, August 16, 1963 (NASA).

ability of flying in space, reentering the atmosphere at hypersonic speeds like a ballistic capsule and still retain enough lift and maneuverability to perform a controlled landing on a runway. Along with other ongoing research, the lifting-body program was designed to answer some of these basic questions, and they would be directly related to today's Shuttle program.

Dr. Alfred Eggers conceived the idea of using a lifting body[1] as a method of reentry from space in 1957, when he was an assistant director at NASA's Ames Research Center at Moffett Field, California.[2] The lifting-body program was designed to test the new concept that a low lift-to-drag vehicle could be safely maneuvered and landed safely on a runway.

The first lifting body was a makeshift plywood and tubular-frame craft weighing 1,200 pounds, built by NASA on a $30,000 shoestring budget.[3] In fact, it had no official budget at all. Funds were found (read "skimmed off the top") from other programs, and the personnel-hours to build the craft were for the most part donated. It was called the M2-F1 (Manned, Modification 2, Fuselage Number 1), and it looked like half of a nose cone with fins, but it did fly. Gus Briegleb, whose occupation was constructing sailplanes out of Mirage Dry Lake, California, built its plywood shell.[4] The tubular steel frame was built at the Dryden Flight Research Center, Edwards Air Force Base in California.

At first, the M2-F1 was towed across Rogers Dry Lake Bed at Edwards at about 120 mph by a Pontiac convertible decked out in NASA markings.[5] The car had been souped-up by racecar driver Mickey Thompson for test runs beginning in 1963. It was first towed unmanned, but later the wooden craft would be piloted by Milton Thompson[6] (no relation to Mickey), a veteran research pilot for NASA, who would later become Chief Engineer at NASA's Dryden Flight Center at Edwards.[7] Thompson had flown the X-15 space plane for NASA and had become very interested in the lifting-body program. The lifting body would go nowhere near the 214,000 feet and 3,700 mph he had flown in the X-15, but it was a pure research aircraft and an entirely new and exciting idea. Thompson jumped at the chance to fly this one-of-a-kind and entirely new type of aircraft.

As for the X-15, Shuttle Commander John Young would honor its achievements by stating, "The lift-to-drag ratio of the Space Shuttle is almost identical to that of the X-15. They were very similar programs and there was a great deal of feedback from the X-15 to the Shuttle. It really paid off."[8]

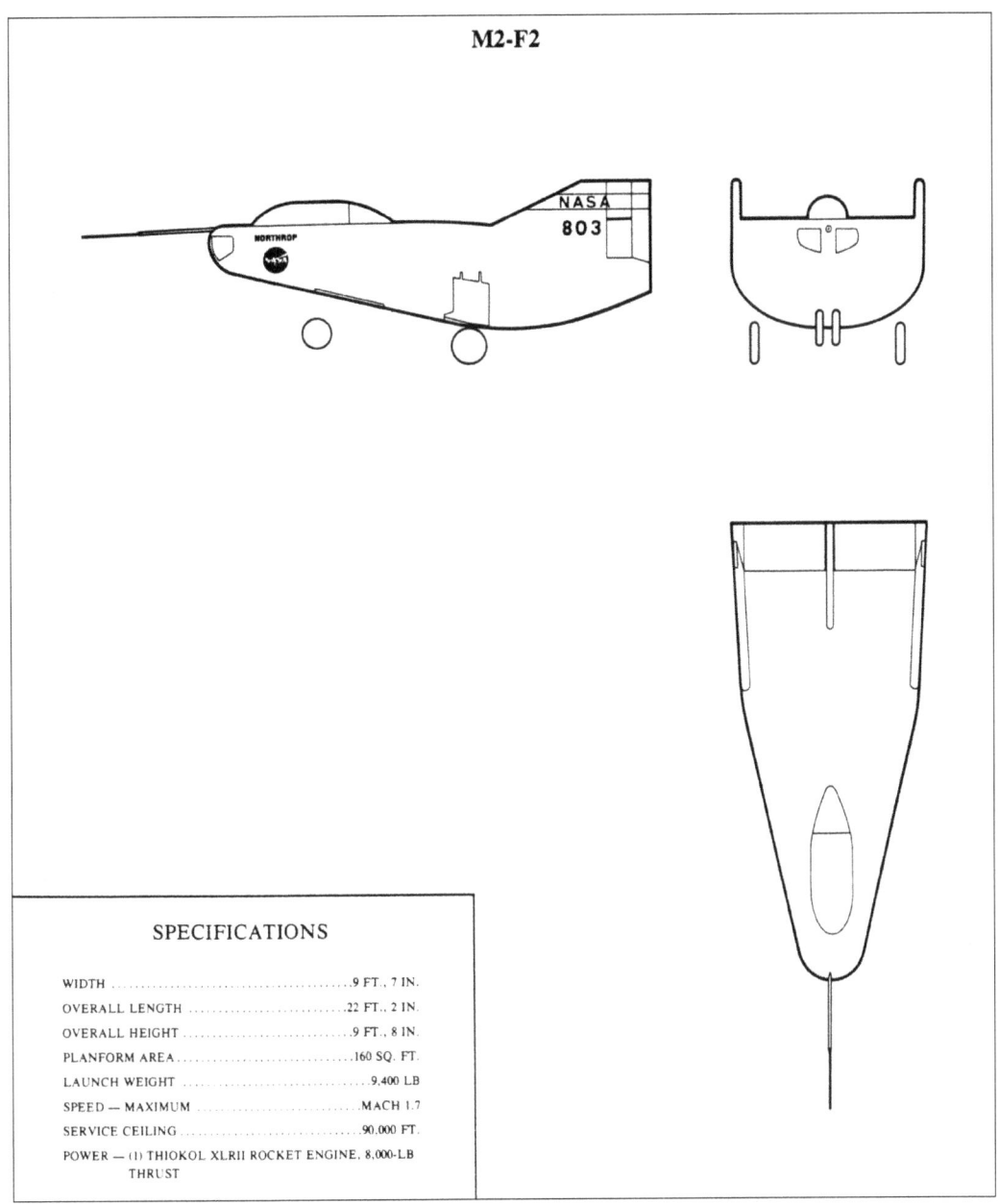

M2-F2, Lifting body research aircraft, built by Northrop Aircraft (NASA).

Later, the M2-F1 was towed to 12,000 feet by a C-47 cargo plane known as a Goony bird[9] before being cut loose from its 1000-foot tow cable for a no-power, glided landing at 80 miles per hour, again with Thompson at the controls.[10] The date was August 16, 1963, and a lifting body had made its first successful manned flight. These flights would last around two minutes as the craft dropped out of the sky like a flying brick with a forward speed of up to 120 miles per hour. Nearly 100 flights were made in the series.

The M2-F1 was later flown by Jerauld Gentry, a test pilot for the USAF,[11] who actually flew the M2-F1 upside down twice, not because he wanted to, along with eight other test pilots, who, for the most part, kept the craft right side up. But as Gentry's flight showed, it was not always as easy to control this new craft. After these tests were successfully completed, NASA would give its official (read "funded") approval for construction and testing of an all-metal lifting body to be designated the M2-F2. In all, the little white M2-F1 logged over 500 flights towed by car and plane, thereby proving that the lifting-body concept could fly, even though the craft had no wings. The feasibility of lifting bodies to successfully maneuver and land had been tested. Later models in the M2 series would be powered by the XLR-11 rocket engine.[12] This was the same engine which had powered the famous Bell X-1 to its historic breaking of the sound barrier in 1947, piloted by Captain Chuck Yeager. The M2-F1 can now be found displayed at the Dryden Research Center, Edwards Air Force Base, California.[13]

Lifting-Body Test Flights

Northrop Aircraft Corporation based in Hawthorne, California, in mid 1964 was selected to build the new all-metal lifting body along with another design designated the HL-10 (Horizontal Landing, 10th Concept). These craft were designed for hypersonic speeds, but would not be tested to those levels because of limited heat-dissipation capabilities built into the test vehicles. High heat during flight was not part of the test program, but would become critical to future Shuttle flights.

The M2-F2 was an Ames Research Center design while the HL-10 was designed by Langley Research Center. Both are NASA facilities. The cost for construction of the M2-F2 and the HL-10 was $1.8 million.[14] It was quite an increase over the M2-F1, but very small as experimental aircraft costs go. Both would be dropped from the wing bomb rack of a B-52 bomber from 45,000 feet to further test the capabilities of the new design. However, before these new craft could fly they would be tested in the wind tunnel at NASA's Ames Research Center. Selected for the growing project were Milt Thompson, Jerauld Gentry, Bruce Peterson, and Donald Sorely, all veteran research pilots who saw the possibility of one day flying into space on a lifting-body spacecraft.

The main objectives of the new lifting-body programs were to evaluate the handling characteristics of this new type of aircraft, and what was discovered would be directly related to future Shuttle developments. In this test series, transonic and supersonic flight would be investigated as it related to pilot-vehicle compatibility. Simply put: could a pilot control this new type of vehicle or would it go out of control and crash? This was not unlike NASA's approach to the manned Mercury program, which was designed and tested not only to rate the capsule but for the capsule to rate the man.

The twenty-two-foot-long M2-F2 rolled out of its Northrop plant in Hawthorne on June 15,

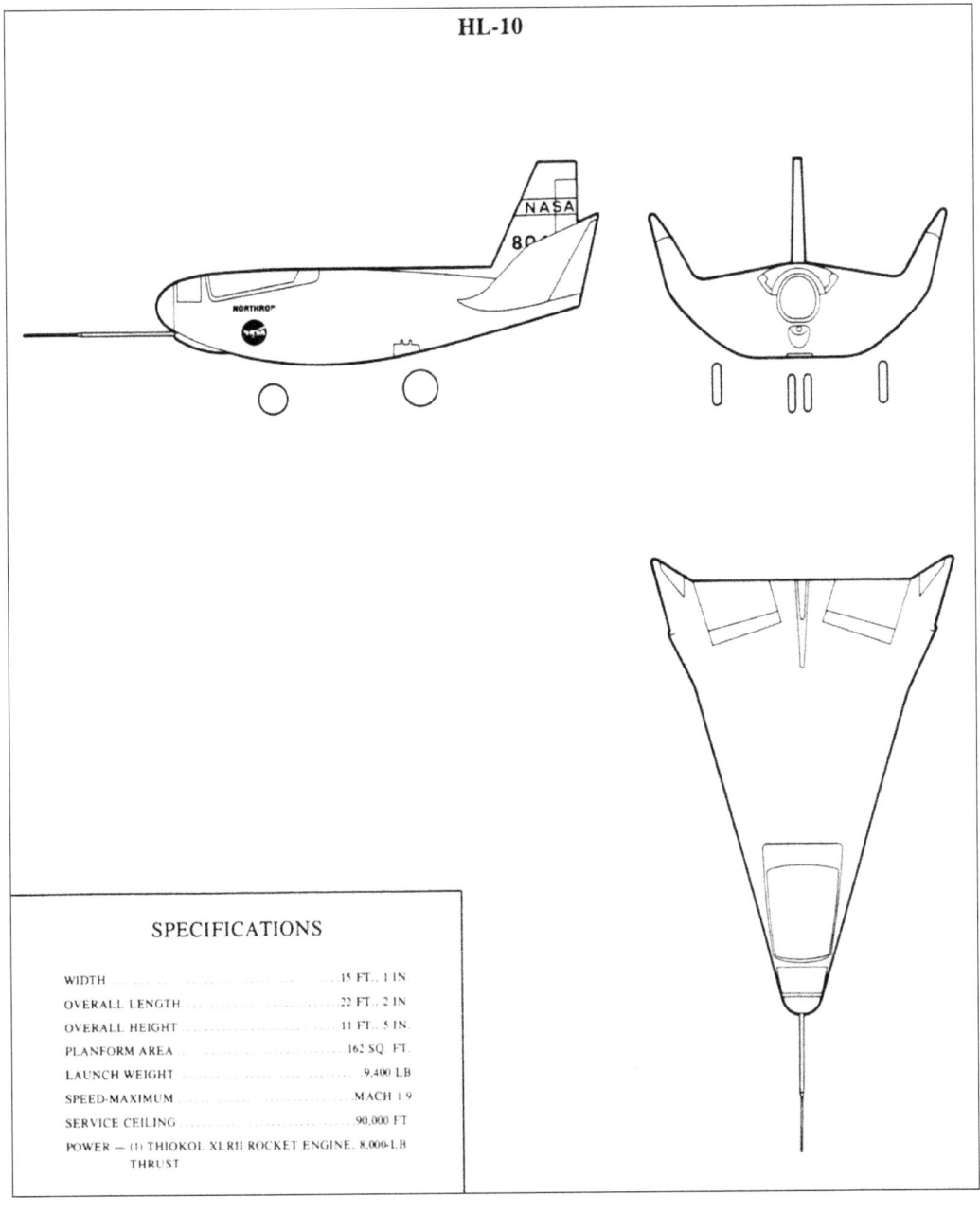

Left and above: HL-10, Lifting body research aircraft, built by Northrop Aircraft (NASA).

1965, weighing 4,620 pounds and lovingly called the "flying brick yard."[15] It made its first manned test flight on July 12, 1966, piloted by Milt Thompson. Dropped from a B-52 bomber, the same aircraft used to launch the X-15 at nearly 45,000 feet, it flew to a top speed of 452 miles per hour on this first test run.[16] This was an unpowered test, which showed just how tricky flying a lifting-body craft could be. As Thompson was completing a 90-degree turn toward the landing strip for his final approach, the M2-F2 began to oscillate in what is called a pilot-induced oscillation (PIO). Although the movements were quite dramatic Thompson soon brought the round-bottom, flat-top, half-cone craft back under his control for a very smooth "dead stick" landing at 200 mph on the floor of Rogers Dry Lake bed. Although the drop tests were very brief, the wait in the lifting body was at times long and hot. Milt Thompson, when asked what was the most tiring part of the flight, answered, "The 45 minutes I sat in the M2-F2 at the end of the pylon waiting for the B-52 to reach launch altitude."[17] It was a long wait for a wild ride.

While the M2-F2 was being fitted with its rocket engine for the powered phase of the test program, the HL-10 was being rolled out of the hanger on January 18, 1966, ready for its first unpowered drop test. Even though these were unpowered tests, both the HL-10 and the M2-F2 were equipped with small rockets in the event that some power was required for extending the landing approach. It was a safety feature that was never required, but was nice to have on hand just in case. Piloted by Bruce Peterson on December 22, 1966, the HL-10 also developed stability problems almost immediately upon being released from its B-52 carrier.[18] The HL-10's control surfaces did not respond as expected to the commands inputted by the pilot. However, Peterson was able to develop enough control to manage a bumpy but safe landing on the lakebed. The HL-10 was then grounded until major modifications could be made to its control surfaces.

The next flight by Bruce Peterson would be one he would never forget. It was a flight many have seen on television. On May 10, 1967, the sixteenth and last unpowered flight of the M2-F2[19] was scheduled. The lifting body had again developed violent stability problems during its approach to the landing strip which took a great deal of effort by the pilot to suppress. These were lateral oscillations, which proved difficult to overcome. In a test run of around three minutes from drop to landing, everything develops with great speed. By the time that some control had been established, Peterson was off his intended runway position and near an observation helicopter, or at least he felt he was. With no ground markings to judge his distance to the desert floor, Peterson lowered the landing gear a half second too late. In fact the gears were coming down when he hit the dry lakebed at 250 miles per hour.[20]

Landing without gear caused the M2-F2 to bounce eighty feet down the landing area and then roll over six times. The M2-F2 was badly damaged and so was Bruce Peterson, who would spend the next eighteen months recovering in a hospital. He would lose an eye; however, that would not stop him from flying again. This is the crash landing shown at the start of *The Six Million Dollar Man*,[21] the seventies television series, only in this case the pilot and the craft were both real. It was the type of crash which could be expected to occur if a Shuttle failed to deploy its landing gear on time during one of its landings. Contrary to published NASA reports, live video footage of landings has shown that landing-gear deployments have been late several times on Shuttle landings.

Due to the crash and the craft's demonstrated stability problems, the M2-F2 was returned to Northrop where it was completely disassembled and rebuilt into the M2-F3 configuration, but this time it would have a center third vertical fin added for greater vertical stability.[22] This would improve its overall control characteristics and help remove most of the lateral control problems. To gain research data on a new control system, a reaction-jet control system was also installed. This system is similar to ones used on orbiting spacecraft such as the Shuttle, first tested on the M2-F3. During the rebuild, attention was again focused on the now-ready HL-10.

After modifications were finished and eleven successful unpowered drop tests were completed to test its stability and control, the HL-10 was ready for its first powered flight. On October 23, 1968, the HL-10 was dropped from its B-52 carrier aircraft, but trouble soon developed.

One. In the Beginning   11

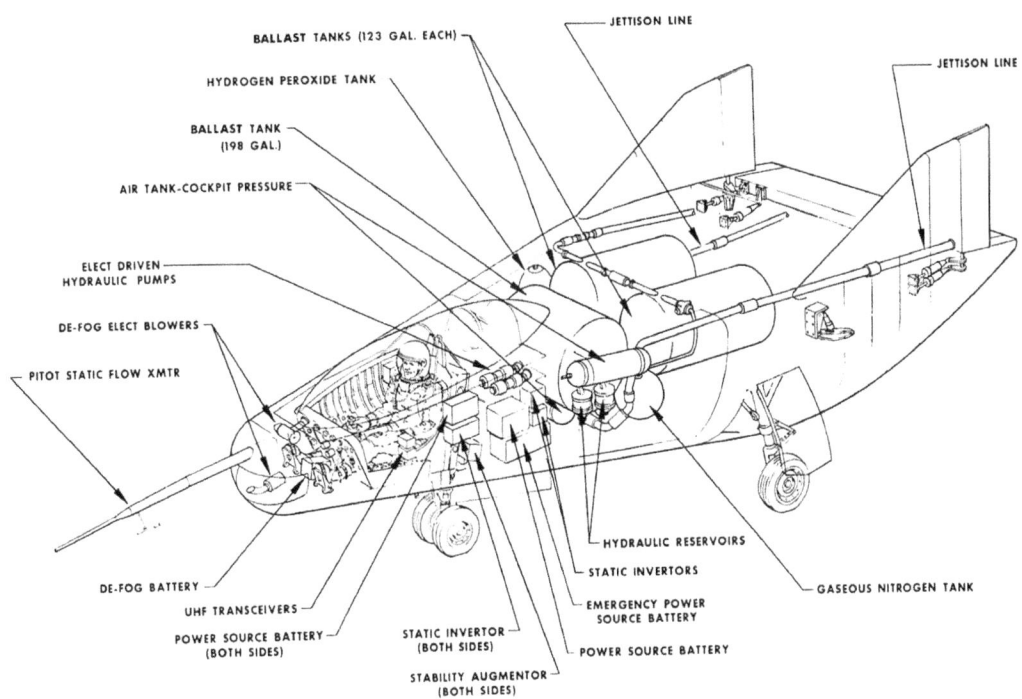

M2-F2, Lifting body research aircraft, built by Northrop Aircraft (NASA).

Only one of its four rockets would fire, so the test had to be abandoned. Pilot Jerauld Gentry jettisoned his fuel and landed on the dry lakebed. This failure would not stop the program.[23] The HL-10 went on to make its first successful rocket-powered flight on November 13, 1968, piloted by John Manke. By the time it made its fourth rocket-powered flight, Manke had pushed the HL-10 past the sound barrier to 724 miles per hour. And by October 1969 the HL-10 had reached an altitude of 79,000 feet and speeds of mach 1.52 (1003 mph) in powered flight.[24] It was not bad for a "flying brick," and even though the term "Shuttle" would not be spoken for a few years to come, these tests would become the bedrock of its overall development.

During a test on February 18, 1970, the HL-10 was able to reach an even more impressive speed of 1,228 mph (mach 1.86),[25] during a test flight lasting seven minutes, flown by Air Force test pilot Major Peter Hoag. This was the fastest speed ever obtained by a lifting body. On February 27, 1970, only nine days after making the speed record, the HL-10, piloted by William Dana, flew to a record 90,030 feet,[26] becoming the highest ever flown by a lifting body. By this time it was evident that lifting-body designs could fly under pilot control and perform pinpoint landings. The HL-10 was flown a total of thirty-seven times and is now also displayed at the Dryden Research Center's main entrance.[27]

On June 2, 1970, the reconfigured M2-F3 made its first flight with pilot William Dana on board.[28] It was an unpowered glide test to examine the modifications made since its crash as the M2-F2. This flight showed that much better lateral control had been achieved, and it was deemed a success. The first powered flight of the newly modified M2-F3 came on November 25, 1970. At 53,000 feet, pilot Dana fired three of the four XLR-11 rocket-engine chambers and pushed his craft to Mach 0.8.

By August 25, 1971, the M2-F3 was ready for its first supersonic test flight. Launched from a B-52 Stratofortress at 42,000 feet, the M2-F3 fired its rockets and reached a speed of 699 mph

Mercury Redstone (MR-3) Launch (NASA).

and an altitude of 67,000 feet before Dana shut down the engines.[29] The seven-minute flight ended with a glided landing at Rogers Dry Lake at Edwards Air Force Base, using a new fly-by-wire system for control. This system is similar to the one now used by Space Shuttles as they configure from space flight to aircraft control surfaces.

In all, the M2-F3 flew twenty-seven missions, reaching a rocket-powered top speed of 1,064 mph (mach 1.6).[30] On its final flight, pilot John Manke pushed the craft to its highest altitude of 71,500 feet on December 21, 1972.[31] Two years after finishing its work, the M2-F3 was placed on permanent display in the National Air and Space Museum in Washington, D.C. It is now part of the test-aircraft series on display, which includes the "test bed" Space Shuttle *Enterprise*.[32]

Into the Wild Blue Yonder

While NASA was flying its lifting bodies, the air force was developing its own program. Since 1967, the air force had been developing and testing the Martin-built X-24A lifting body, also at Edwards Air Force Base. This vehicle was built to fly to 100,000 feet at mach 2. It looked like a teardrop with three vertical fins in the rear. It weighed 6,270 pounds and was twenty-four feet long and fourteen feet wide. The air force knew that eventually military astronauts would fly into space, and this test program was designed to examine the type of craft which could be expected to fly them there and back safely to earth. It must be remembered that at the time the United States was deep into the Cold War with the Soviet Union. U.S.-Soviet space competition was in the forefront of the nation's strategic plans, and the air force felt a great need to be on the cutting edge of this superpower struggle for supremacy in space. The air force wanted a space fighter.

The X-24A flew its first glided, unpowered flight on April 17, 1969, piloted by Air Force Major Jerauld Gentry. This same pilot would test the new craft during its first rocket powered flight on March 19, 1970. The X-24A would eventually fly as high as 71,400 feet and at speeds of 1,036 mph (mach 1.6).[33] On June 4, 1971 the X-24A flew its twenty-eighth and final test flight with NASA test pilot John Manke at the controls, helping to validate unpowered landing systems developed in the lifting-body program.[34] Once again those systems would in the near future become critical to Shuttle development.

Even before the lifting-body program, the air force was testing reentry vehicles under program names such as START, ASSET and PRIME.[35,36] By 1969, under these programs, the air force had conducted six launches using a modified Thor booster rocket to place a subscale, flatiron-shaped spacecraft into a suborbital trajectory.[37] The subscale so-called "test bed" looked very much like the future X-24B. These flights concentrated on reentry trajectories and the effects of reentry on heat shields, and were launched from the Eastern Test Range at Kennedy. During the same period, three flights were made from the Western Test Range at Vandenberg Air Force Base in California using Atlas rockets.

On August 1, 1973, the air force tested its new X-24B for the first time. To save costs, the air force had sent the X-24A back to Martin for modification and reconstruction into the X-24B.[38] This design looked like a flying flatiron with a rounded top and a pointed nose. Piloted by John Manke, the X-24B was dropped from a B-52 at 40,000 feet. Four minutes later it landed at 200 mph on Edwards' dry lakebed. It would fly its first powered flight on November 15, 1973, also piloted by Manke.[39] This craft was thought to be the forerunner or prototype of the Space Shuttle, not only for its appearance, but also for its handling qualities, which were close to the Shuttle on landing. Like many of the lifting bodies, the X-24B landed on the same lakebed landing strips the Shuttle would use in the years to come.[40]

During 1974 and '75, the X-24B flew a test program which, along with other programs, was used in the direct development of the Space Shuttle. Aerodynamically it was the closest to the Shuttle of all lifting-body programs, therefore the data was of great importance to its designers. One of the pilots assigned to the program was future *Challenger* commander Francis R. Scobee, who was test flying for the air force at the time. The X-24B flew at mach 1.76 and 74,000 feet, which was as fast as it would fly. First piloted by John Manke and then by Air Force Major Mike Love,[41] the X-24B made two pinpoint landings on the main concrete runway at Edwards.[42] These were the final objectives of the program and helped clear the way for the Space Shuttle to land on an airstrip without power. When the X-24B finished its test series in 1975, the lifting-body programs were successfully completed.[43]

After the lifting bodies had shown continued success with drop tests, NASA began to have confidence in the ability of a Space Shuttle to land on a runway without power. These test flights were a major stepping-stone to the Space Shuttle, showing it too could be flown and landed safely without power on a runway.[44] Originally the agency was designing the Shuttle with at least two jet engines which would give the Shuttle a go-around capability during landing in the event that problems occurred. With proven runway landings and the need for engines eliminated, NASA was able to greatly increase the cargo capability of its ongoing designs.[45] The Shuttle however, would still become a compromised spacecraft in its design. It blended the knowledge of reentry tests, aircraft technology, and the experience gained from the lifting-body flights into the Shuttles we see today.

Building on the successes gained in lifting bodies, NASA's Langley Research Center began development of a new project, which in reality would have been a mini–Shuttle. The center was working on the HL-20, which they envisioned as an orbital taxi to be launched on an expendable booster, then reenter to land on a standard runway. Even legendary moon-rocket designer Werner von Braun liked the idea, and suggested that two of the new HL-20s or the old HL-10s be orbited on a single Saturn 1B.[46] The plan called for the first HL-20 to reenter the atmosphere unmanned, to test the technology. If the test were successful, the second lifting body would be piloted to a glided landing at Edwards. But the funds were never budgeted and the program never got past the planning stage. Piloted reentry of a spacecraft to a controlled landing on a runway would have to wait for the Space Shuttle. However, with the two Shuttle disasters, the HL-20 concept is now getting a major new look as a possible replacement for the aging and deadly Shuttle system.

A Reusable Manned Spacecraft

> *"It is the view of the* Columbia *Accident Investigation Board that the* Columbia *accident is not a random event, but rather a product of the Space Shuttle program's history and current management process."*

With all of NASA's manned political chips placed firmly on the race to the moon, it was not surprising to find the agency groping for a new direction for its human space flight capability after the final Apollo moon landing in December 1972. A bit of loose talk within NASA concerning a manned Mars flight in the near future was going nowhere. There were three upcoming manned missions to Skylab, NASA's new orbital space station, but that was about all. The station had been developed and built using Apollo-driven technology. By converting the third stage of a Saturn V moon rocket, NASA was able to build and fly an earth-orbiting space laboratory for scientific research. This was a small step forward, using the catchphrase of "Apollo Applications," but NASA knew that it was to be a short lived program. In the age of big-budget Apollo, NASA even had the money to build a backup space station. The second Skylab was completed but never flown and is now on display at the Smithsonian Air and Space Museum in Washington, D.C.

When Skylab was launched on May 15, 1973, it was the final flight of the powerful Saturn V moon rocket and it was nearly a failure. During launch, a meteorite shield and solar panel were both ripped off by aerodynamic forces. This caused the can-do space agency to quickly develop a rescue plan for the now greatly underpowered and overheated laboratory. In the heady post–Apollo-moon-landing days, this rescue of Skylab using the first crew to occupy it as a repair team showed NASA at its best and for a while put NASA back on the front pages of the world's press. It should be noted however, that if the lab had failed to reach orbit the space agency did have a backup lab and one last Saturn V ready to go.

NASA had wanted to place a new, larger, permanently manned space station into orbit, but the mood in Washington led to a cut in funding, not an increase. The general public was simply not interested. In this light, a manned return to the moon for longer stays, or a U.S. manned flight to Mars were totally out of the question. The United States would be confined to earth orbit for more than thirty years. There had even been a good deal of talk about ending manned flights altogether. If NASA were to continue with a manned space program, the agency would need to push

Launch of Skylab on May 15, 1973, America's first manned space station (NASA).

hard for the Shuttle space plane. It was the only show in town and NASA knew it. The agency needed to find a way to sell it to the Congress and the American people. These were people who had already become bored with moon flights and astronauts, and whose focus and concerns were on a war a long way away. It would be a hard sell.

Those who believed in the Space Shuttle as the answer to America's future manned space development would have liked to feel that the building and flying of the Shuttle was on a firm economic foundation. Nothing could be further from the truth. As with many of our space projects,

Skylab in orbit with missing solar panel and new solar shields (NASA).

for political and military reasons it was never going to be profitable. This was something NASA always knew and said little about. During this early stage of America's space program, the nation went into space for reasons beyond mere return on investment dollars. It was to build a foundation for the future. Space flight is always pure research, and the investment is always high.

Even though the Soviet Union had been soundly beaten to the moon, they were still a very powerful force on the world space scene. They were also, in most of the world's view, not very far behind the United States. In one technical area, in fact, the Soviets were clearly ahead of the U.S., and were to dominate the use of small, very risky, but effective space stations for many years. With the Soviets going for longer and longer stays on board their Salyut space stations, NASA would need to jump ahead to a totally new technology to have any chance at all for funding.

The space agency would also need the support of the military branches, which were looking for more and more ways to access space. The military had a larger post–Apollo space budget than the civilian space agency, and getting them on board the Shuttle program would become critical to its success since it was in many ways too costly for NASA alone. Some inside the civilian agency felt that if one or two Shuttles were assigned to the air force, the situation would cause NASA to lose control of the entire program. With operational control of half the Shuttle fleet, its own launch site in the future on the west coast, and use of two sites on the east coast, it looked to some that loss of control was a distinct possibility somewhere down the road. Added to this, the possibility that the air force would eventually build its own control center in Colorado made for a very nervous NASA. Both NASA and the Pentagon would find compromises; both knew that high flight costs and political realities would push them toward a joint, reusable manned system if space was to continue to be accessible.

In September 1969, only two months after the first manned lunar landing of Apollo 11, the Presidential Space Task Group[47] named by President Nixon and chaired by Vice President Spiro Agnew, recommended that a Space Shuttle be developed for low earth orbit. It also recommended that the Shuttle be fully reusable. It was to carry many types of payloads into low orbit as a "space truck" with a payload bay large enough to carry the USAF KH-11 reconnaissance satellite into orbit. Along with this so-called space truck, an upper-stage booster would be developed which could be carried on board and used to place payloads into higher geostationary orbits. Later, this upper stage was to successfully propel unmanned explorers to Venus and Jupiter.

Earth- and lunar-orbiting space laboratories, starting with twelve-person crews, were also part of the recommendation.[48] These space stations would be permanently manned, at least the one orbiting the earth, eventually by forty to fifty astronauts from the United States, with the possibility of room for astronauts from other "friendly" nations.[49] An even larger 100-person station in earth orbit was also planned. Included in two of the options proposed was a manned flight to Mars.[50] Using the Apollo technologies developed to land a man on the moon, and space station technologies then being developed, it was felt that all of these projects could be completed in sixteen years (by 1985). The Shuttle was only one small portion of an overall plan, but it would be the only area to see the light of day for nearly thirty years. The space station would become the second project to see flight, yet as of this writing it is only three-fourths completed, and Mars is only a dream.

After Apollo there was almost zero interest in long-range manned space proposals, not only by the general public but by most of Congress. The giant Saturn-V rocket program had been canceled. By then two of the final moon flights (Apollo 18 and 19) had also been canceled, even though most of the flight hardware had already been built. The general view at the time was that this was a grand waste of taxpayer's money, and the Apollo Applications Program would see only one Skylab space station launch before it too was ended. Post-Apollo/Skylab NASA would become an organization looking for a mission for its human space flight capability, and the options were becoming very limited. The political realities at the time guaranteed that manned Mars flights and a group of space stations were out of the question.[51] There were no guarantees that the United States would even continue sending astronauts into space. This possibility of disbanding the human space flight office could easily have happened. NASA would need to create a reason to build and fly shuttles, because there would be no space station to "shuttle" to.[52]

By late 1969, however, while NASA was preparing to send its second Apollo mission to the moon, Apollo 12, its office of Advanced Research and Technology was inviting possible contractors from the United States and Western Europe to a Shuttle conference. The conference was held in the Smithsonian's Museum of Natural History in Washington, D.C., and focused for two days in October on Shuttle designs.

The designs were to be simple and straightforward. The new Space Shuttle would be piggybacked from a larger aircraft flown from a runway or launched on the back of a carrier rocket or from a launch pad. It would go into low earth orbit, deploy payloads or dock with a space station and return to earth, piloted to a runway and landing like an aircraft. Primary to its design would be the requirement that the Shuttle be completely reusable. It would return to its hangar for a short overhaul, be refueled, and be ready to fly again. It was hoped that with many launches payload costs could be cut from $1000 a pound to as little as $100[53] and perhaps the system might someday even turn a profit. The Shuttle also had to be designed to fly at least 100 missions. But it would prove to be far too costly to fly cost effectively. Such is the price of developing technologies. In the end, the cost of a pound delivered into space on the Shuttle would reach $10,000. There would be no cost savings in the Shuttle program.

After the Space Task Group recommended development of a reusable space transportation system, President Nixon, in March 1970,[54] approved a two-year feasibility study. One method NASA used to push the Shuttle funding and sell it to the Nixon administration was to cut three of the last Apollo moon flights and end production of the Saturn V booster. The Shuttle was to be a fully reusable space/aircraft, but soon NASA found that changes would have to be made to preliminary concepts. This would not be an easy task, and NASA managers struggled to come up with a design.[55]

Several aerospace companies developed preliminary ideas about what the Shuttle should look like. Among them were Martin Marietta, Lockheed Aircraft Corporation, Rockwell, General Dynamics, and McDonnell Douglas. Most of the designs proposed a piggyback configuration with a larger aircraft used to deploy a smaller Shuttle to a high altitude which would allow it to fly rocket-powered into space.[56] Both craft would land on runways, the carrier after it had taken the Shuttle up to 31 miles altitude and the Shuttle after it had made a trip into space.[57]

With the new, larger space station now gone from the plan, support in Congress and from the general public would not be easy for a Shuttle that had nowhere to go other than simply into earth orbit. (Been there, done that.) It was felt that neither would accept the Shuttle as a way to open new areas of industrial use of space. It was simply too early to press for this view and even NASA did not know how to sell that approach. Private companies have bottom lines to worry about. However, in the years to come NASA foresaw the industrialization of space. NASA also knew that despite the tremendous success of the moon-landing program, the public and Congress were starting to become bored with space. NASA and the administrations of Kennedy, Johnson and Nixon had sold the moon flights as a race with the Soviet Union. In the minds of a great many people after the lunar landings, the race and the manned space program were over.

The only way to sell the Space Shuttle as the new manned space vehicle was to market the plan as an all-purpose "space truck" and attempt to show that it was economically viable. It would be a vehicle that could do almost anything the users could think of, from deploying payloads into orbit, doing repairs and recovering damaged satellites, to placing military men and equipment into space. It would be a little of something for everyone. NASA could not afford to forget the military in this project, so they dangled the possibility of sending complete military crews and payloads into space. It was a prospect that Defense Department officials had long dreamed of and one they were more than willing to pay for.

NASA planners had thought that the cheapest operational Shuttle method would be a two-aircraft Shuttle system named Virtus,[58] but they soon discovered it would be the most expensive to develop and build.[59] Later, as smaller and smaller budgets for the Shuttle were approved, NASA would find that these deep cuts extended the development and eventually the costs. In the long run, real costs would begin to approach the projected costs of the two-aircraft Shuttle. NASA would not be able to save any money on its Shuttle development program.

The two-aircraft piggyback system was thought by some NASA engineers to be able to place 12.5 tons of payload into earth orbit at a cost of around $5 million per flight. This was a very low cost per flight ratio, but the study soon became moot when NASA found no support for this type of system. Instead, the agency would have to find a low-cost compromise for the Shuttle's construction. In May 1971, NASA was informed that only $5 billion could be spent on Shuttle development over the next 5 years. By January 1972 that figure had become $5.5 billion over six years.[60] The fully reusable two-stage system was gone, and NASA spent the next six months trying to find a new system.

The compromise that the space agency was forced into accepting focused on a fully reusable Shuttle which would be attached to a main fuel tank. Extra rockets on the side of the tank, of either liquid- or solid-fuel design, would help lift the craft for the first two minutes of powered flight before being separated and dropped by parachute into the ocean for recovery and reuse. The only part of the stack which was not reusable would be the main fuel tank which would, after use, be burned up by the earth's atmosphere as it reentered from its suborbital trajectory. Later, NASA would develop ideas to use these tanks in some type of low earth-orbital lab, but the project was found to be unsound, and very costly.

Shuttle Approved

On January 5, 1972, the Nixon administration issued the order which allowed NASA to proceed with the Space Shuttle.[61] The administration would sell the idea to the American people that the Shuttle was going to "revolutionize space transportation." To most people this meant that costs to place payloads into space would be greatly reduced, but this promise was soon lost. For President Nixon the program promised political benefits, as key states he needed for reelection would enjoy increased employment.[62] Funding had been approved for the design and construction of three spacecraft, all prototypes. One would be used for static ground tests, which would be followed up with two others built for the initial test flight series. The Space Shuttle Transportation System (SSTS) was on its way, but NASA management realized early on that each year would bring budget fights with a reluctant Congress. The only way to win these fights would be to maintain well-publicized progress with their program. This also meant that the test-flight series would be a very short and well-publicized one. It would also need to be reported as a complete success, but technical goals would not always be met. And at least two of these areas which were never fully successful, O-rings and debris hits, would prove deadly.

By the time funding had been approved, NASA had already begun preliminary engineering and financial feasibility studies. Two major aerospace companies were already in competition for the expected multibillion dollar contracts: Rockwell and McDonald Douglas. During a news conference to discuss the newest NASA manned program on January 10, 1972, in Washington, D.C., Dale Myers stated, "Our final conclusions were that we should have a fully reusable orbiter, a hydrogen tank that is discarded in orbit, and a choice between recoverable liquid-fuel or solid-fuel booster rockets."[63] The contract scramble was on for the only manned vehicle NASA would order for the next thirty plus years.

The original design had also placed two jet engines in bays under the Shuttle, which would have allowed the Shuttle a "go-around capability" during landing. It was a safety edge, but one which was very costly and added payload weight. It was also envisioned that the Shuttle could fly with four jet engines from California to the Kennedy Launch Center in Florida under its own power. These ideas were quickly abandoned when the weight of the engines and fuel for them were taken into consideration. Also taken into consideration were the successful drop tests at Edwards by lifting bodies which showed that a no-power, so-called dead-stick landing could be safely made by highly trained pilots.

Adding to NASA's confidence in dead-stick landings was a series of tests conducted at NASA's Dryden Flight Research Center. These tests were conducted using a Convair 990 to simulate an orbiter on descent from 40,000 feet. The flights were flown with pilots as well as with an autopilot, and proved successful in the summer of 1972.

The House Committee on Science and Astronautics also released its report on the Shuttle in early 1972. The report came to five conclusions, which were tantamount to a glowing endorsement of the Shuttle program[64]:

1. The technology and fundamental resources exist to successfully develop the configuration chosen by the National Aeronautics and Space Administration for the Earth Orbital Shuttle.
2. Within reasonable developmental risk, it appears that the current Shuttle design will allow development total cost to stay within $5.150 billion, as projected by NASA, and, allows NASA to achieve an operational cost of $10.5 million per flight.
3. The success in meeting the operational cost per flight for the Shuttle is particularly sensitive to the cost of the propellant tanks and acceptable recoverable or refurbishable cost of the Solid Rocket Boosters.
4. The development of the space tug is of key importance in gaining full utility of the Shuttle.
5. If the nation is to realize the full benefits of near space in terms of scientific exploration, practical application, and national security, the development of a low-cost Space Shuttle system is essential.

In 2003, the *Columbia* accident board would state that the Shuttle program "never met any of its original requirements for reliability, cost, ease of turnaround, maintainability, or, regrettably, safety." NASA managers knew this but had continued to push aside any concerns.

NASA would soon find that the development of the Shuttle was not going to be a simple continuation of older technology. The agency would also see the $10.45 million per flight estimate quickly fade along with any development of a space tug. For then NASA Administrator James Fletcher, cost was not a concern. "For the U.S. not to be in space, while others do have men in space, is unthinkable, and a position which America cannot accept."[65]

On March 31, 1972,[66] North American Rockwell's Rocketdyne Division was given the contract for the design and development of the Shuttle's main engines.[67] These would be the first space flight engines designed to be reused dozens of times. As a warning about expected developmental problems when designing a new era of rocket engines, NASA Administrator Dr. Robert Frosch stated, "A Shuttle engine is not like a car engine where something could malfunction and the car would still run. If a part failed in a Shuttle, the engine would blow up or simply melt." It would not be an easy or inexpensive engineering task.

As early orbiter designs were developed, NASA decided to place escape rockets on the Shuttle to pull it away from the external tank, in the event of an emergency. However, computer-simulated studies soon showed that the Shuttle could not stand the stresses placed on it by an abrupt change in flight direction, which would occur if rockets were used. This problem was dramatically proven when *Challenger* was destroyed not by an explosion, but by a dramatic change in direction and aerodynamic loads. Later, those same aerodynamic loads would also help destroy *Columbia*.

By late 1972, that escape plan was scrapped but a new one was quickly developed. The Shuttle would have ejection seats, armed during the first four test flights. However, this system would also have its limitations. They could not be used when the Shuttle was on the pad because the crew would have to eject sideways, parallel to the ground. There would be no time to allow a parachute to open. After the first four test flights, the ejection seats were deactivated and then removed. There would be no way to eject from a flying Shuttle after the test series was completed. The only way back safely would be to land on a Shuttle runway or emergency airstrip. The question then asked was whether NASA could build a dependable enough system without the need for some type of escape system. NASA said yes, but flight history has said no.

There were in fact never any real plans to build an escape capsule into the Shuttle, which could have pulled an entire crew to safety in the event of an accident. Such a system would have added a great deal of weight, which would have cut back on the amount of payload a Shuttle could carry. (The B-1 bomber has just such a system.) Also, NASA simply did not have any money budgeted for the extra costs needed for its development. It is interesting to note that after the *Challenger* accident, NASA would be able to find the funds for a modified low altitude escape system. The NASA administrator at the time, Dr. James Fletcher, made the decision that it would cost too much for the development of what was called a "crew compartment escape system," an estimated $292 million.[68] Funding once again was the critical problem. If the space agency was going to keep the program funded it would have to take risks, and this meant that crew safety took a back seat. It is something the space agency does not like to publicize, but it is a reality continued to this day. The *Columbia* accident report would state, "For the past 30 years, the Space Shuttle Program has been NASA's single most expensive activity, and of all NASA's efforts, that program has been the hardest hit by the budget constraints of the past decade. Throughout the decade, the Shuttle Program has had to function within an increasingly constrained budget. Both the Shuttle budget and workforce have been reduced by over 40 percent during the past decade."

This would also be the first time a manned spacecraft would be tested without an unmanned flight beforehand. It would fly its first mission with a two-man crew. NASA knew it was taking a great risk, as did the astronauts, but the agency had great confidence in the craft and the highly trained crews who would be flying it. What the agency did not greatly publicize was the fact that there were many situations during the launch phase that even the best pilots could not survive. These were called "black areas" in the launch profile, including the first two minutes the Shuttle was on the solids.

During April 1972, Congress approved a space budget[69] which included funding for the Shuttle. While walking on the moon and setting up the American flag during their Apollo 16 mission on April 20, John Young and Charlie Duke were informed by Mission Control that the House had passed the Space Budget 277-60 the day before.[70] John Young, who nine years later would command the first Shuttle test flight, replied, "The country needs that Shuttle mighty bad, you'll see."[71] By the time NASA launched the Apollo 17 mission in December 1972, the space agency had to pay the television networks to cover the moon walks live.[72] However, work was well on the way developing the Space Shuttle.[73]

On July 26, 1972,[74] North American Rockwell Corporation and its Space Transportation and Systems Group (which would later be renamed the Space Division) was selected as prime contractor for the Space Shuttle Transportation System (SSTS) project.[75] NASA then selected Rockwell's Rocketdyne Division for the Shuttle Main Engine production contract on August 16, 1972. The contract called for production of twenty-seven engines through June 30, 1979.

With no space station for the Shuttle to dock with, missions would be proposed that would make the system a semicommercial success. The new Space Transportation System would become a platform for launching commercial satellites and military payloads from its cargo bay. However, these missions would not satisfy the science community who wanted to expand beyond the three manned Skylab flights then scheduled for 1973. The science community had lost the second Skylab and was pushing hard for some type of laboratory, which would work in concert with the new Shuttle program.

With this goal in mind, Marshall Space Flight Center proposed a Space Shuttle Science Module. It would be a small space-research laboratory, which would be able to perform short-term research while being housed in the Shuttle's cargo bay. NASA already knew there was no money for a larger permanent space station, at least not at the time. But perhaps a way could be found to fund the smaller and less expensive lab. It would also be a method of keeping Marshall budgeted with manned space-flight projects. The engineers and managers at Marshall were well aware of the Congress's views toward ending their operation. They could become the first major NASA research center to be scrapped during the post–Apollo funding period.

Marshall called their 52-foot-long lab the "RAM"—Research and Applications Module.[76] It was envisioned that on later flights this module would be set free of the Shuttle. Later, it would be picked up and again placed in the cargo bay to be brought back to earth. More advanced modules could be left in space as a possible beginning to a much larger lab, perhaps the start of a real space station. Although not generally stated, some workers at Marshall felt that the lab could become a backdoor approach to a permanently manned space station.

However, NASA could not find funding for even this modest project, so the agency looked to Western Europe to fund the project now known as Spacelab.[77] Meanwhile, in December 1972, NASA was flying its last manned flight to the moon with Apollo 17. On board that flight were three astronauts flying an Apollo spacecraft with a lunar-landing module named *Challenger* in tow. Later, as an experiment to test the moonquake detection instruments left on the surface, the lunar module *Challenger* would be sent crashing onto the lunar surface near Taurus-Littrow valley.

On June 27, 1973, an eight-month study prepared by the General Accounting Office was released.[78] This study, requested by Senator Walter Mondale, questioned whether or not the Space Shuttle project was economically justified.[79] Mondale saw this as a potential political lever he could use. NASA should never have tried to show that a new system, such as the Shuttle, could ever be fully cost effective. This new manned space vehicle, would have commercial applications, but it would always be experimental. NASA would soon find itself designing a vehicle that would attempt to be all things to all people.

Mondale was never a friend of NASA, and had tried several times to end the program. This would not be his last attempt. Senator Mondale simply did not want an American manned presence in space and no program was ever going to pass his muster. Even with the Soviets pressing forward with their earth orbital stations, Mondale felt other areas should be funded to the detriment of America's space program.[80]

When the Apollo One (Apollo 204) crew of Grissom, White and Chaffee were killed in January 1967 during an Apollo launch-pad fire inside their spacecraft, it was Mondale who would use this disaster for political gain. Mondale told then NASA chief James Webb, who would lead the space agency from 1961 to 1968, "I intend to ride this for every nickel's worth of political power I can get out of it. I don't give a hoot in hell about the space program or about your future."[81] Other senators such as William Proxmire and Jacob Javits warned that the costs would be too great and that sending men into space would achieve nothing. Many felt that no matter what the cost, unmanned craft could do the job just as well. The Shuttle was even called a "manned space extravaganza." However, this was exploration and research, which is always an unknown. Even Columbus had a hard time justifying his exploration as cost effective although he thought he knew where he was going. History would show that China was just a bit farther than he had anticipated, not to mention being blocked by a continent.

Constant underfunding of the Shuttle development program by Congress would keep pushing the Shuttle schedule back and Shuttle safety to lower levels. Also, instead of keeping engineers on the program to monitor developments after designs were completed, engineers were transferred to other underfunded areas. This was a new and potentially dangerous situation, which caused many problems as the new system went online. It continues to be a problem to this day as Shuttle development staff numbers fall, despite two Shuttle disasters which have claimed fourteen lives. Will it take a third disaster to learn this lesson?

NASA's Willis H. Shapley, then Deputy Associate Administrator, told the U.S. House of Representatives in testimony on June 26, 1973, "In our view, the GAO (General Accounting Office) review has not found any substantial reasons to question the decision that has been made to proceed with the Space Shuttle." It should also be noted that most Congressional support for the Shuttle at this point did not relate to scientific or engineering developments. Most support on Capital Hill came from members scrambling for Shuttle funding that might be spent in home districts.

Seeing an opportunity in NASA's funding problems, the European Space Agency (ESA) jumped at the chance to develop and build the orbiting lab that Congress would not fund. It would be a method of placing European astronauts into space, years sooner than they were then planning, and on a much grander scale. Support may have been lacking in the United States, but Western Europe saw this as a great opportunity to enter space cost effectively. They were happy to be riding an American Space Shuttle.

In August of 1973, ESA and NASA signed an agreement that would become the basis for the Spacelab project. At the time the European group was simply an eleven-nation European consortium. It would not be formally ESA until April 1975. At that time Spacelab was put under the control of Europe's own space agency. According to the agreement, the first Spacelab built would be shared by ESA and the United States. Later, NASA would purchase a second module for its own scientific projects. With this agreement, NASA felt that it was well on the road to success with its new program, so much so that space agency officials were looking forward to operating sixty flights a year by 1986. However, even in 1973 this vision was wildly blurred. There was simply no way the contractors for the external tanks and solid rockets could have kept up with that flight pace, not to mention the amount of operational stress that pace would have caused on workers at Cape Kennedy. There were only so many facilities and personnel to do the job, and there was only one launch pad at the time. Operational reality was not the order of the day at NASA in the early 1970s, at least as far as the Shuttle program was concerned. It was a "can't fail" attitude with no real reason for believing it could happen. The bottom line was that these launch rates were simply wide guesses not based in reality.

On August 16, 1973, Martin Marietta was given the contract to build the Shuttle's huge external tank, and production began in June 1974.[82] It would be one of the largest fuel tanks ever developed for space flight. It would be the penetration of one of these tanks by a flame from a field joint of a solid rocket which would lead to the destruction of that tank and the Shuttle *Challenger*. It would also be insulation from one of these tanks ripping loose during launch which would contribute to the loss of *Columbia*. In fact, post–*Columbia* disaster studies would show that 80 percent of the seventy-nine missions that had imagery available indicated that foam had been lost from the external tanks during launch, and that many hits had been made by those falling pieces on the Shuttles and their thermal protective systems (TPS). In fact, not one mission has ever flown without some damage to the TPS.

During this period, NASA was undergoing a great deal of criticism by the press from many different angles. Some newspaper reports called on NASA to be cut back, while others asked why the space agency had "slowed down." The answer was funding. In a November 27, 1977, *Los*

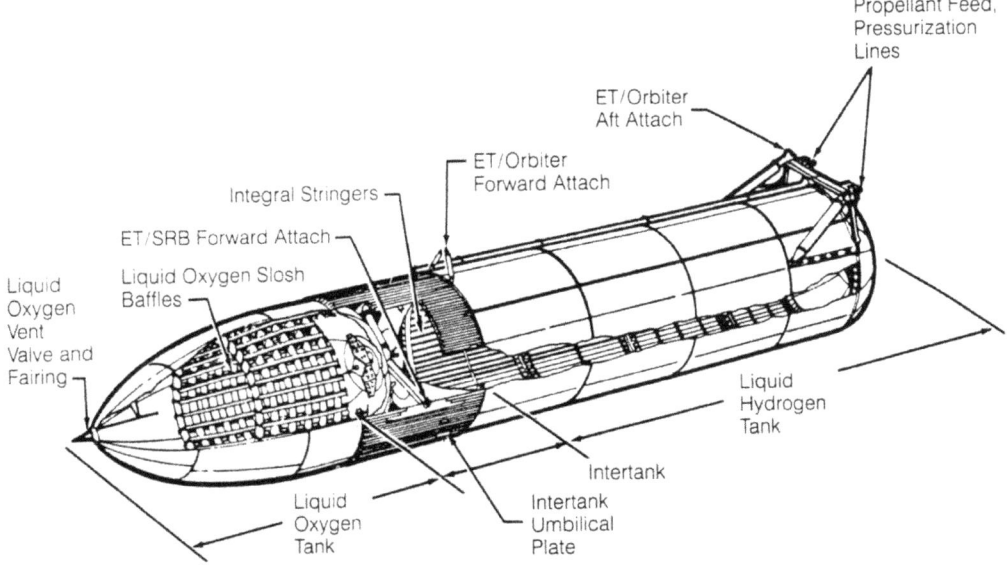

Space Shuttle external tank (graphics from *Challenger* report, page 41).

Angeles Times article, it was reported that the agency was "beginning to show signs of aging. Where it was once the moon ... and then the stars, now it's technology transfer with programs that are ... just to satisfy the White House and Congress." The article further stated, "NASA is on the verge of tarnishing its image as a purely civilian agency because, as operator of the Space Shuttle program, it will be launching a variety of super-secret military and intelligence satellites for the Pentagon and the intelligence community."

NASA, it would seem, could not win, as it and the military would be working together. In the background, however, NASA was indeed having problems controlling its ever more powerful American space allies. The air force and the Defense Department wanted full operational control over the space agency and the Shuttle. Power plays and secrecy were rife and old hands at NASA were becoming concerned. It was a relationship that would not see the light of press reports for many years, but would be addressed by the *Columbia* report.

To add a high level of possibility to ending United States human space flight, President Jimmy Carter in early 1977 directed Dr. Robert A. Frosch, then NASA administrator, to find out if the Shuttle program should be shut down completely before it even flew its first flight. Old hands at NASA knew that Vice President Mondale was working behind the scenes as hard as he could to cut any NASA funding, and the Shuttle was a large target. Carter also wanted to know if the Shuttle could be flown without a crew. He was told that it would be possible but that it would be very costly. And what would be the point of an unmanned Shuttle?

During this time, and on up through the actual first launch and beyond, NASA was running into problems with the Shuttle main engines. They were proving to be a massive engineering task. NASA originally felt the new engines would simply be an upgraded version of the tested and proven second-stage engines used by the massive Saturn V rocket, used to propel American astronauts to the moon. The agency was to find however, that this would be a whole new set of engineering problems never before seen in the space business. Adding to the problem was the fact that not enough funds were being allotted for the development of the main engines. NASA was forced to test them "all up," which meant they would be tested only as a complete unit, and continuing cost overruns were always a problem. Flying safely into space cannot be done on the cheap.

Under the original contract, Rockwell built what was called a Main Propulsion Test Article (MPTA-098). It consisted of a full-size aft orbiter fuselage; a simulated orbiter mid-fuselage, and a complete main-engine thrust structure. The thrust structure consisted of all main propulsion-system plumbing and a full electrical system.

Shuttle Testing Begins

On June 24, 1977, the MPTA-098 was shipped to the National Space Technology Laboratory. There it was mated with a simulated external tank known as the Main Propulsion Test Article External Tank or MPTA-ET.[83] When fully integrated, these two units would be used for testing the Shuttle's newly developed main engines. NASA soon found that this engine would indeed be a whole new technology never tried before. This was, after all, the first time anyone had attempted to design a rocket to be used up to fifty times. Before this program, all rocket engines were a use-once, throwaway affair. This engine would also be required to operate at higher pressures and for longer periods of time then ever before attempted. This was something that NASA was not ready for, and the space agency would take years to solve its problems.

The agency and its contractors found that with very high pressures being encountered, many failures were occurring. Some of these were ruptures in the fuel lines, cracked turbine blades, and failed seals. At times these and other failures not only caused the engines to shut down, but resulted in explosions[84] which not only destroyed the engines but parts of the test equipment.[85] During all of 1978 not a single engine test was successful, and NASA had hoped to be flying in 1979. In all, four separate engines failed and were damaged along with a turbo pump.[86] At least four months were lost in downtime while the engines were modified and repaired at a cost of nearly $21 million. By the end of the year NASA knew its main Shuttle engine was in real trouble. NASA had assumed that it would be able to expand the main engine capability to deliver 109 percent of rated thrust, but so far was unable to even perform at 100 percent. Eventually, NASA did solve the problems, and today these engines stand as the most efficient and most technologically advanced rocket engines ever designed. They develop more usable thrust per pound of fuel than any other engine. But the road was long and costly, as NASA's "all up" program showed the agency that up-front cost projections, properly funded, were the best way to do development.

On July 2, 1979, the MPTA-098 suffered major structural damage in an explosion when a fuel valve fractured. It would not be until December 17, 1979, that a successful complete static test firing would be accomplished. On that date all three main engines were run up to 100 percent of rated thrust for 554 seconds.[87] It was a major milestone in the development of the world's first reusable space-flight rocket engine. As testimony to their strength and ability, the three main engines from *Challenger* can be seen in flight videos to still be operating, even after the Shuttle was destroyed.

The final major Shuttle contract was signed on June 27, 1974,[88] with the Wasatch Division of Thiokol Chemical Corporation. This contract would become one of life and death. Thiokol was named to design and build the Shuttle's solid rocket boosters (SRBs). The solids would be called upon to deliver almost 80 percent of the thrust required at liftoff. It would be the failure of a field joint on one of these rockets which would doom the *Challenger* and her seven-astronaut crew. And even though NASA knew that operational costs for solids would exceed liquid-fuel rockets, the agency went with solid rockets for the first time on a manned space vehicle.

On March 1, 1975, the first full-scale mock-up of the Space Shuttle was unveiled at Rockwell's Downey, California, facility. This test mock-up was used as a size check for future operational

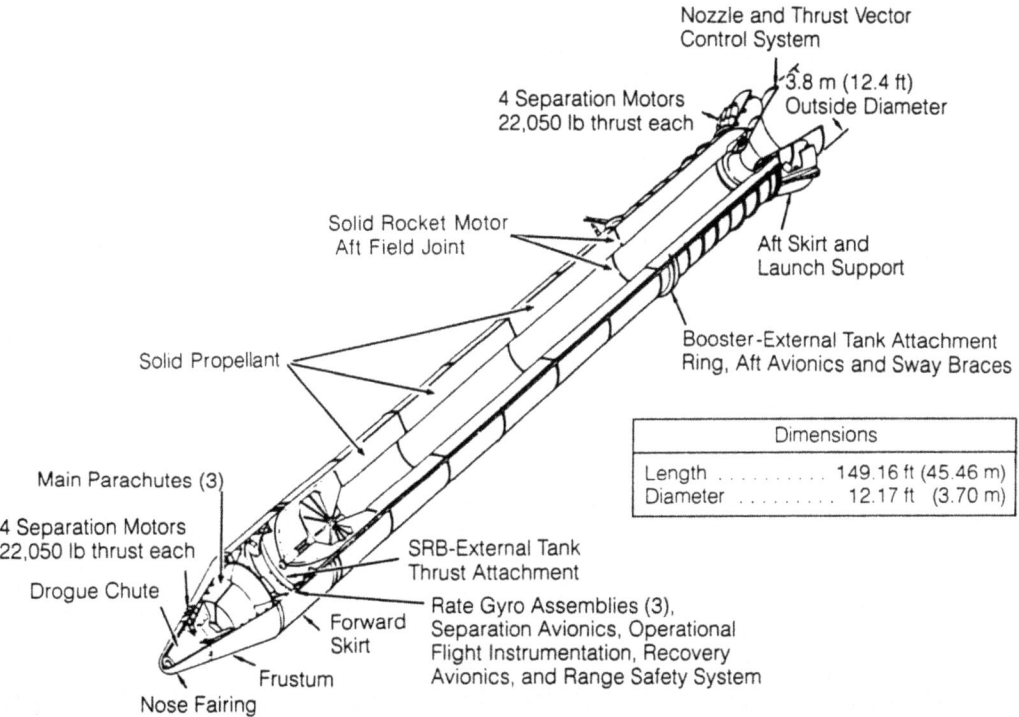

Space Shuttle solid rocket (graphics from *Challenger* report, page 56).

Shuttles. To the press and public it was the first chance to view the massive size of this latest manned spacecraft being built.

Another vexing problem for the space agency at the time was the development of the thermal protection system, which would be the only protection the Shuttle would have from the searing heat of reentry, which would range up to 2,750 degrees Fahrenheit.[89] Before the Shuttle, all spacecraft relied on a heat shield, which ablated, or burned away, as the craft entered the earth's atmosphere. As the shield wore away, pieces flew off in a dazzling display not unlike a comet, taking the heat with it. This was just fine for your run-of-the-mill, use-once-and-park-it-in-a-museum spacecraft, but not for the Space Shuttle.

The Shuttles were being designed for 100 missions each, and replacing the heat shield after every mission was simply not feasible, and far too costly to even consider.[90] A method would have to be found which could protect the Shuttle and at the same time be usable many times over, as well as being a permanent part of the vehicle. It would also have to be lightweight. The Apollo heat shield had weighed 3.9 pounds per square foot, but weight restrictions demanded that the Shuttle's thermal system weigh in at no more than 1.7 pounds per square foot. This problem was solved by a new ceramic tile system, which could hold in the heat of reentry without requiring it to be ablated away or passing it on to the Shuttle's skin.[91] These tiles were lightweight and very heat resistant, and there were 31,000[92] of them of varying thickness and shapes to be glued on to each Shuttle, but they were very subject to impact damage. Their thickness depended on how much heat they would have to endure. Later, a system of thermal blankets would be developed and used on areas of the Shuttle which were subjected to lower heating, but no system has yet been developed to replace the tiles on the truly white-hot areas on the underside and leading edges of the Shuttle.[93]

NASA had originally hoped the new Shuttle would be ready in time to attempt a rescue of the aging Skylab orbiting workshop. Since it had been last manned in 1974, the 85-ton lab had been orbiting the earth uncontrolled and would eventually reenter the atmosphere. NASA feared that large parts of Skylab would survive reentry and pose a danger, albeit very small, to individuals on the ground. In truth, the odds of anyone actually being struck by a piece of Skylab were astronomically small, but the space agency was obligated to ensure that they were prepared just in case. The media were watching this situation very closely as they did with both the *Challenger* and *Columbia* disasters.

Plans were developed in early 1978 to fly a Shuttle crew on one of the early missions (some sources say second[94]) to dock with the lab and deliver a rocket package. This package would be used to boost the lab into a higher and thus longer-lasting orbit, or used to direct the failing Skylab toward an ocean impact, far away from any inhabited areas.[95]

By radio commands, NASA was able to regain control over Skylab and add a few months of life to its orbit by directing the small end forward, which created less drag from the earth's upper atmosphere.[96] However, it was not enough to save the Skylab, as the sun's heat expanded the upper atmosphere, causing more drag than had been expected. By December 1978, it had become clear that America's orbiting workshop would soon reenter and burn up before a Shuttle would be ready to fly. The rescue effort was abandoned and NASA could only direct as best it could the huge derelict to a July 12, 1979, reentry into the earth's atmosphere.[97] And, despite the press hoo-ha about possible injuries and damage, no one was hurt as Skylab came to its fiery end in a 4000-mile debris field over the Indian Ocean and the southwest corner of Australia. It would be more than twenty years before the United States was ready to once again begin placing a fully operational space station into orbit. By 1999, the only space station the Shuttle had docked with was Russian. By 2003, the United States was well on the way to building their long-awaited space station, but that work would be halted for years by the *Columbia* disaster, and grounded once again after only one flight by *Discovery* in July 2005. The flight of *Discovery* would prove that the problems which had brought down *Columbia* had not been solved.

In the fall of 1978, during the Carter administration, NASA's planned (or rather hoped-for) seven-orbiter fleet was cut to six and then again to five. It would be under Mondale's pressure that the fleet was cut to just four orbiters, the minimum required number for operations then being planned. If problems continued in the development, there was a possibility that the number would be cut even more. With Mondale as vice president, NASA had one of its worst enemies in the White House, a heartbeat away from the presidency and the possible end of the Shuttle program, as well as the effective end of America's manned space efforts.

On January 29, 1979, despite pressure from many sides, NASA placed the contract with Rockwell to construct Shuttles *Discovery* and *Atlantis*, and upgrade *Columbia*.[98] The contract was also to modify *Challenger* from the static test article STA-099, to a fully operational orbiter. NASA and the country would have to settle for a minimum four-orbiter fleet.

By early 1979, NASA was starting to feel pressure from Congress to fly. With engine and tile problems starting to work themselves out, NASA announced a launch date of November 9, 1979,[99] for its first Shuttle test flight (STS-1). It was a launch date no one in NASA felt they would meet, and yet the announcement came. But with pressure from the White House, Congress and the military, NASA felt obligated to put a date on the table. It was this type of pressure, which would be felt time and time again before and after the *Challenger* disaster. Most of that pressure was to finish the International Space Station. During the heady days of the Apollo program, NASA was more than a great agency, they were also very lucky. They were engaged in a dangerous and wonderful adventure, but they kept in mind, especially after the loss of the Apollo 1 crew, the many dangers the crews would face. When the Shuttle came along NASA seemed to forget those dangers. NASA is an example of the wisdom of the advice that one should never read or believe ones own press releases. As far as the technology is concerned, "Trust, but verify."

Enterprise OV-101

Three years earlier, as NASA was struggling with many engineering problems, the agency rolled out its first Space Shuttle with as much fanfare as the agency could muster. Designated OV-101 (for Orbital Vehicle), the new Shuttle was towed out of Rockwell's Plant 42 in Palmdale, California, on September 17, 1976.[100,101] OV-101 was originally to be named *Constitution*, in honor of the U.S. Constitution's Bicentennial, but Star Trek fans across the nation and from around the world began a letter-writing campaign to the White House. At their urging, President Ford changed his mind and *Enterprise* became the name of the first Shuttle.[102]

Later, *Star Trek* fans would be disappointed to learn that the *Enterprise* would only be used to test the Shuttle's systems within the earth's atmosphere. Originally OV-101 was to be used for testing and then rebuilt as an earth-orbiting Shuttle, but it was found that the costs were nearly as much as building a new one; and more to the point, the test bed which would become *Challenger* OV-099 was much closer to flight specifications. *Challenger* would fly in space, but *Enterprise* would not. *Enterprise* would, however, fly piggyback test flights for many hours, as well as approach and landing tests for a total of 20 minutes and 57 seconds, well within the earth's atmosphere, opening the way for Shuttle flights into space.[103]

After rollout, *Enterprise* was transported overland to Edwards Air Force Base, 36 miles away, for its full series of approach and landing tests.[104] For the tests on February 25, 1976, two NASA astronaut crews were selected. Fred Haise and Gordon Fullerton were one crew, and Joe Engle and Richard Truly the other.[105] All but Fred Haise would eventually fly the Shuttle *Columbia* on test flights into space. Haise, who had flown to the moon on the ill-fated Apollo 13 mission, would leave the program before orbital flights began.

Manager for the approach and landing tests was astronaut Donald K. Slayton, who had flown on the Apollo Soyuz Test Project (ASTP) mission in 1975 as docking module pilot. Slayton had been one of the original Mercury 7 astronauts, but had not flown at the time because NASA doctors had thought that his occasionally irregular heartbeat might cause a problem for him in space. The doctors were wrong. The ASTP flight meant little technically however, since it was to test a rescue of another nation's astronauts, even though NASA had no more Apollo hardware to use for any rescues. This propaganda flight did succeed in allowing the Soviets a close view of America's moon-flight technology. The flight also gave Slayton his long-awaited chance to fly in space.

John Manke, who had flown the M2-F2, HL-10 and the X-24B lifting bodies, was now Chief of Flight Operations at Dryden Flight Research Center, and responsible for these tests. He would be able to see the results of his and other research pilots' efforts come to fruition on the same runway he had used only a few years earlier. The lifting-body program had come full circle, as Shuttles began their test flights.

The first tests were simple taxi tests (if anything can be called simple when a spacecraft is tested), with the *Enterprise* mounted atop its Boeing 747 carrier aircraft.[106] Three such tests were conducted on February 15, 1977. Later, still mounted on the 747, the Shuttle was carried on five captive flights without a crew on board, so that NASA could assess the structural integrity and the handling qualities of the two mated craft. Next were three manned captive flights with astronaut crews operating the *Enterprise*'s flight controls, while it remained firmly bolted to the 747. The 747 pilots selected for the tests were Thomas McMurty and Fitzhugh Fulton, Jr.[107]

The unmanned "captive-inactive flights"[108] concluded on March 2:

- February 18, 1977 (16,000 feet, up to 287 mph, 2 hours 5 minutes)
- February 22, 1977 (22,600 feet, up to 328 mph, 3 hours 13 minutes)
- February 25, 1977 (26,600 feet, up to 425 mph, 2 hours 28 minutes)
- February 28, 1977 (28,565 feet, up to 425 mph, 2 hours 11 minutes)
- March 2, 1977 (30,000 feet, up to 474 mph, 1 hour 39 minutes)

Because of the success of these flights, a sixth captive-inactive flight test was cancelled by Slayton.[109]

The manned "captive-active flights"[110] all took place in June and July:

- June 18, 1977 (14,970 feet, up to 208 mph, 55 minutes 46 seconds, Haise and Fullerton)
- June 28, 1977 (22,030 feet, up to 310 mph, 62 minutes, Engle and Truly)
- July 26, 1977 (27,992 feet, up to 312 mph, 59 minutes 53 seconds, Haise and Fullerton)

Finally, the *Enterprise* was flown on five unpowered, dropped free flights. After being released from its carrier aircraft, the Shuttle *Enterprise* was maneuvered to landings at Edwards Air Force Base. On its first four landings the Shuttle came down on the dry lakebed, the same lakebed used by the X-15 and lifting bodies.[111] However, on its fifth and final flight, the Shuttle landed on Edward's main concrete runway simulating a return from space. These tests verified the Shuttle's pilot-guided approach and landing abilities and verified the orbiter's subsonic airworthiness. This was the last time the *Enterprise* would ever be flown by itself, but it completed a critical test phase, which would do much to qualify the Shuttle system for human space flight, then a little more than three years in the future.[112]

The dropped free flights were brief:

- August 12, 1977 (Tail Cone On, 27,000 feet, 5 minutes 21 seconds, Haise and Fullerton)
- September 13, 1977 (Tail Cone On, 24,000 feet, 5 minutes 28 seconds, Engle and Truly)
- Sept. 23, 1977 (Tail Cone On, 24,000 feet, 5 minutes 34 seconds, Haise and Fullerton)
- October 12, 1977 (Tail Cone Off, 20,536 feet, 2 minutes 32 seconds, Engle and Truly)
- October 26, 1977 (Tail Cone Off, 17,000 feet, 2 minutes 2 seconds, Haise and Fullerton)

After the drop tests and other tests were completed, in April 1979 the *Enterprise* was flown to Kennedy Space Center where it was mated to an external tank and solid rockets.[113] The stack was then rolled out to launch complex 39A to serve as a practice and launch-complex fit-check verification Shuttle, which represented the Shuttles which would take astronauts to space.[114] The *Enterprise* would be the only Shuttle to be placed on both launch pads at Kennedy and SLC-6 at Vandenberg Air Force Base in California, even though it would never fly in space. It is now on display at Dulles Airport, in Washington, D.C., and has been part of the Smithsonian Institution's Air and Space Museum since November 18, 1985.[115]

The March to a Launch

Attempting to meet its stated November 1979 launch date, NASA began to push itself to finish the massive tile-installation project on *Columbia*.[116] The installation would take around 670,000 hours, or some 76.5 years of labor time to finish. Rockwell was averaging less than two tiles per worker per week (1.8) and there was no incentive to finish the job any faster.[117] After all, many who worked on the project felt with good reason that they would be laid off when the job was complete. In fact, it was found later, but not very well published, that deliberate sabotage had occurred on the project.[118]

Rockwell had developed a "good citizen program" in which they had hired ex-convicts to install thermal tiles on a part-time basis. During a random quality control inspection it was found that small holes were being punched into many of the tiles already on *Columbia*. It was later learned that some of the "good citizens" had punctured the tiles with hand-held sharp objects in

Shuttle *Enterprise* crawls towards launch pad 39A at the Kennedy Launch Center (NASA).

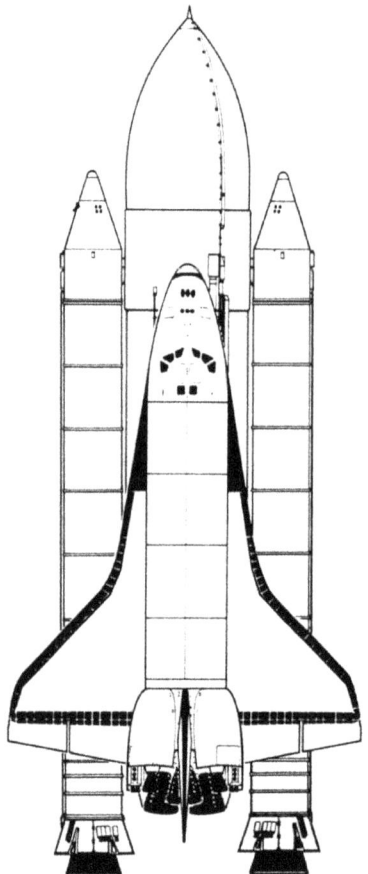 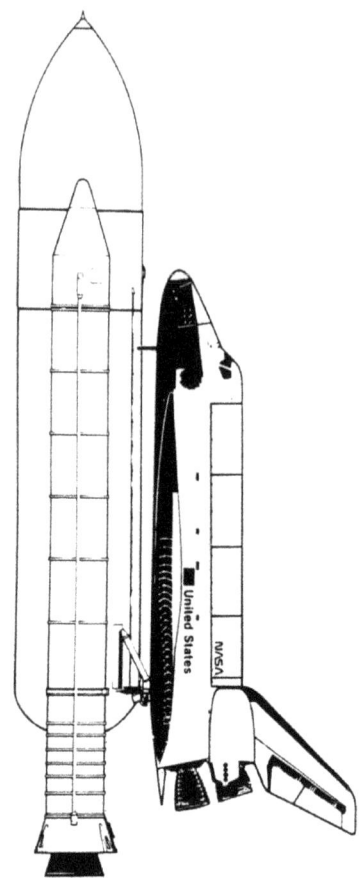

Artist's drawing depicts Space Shuttle stacked for launch in view from dorsal side of Orbiter (left) and from the left side of stack.

Drawings of the stacked Space Shuttle ready for launch (graphics from *Challenger* report, page 3).

order to extend their period of employment with Rockwell.[119] It was time for NASA to go to a low-paid college-student workforce and leave the so-called "good citizens" behind.

In the hope that installation could be accomplished faster and perhaps safer at the Kennedy Space Center, NASA made plans to fly the Shuttle *Columbia* to the Cape. At the time only 23,200 of the tiles had been installed, so to protect the exposed areas, nearly 10,000 dummy tiles were used to fill in the empty areas. These dummy tiles were held in place by adhesive tape. This proved to be a mistake. When *Columbia*, sitting on top of its 747 carrier, rolled down the runway at Dryden for a test flight of the tiles, thousands of dummy ones simply fell off. It was said to look like confetti.[120] Along with some 4,800 dummy tiles, 100 "permanent" ones came flying off.[121] Clearly the tape solution was not what NASA was looking for. The next solution was to glue the corners of the temporary tiles in place. This would hold them for the flight to Florida from California. Later they would be removed and replaced with the thermal tiles. But with 100 permanent tiles lost, NASA managers needed to develop a method to pull each tile to test its ability to hold and not come loose during the stresses of space flight.[122]

On March 20, 1979, the 747 carrying *Columbia* flew to Kennedy Space Center.[123] The flight would not be a direct one however, as the 747 flew a path intended to keep it away from rainstorms.[124] Any kind of rain could easily damage the tiles. Waiting out a rain storm in Texas, a

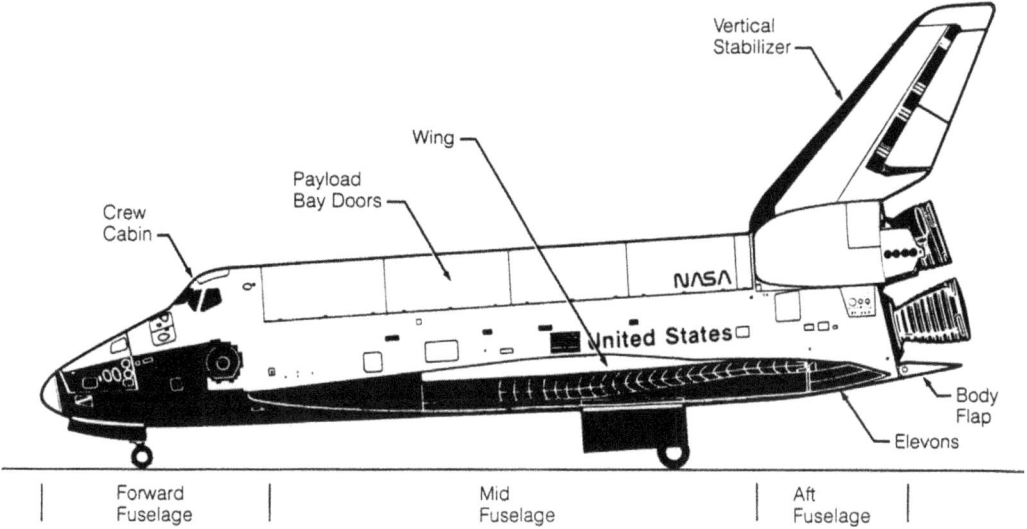

Sketch of Space Shuttle Orbiter in the landing configuration viewed from -Y position identifies aerodynamic flight surfaces.

Space Shuttle side view (graphics from *Challenger* report, page 47).

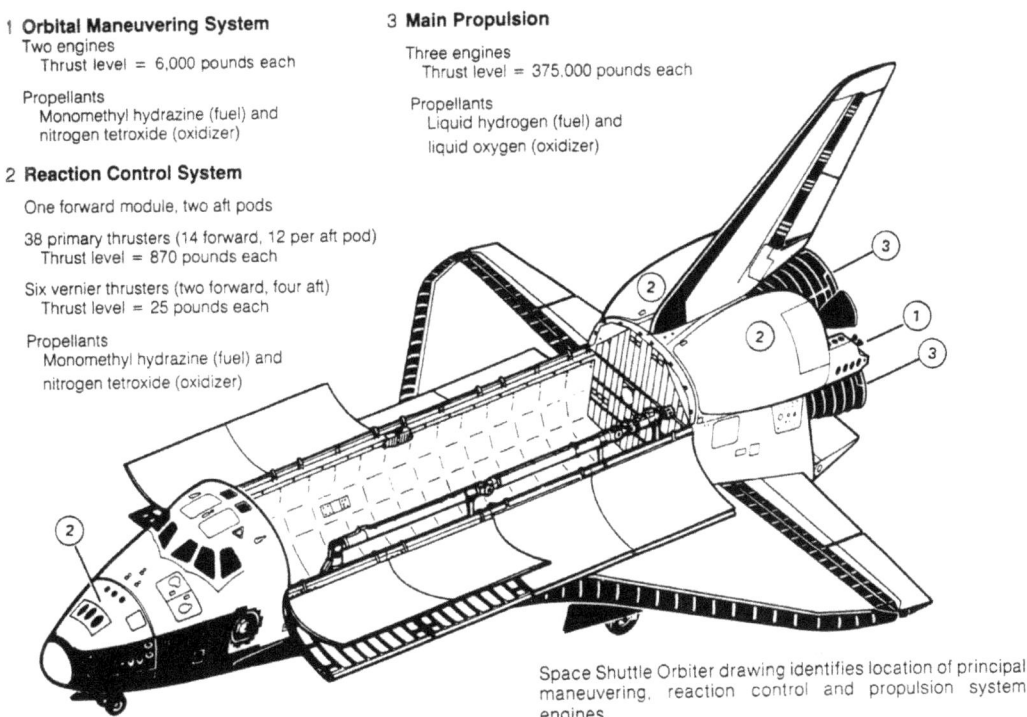

Space Shuttle Orbiter drawing identifies location of principal maneuvering, reaction control and propulsion system engines.

Space Shuttle 3/4 view (graphics from *Challenger* report, page 46).

flight of only a few hours turned into a trip of four days. As *Columbia* arrived at Kennedy on March 25, the Shuttle looked very much like a patched-up old warhorse and it had not yet flown in space.[125] At the Space Center, the news media were less interested in recording *Columbia*'s arrival, and more on how soon the Shuttle would fly into space. This was a question that even the Shuttle managers could not answer.[126] The Shuttle was behind schedule and that meant more costs.

Space Shuttle *Columbia* is flown piggyback to the Kennedy Launch Center (NASA).

Clearly NASA was beginning to feel the pressure to "put up a bird." The media were unaware of just how difficult a job developing the Shuttle system was. All the press would have needed to do was to take a hard look at the wide range of people working on the tiles, from unemployed crop pickers to college students on summer vacation, to see just one of the many problems facing NASA. Everyone at the space agency was still learning how to process the first reusable spacecraft, and the Shuttle system was being pushed too hard and too fast.

The Defense Department was also becoming concerned with the delays. The Shuttle was correctly viewed by the Pentagon as a national defense resource which would soon be required to place heavy, top-secret payloads into orbit. These critical payloads could not be put on hold indefinitely, and because the Pentagon had so highly prioritized the launch of the Shuttle it wanted more than a press conference. The Pentagon wanted results, and soon.

By the summer of 1980, after nine postponements of its first flight, NASA Administrator Robert Frosch was ready to announce a launch date of March 1981.[127] It was a date NASA had every intention of keeping. Frosch was convinced that problems with the Shuttle main engines and the heat tiles falling off would be solved by that time. By November, NASA managers announced that the tile problem had indeed been solved and that *Columbia*'s heat shield had been, for the most part, completed. The only area left was to fill in narrow gaps between some of the tiles, and that could be completed while other Shuttle processing was in the works.

On November 21, 1980,[128] the first operational Space Shuttle, *Columbia*, was rolled out of the OPF (Orbital Processing Facility) and placed in the VAB (Vehicle Assembly Building). Inside the VAB, *Columbia* was mated to its external tank and twin solid-rocket boosters.[129] On December 29, *Columbia*, now fully integrated and mounted on its mobile launch platform, began its three-mile crawl to launch pad 39A.[130] It was the same pad which had launched astronauts to the moon in the late 1960s and early '70s, and had been tested for fit and given operational checks using *Enterprise* only twenty months earlier. The pad was ready, but were the Shuttles? *Columbia* would be the first "space-rated" orbiter, and as such was built slightly differently than the rest of the fleet.

Crew for the first Shuttle test flight on *Columbia* STS-1, Shuttle Commander John W. Young and Shuttle Pilot Robert L. Crippen (NASA).

It was heavier and so would be judged insufficiently cost effective to fly payloads to the Space Station.

The trip to the pad took nearly six hours, and then the six-million-pound assembly was placed precisely where it needed to be.[131] This was the first time a fully operational Shuttle was on the launch pad. It had been more than just a slow crawl to the pad. It had been the accumulation of more than eight years of work after Rockwell had gotten the contract. *Columbia* now stood ready to again place Americans in space. However, to clear the way, the engines would have to be fully certified. On January 17, 1981, only three months before the first flight of *Columbia*, the main propulsion test article (MPTA-098) successfully completed a 625-second firing. The three flight-rated engines from *Columbia* were then removed and installed on MPTA-098 where they completed a 520-second test firing. Finally, the engines were shipped back to Kennedy and installed on *Columbia* where they performed a 20-second flight readiness firing (FRF).[132] *Columbia* was now cleared to launch and Americans were once again ready to view the earth from orbit, this time from a new generation of manned spacecraft.[133] "Upon the conclusion of the readiness firing, steps were taken to repair a small portion of the External Tank's super light ablator insulation which became debonded during a tanking test of the orbiter's super cold liquid oxygen and liquid hydrogen propellants in January."[134] This was the first red flag on the insulation problem.

On March 21, technician John Bjornstad was killed and another was critically injured when they entered the Shuttle engine compartment and were overcome by nitrogen gas. The gas was thought to have been pumped out before they entered *Columbia*.[135] Days later the second man, Forrest Cole, would die.

On April 12, 1981, Space Shuttle *Columbia*, commanded by John Young (veteran of Gemini and Apollo) and piloted by Robert Crippen (from the old MOL project), would leap from the pad on its maiden voyage. NASA was so worried about the new thermal tiles coming off that the crew turned the Shuttle underbelly in a direction which allowed the KH-11 spy satellite a chance to photograph the tiles.[136] It was a classified and as-yet-unpublished test, but the result has been referred to as "remarkably successful" by one NASA manager.[137] The bottom line was that the tiles had worked, though not as well as NASA wanted. *Columbia* had lost 16 tiles[138] and 148 others had been damaged during the flight. *Columbia* had taken over 300 debris hits where none was the expected standard. This flight was designated STS-1 (Space Transportation System 1) and America was back in the human space flight business. It should be noted that when damage was suspected on the tiles on *Columbia*'s final mission, NASA managers considered and then discounted asking for photos to be taken of the tiles while the *Columbia* was in orbit. Since history would show that NASA has never flown a mission without some type of damage to the tiles, it was not a good call. In fact, it was a fatal one.

After its first four flights, the Shuttle program was called operational, but it will never truly live up to that designation. Each flight of a Shuttle is an experiment. Perhaps in the future a second- or third-generation Shuttle will be built using the knowledge gained by today's Space Shuttle, and it may even be profitable. The *Columbia* accident board put it this way: "Throughout the history of the program, a gap has persisted between the rhetoric NASA has used to market the Space Shuttle and operational reality, leading to an enduring image of the Shuttle as capable of safety and routinely carrying out missions with little risk." The Shuttle has yet to live up to NASA's press reports on safety and reliability.

Challenger OV-099

Space Shuttle *Challenger* began its life as STA-099. It was a high-fidelity "structural test article," never meant to someday fly in space. Named after an American navel research vessel which sailed in both the Atlantic and Pacific in the 1870s, *Challenger* was to become the second operational orbiter when it was launched on its first flight, designated STS-6, on April 4, 1983.

The test article airframe was completed at the Rockwell plant and delivered to Lockheed for tests on February 4, 1978. Due to the lightweight design, nearly all components of the frame would be required to withstand large structural stresses. These stresses would be difficult to simulate to the levels required by NASA using computers available at the time, so direct testing was conducted.

STA-099 was put through an 11-month program of vibration testing in a 43-ton rig designed uniquely for Space Shuttle tests.[139] This rig was built with 256 hydraulic jacks, which could distribute flight-simulated stress loads to more than 836 load-application points on the Shuttle. This intensive testing could simulate all expected stresses the orbiter could normally expect to undergo during launch, ascent, orbit operations, reentry, and landing, all under computer control. To simulate the thrust of the three Shuttle main engines, three one-million-pound-force hydraulic cylinders were used. Adding to that were tests which would simulate the heating and thermal conditions of space flight. STA-099 came through with flying colors. Indeed, it was stronger than designers had expected.

The original contract had called for the construction of two static test articles and two flight test vehicles, designated *Enterprise* OV-101 and *Columbia* OV-102. But in 1978, NASA decided not to convert *Enterprise* to a fully operational Space Shuttle, which left only *Columbia* as space-flight capable. With this in mind, NASA awarded Rockwell a supplemental contract on January 29, 1979, to convert STA-099 from its test configuration to a fully operational orbiter. This orbiter became known as OV-099 or, as most will remember her, *Challenger*.[140]

On November 7, 1979, STA-099 was returned to Rockwell to begin its conversion. The conversion would be easier than the one originally planned for the *Enterprise*, but it was still a major undertaking. It would involve a major disassembly of the craft.

Here are some of the *Challenger*'s construction milestones[141]:

- January 5, 1979 (Contract awarded)
- January 28, 1979 (Structural assembly of crew module began)
- November 3, 1980 (Final assembly started)
- October 23, 1981 (Final assembly completed)

Launch of Space Shuttle *Challenger* on STS-6, April 4, 1983 (NASA).

- June 30, 1982 (Rollout from Palmdale)
- July 5, 1982 (Delivery to Kennedy Space Center)
- December 19, 1982 (Flight readiness firing)

When STA-099 had been built, a simulated crew module had been added to aid in the testing. In order to access this crew module the forward fuselage halves had to be separated. Also, the wings had to be modified and reinforced in response to findings from the structural testing. Finally, two new heads-up displays known as HUDs were installed in the now completely rebuilt cockpit. When the main engines were installed, the new orbiter *Challenger* weighed in at 175,111 pounds, which was 2,889 pounds lighter than its sister ship *Columbia*. This would translate to larger payloads being able to be placed into orbit by *Challenger*, or higher obtainable orbits to include docking with the International Space Station, which *Columbia* would never reach.

Several major components were removed from STA-099 so they could be modified for flight, including the vertical tail assembly shipped to Fairchild in New York in March 1980. The payload bay doors were removed and shipped to Rockwell's plant in Tulsa, Oklahoma, so that the orbiter's space radiators could be put on. The aft fuselage was sent to Downey, California, so that secondary structures required for space flight could be constructed. At that time, due to budget cuts, NASA was even looking at the *Enterprise* as cannibalization material to construct parts of *Challenger* and *Discovery*. Funding was still being cut to the bare bones. By April 27, 1981, news agencies were reporting that *Challenger* had 8,100 of its 30,622 tiles already in place, and that its reinforced wings had been attached to the fuselage. Sitting next to *Challenger* in the next bay was the *Enterprise*, stored after its test flight series for possible parts supply. It would soon be moved, however, to make room for the third Shuttle, *Discovery*, which would be assembled in September of that year.

All of *Challenger*'s tiles were put through a "densification" process, a chemical permeation process which gave them a much greater surface strength. This also made them better able to hold on to the felt that was used to bond them to the skin of the Orbiter. With the problem of temporary tiles falling off due to movement to Kennedy by *Columbia*, it was decided not to bring *Challenger* to the launch site until all of its thermal tiles were in place and checked. It was also fitted with newly developed thermal blankets, which would take the place of thermal tiles in lightly heated areas.

By November 1981, while *Columbia* was on its second test flight, *Challenger* was nearly 80 percent complete at the Rockwell plant, with *Discovery* now at its side in its beginning stages. *Aviation Week and Space Technology* reported in its November 16, 1981, issue that when *Challenger* was completed it would be the first operational Shuttle "with a 100 mission design life capability." It would also have a strengthened mid-body area and thrust structure as well as be the first to have the new heads-up display to aid in landings. The aerospace news media seemed to have forgotten that all orbiters were to have been designed for a 100-mission capability.

On June 30, 1982, *Challenger*, now completed, rolled out for its transportation to the Cape. Paul Weitz, who would command the orbiter on its first flight, accepted a symbolic key to the Shuttle on behalf of the American taxpayers. By August 1982, NASA was trimming the flights it hoped to make by 1985 from forty-four missions to thirty-two. It was an admission that flying Shuttles was going to prove a great deal more complicated than anyone had ever imagined. In 1975, NASA had projected that the turnaround time for Shuttle processing would be only ten days. By the end of 1985, the realities of flight had pushed that estimate to an average of sixty-seven real working days. This, of course, was increasing the costs to fly, which by 1989 had risen to $140 million per flight.

Challenger was rolled to pad 39A on November 30, 1982, in heavy fog, for a flight-readiness firing of its three main engines. It was during this test that a leak was discovered in one of the main engines, which would soon be repaired.[142] On April 4, 1983, *Challenger*'s mission STS-6 was launched from pad 39A at the Kennedy Launch Center on its maiden flight. It was the first of ten flights by *Challenger*, and NASA now had a two–Shuttle fleet, with two more orbiters on the way. During *Challenger*'s launch, which was the first to use a new lightweight external tank and lighter-weight SRB casings, both the left and right SRBs suffered heat contact on the primary O-rings. The design stipulated that no heat was to reach the O-rings. During launch, *Challenger* was giving NASA its second warning of problems which were to bring Shuttle flights to a crashing halt in less than three years.

"I believe I am both dedicated and qualified enough to do whatever NASA calls on me to do."
— Mission Specialist Judith Arlene Resnik

Astronauts Killed While on Active Duty

Name	Age	Date	Cause
Theodore C. Freeman	34	October 31, 1964	T-38 jet trainer hit a large bird
Elliot M. See	38	February 28, 1966	T-38 Jet trainer crashed during a snowstorm
Charles A. Bassett	35		
Virgil I. ("Gus") Grissom	40	January 27, 1967	Fire in the cabin during a ground test in capsule 204, five days before launch, preparing for Apollo 1 flight
Edward H. White II	37		
Roger B. Chaffee Jr.	31		
Edward G. Givens	37	June 6, 1967	Car crashed near Houston, Texas
Russell L. Rogers	41	September 13, 1967	X-20 Pilot killed in jet crash
Clifton C. Williams	35	October 5, 1967	T-38 crashed near Tallahassee, Florida
Michael J. Adams	37	November 15, 1967	X-15 crash over Mojave Desert
Robert H. Lawrence	31	December 8, 1967	Aircraft crash at Edwards Air Force Base
James M. Taylor	39	September 1970	Air crash
Francis R. Scobee	47	January 28, 1986	*Challenger* flight (first crew killed during an American human space flight)
Michael J. Smith	41		
Ellison S. Onizuka	39		
Judith A. Resnik	36		
Ronald E. McNair	36		
Gregory B. Jarvis	42		
S. Christa McAuliffe	37		
Stephen Thorne	32	May 25, 1986	Stunt plane crashed near Santa Fe, Texas
S. David Griggs	49	June 17, 1989	Stunt plane crashed at Earle, Arkansas
Manley L. Carter Jr.	43	April 5, 1991	Crash of Atlantic South-east Airlines
Robert Overmyer	59	March 23, 1996	Crash of test aircraft at Duluth, Minnesota
Rick Husband	45	February 1, 2003	*Columbia* flight (second crew killed during an American human space flight)
William C. McCool	41		
Michael P. Anderson	43		
Laurel Clark	41		
Kalpana Chawla	41		
David Brown	46		
Ilan Ramon	48		

When the Apollo One crew died, NASA informed the world that the fire had killed them instantly. In fact, it had taken nearly three minutes for the crew to die — not by the fire but by asphyxiation. The screams of the crew could be heard on an open circuit and would be forever recorded on tape.

CHAFFEE: Hey...
WHITE: Fire! We've got a fire in the cockpit!
CHAFFEE: We have a bad fire! We're burning up...
UNKNOWN: A scream, then static on the line.

The events were also recorded on film viewed through a window of the command module.[143] That film has yet to be released to the American people who paid for the entire Apollo program.

CHAPTER TWO

A History of Success? The Solids

It's been nearly a quarter of a century and we thought this might happen sometime but we've delayed that day until today.
— Senator John H. Glenn, January 28, 1986

From April 1981 through January 1986, the United States successfully launched twenty-four Space Shuttle missions from the Kennedy Launch Center. By that time NASA had two fully operational launch pads at Kennedy with a third polar launch site dedicated to the Defense Department nearly operational at Vandenberg Air Force Base, California. The agency had flown all four Shuttles and was looking forward to a very productive year of fifteen Shuttle launches for 1986. Then the "accident" occurred.

At first the public saw the destruction of *Challenger* as a simple accident — sad and shocking, but part of the price we would pay if we were going to continue as the preeminent space-going nation. What the public was unwilling to look at during these flights were the many problems NASA was having with the Shuttle program. The press openly reported on them at times, but the public and many within NASA ignored them. The program had been sold as a simple, safe, "truck" into space. In reality the Shuttle would never move beyond being a research spacecraft, pushing the cutting edge of technology. This was the first reusable spacecraft and the most complicated machine ever created. That is the way NASA should have sold it. The public and press forgot that there was risk in every single launch of the Shuttles before and after *Challenger*'s dark passage. We were reminded of that fact when *Columbia* was lost on February 1, 2003.

A Problem with Design

When Dr. Werner von Braun, the visionary German rocket scientist whose Saturn V helped place astronauts on the moon, was told of plans to use a huge solid rocket booster (SRB) as the first stage of NASA's new Space Shuttle, he was shocked. As director of NASA's Marshall Space Flight Center in Huntsville, Alabama, he was very familiar with the history of solid boosters and knew their many dangers. Von Braun knew that solids could not be turned off once they were ignited, therefore they posed a grave danger to the crews who would fly them.[1] He and his staff at Marshall informed NASA's top management, including Dr. Fletcher, the administrator, that when it came to manned flight, solids were far too dangerous.

As an alternative to solids, some at Marshall, as well as others within NASA, tried to push for a liquid-fuel booster system. This type of system had a long history of success in both unmanned and manned space flight, and had the added advantage of being able to be throttled and shut down in the event that problems occurred during the flight. Marshall wanted a pressure-feed, liquid-fuel system which would not only be safer but could develop even more thrust than the SRBs were being designed to deliver. The system would have been cheaper, safer and stronger than solids, but it was rejected. For Fletcher there were "other" considerations.

These concerns were to no avail. In January 1972, Dr. Fletcher convinced President Nixon that not only would the solids work, they would be cost-effective to operate.[2] And yet, no one had ever produced solid rockets this large before and no real data was available. In fact, after the *Challenger* disaster the Commission would find that even the engineers who operated the system did not fully understand the design or its operational capabilities. The president was informed that solid rockets were very reliable (read "safe") in flight even though this information was based solely on the much smaller rockets (one-third as large) used by the unmanned Titan system, rockets never recovered after their flights. When Fletcher met the president and showed him a model of the Shuttle, it had liquid rockets on its sides.

During this period and later, NASA was desperately looking for projects and funding to continue its manned flight operation. Throughout much of the space agency there was a feeling that cutbacks in all areas would cause operations to slow to a crawl. The Shuttle program was seen by many as the only way to keep NASA alive, and to keep the United States in the manned-space-flight business. Indeed, Marshall Space Flight Center had already been targeted by the Office of Management and Budget for possible closure due to budget cuts. With this and other concerns in mind, NASA assigned development and management of the SRB program to Marshall. The NASA center which had fought the hardest to keep solids off of the Shuttle, now found itself in a leadership position for their development. Yet many at Marshall continued to oppose their use then as they do to this day!

Despite their opposition, Marshall took the lead in SRB development, which included a recommendation to seek a single component bid for the solids. Most aerospace programs were divided up so that contracts could be spread around, thereby garnishing greater political support for NASA projects. It was a fact of life for our nation's space agency. But in this case it was felt that more control over quality and safety would be assured if an agreement was made with a single contractor. However, top management, led by Administrator Fletcher, decided to break up the contact for solid-booster development. This contract breakup would keep politicians happy, at least those who got work for their districts, as well as keep more contractors in the flow. Did political expedience surpass safety considerations in the solid-booster program?

Four companies developed bids and proposals for the solid rocket boosters which were to be reused up to twenty times—Lockheed Propulsion Company, United Technologies, Morton Thiokol and

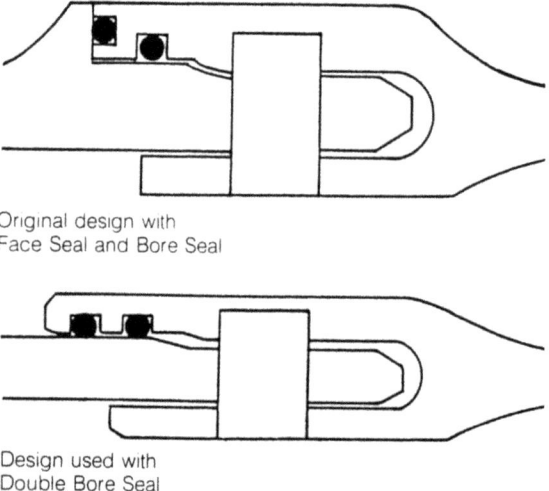

Comparison of original design to design used (graphics from Challenger report, page 121).

Aerojet Solid Propulsion Company.³ At the time only one consideration was paramount within NASA — cost. Safety was not the issue. As in most government contracts, politics was part of the mix.

An evaluation board was formed within NASA to evaluate SRB proposals and bids, and to rate companies on design, development and verification as well as costs and management history for solid rockets. The board found that Morton Thiokol was fourth and last in design, development and verification, well behind the other bidders.⁴ Despite this, Morton Thiokol, a company in Utah, the home state of Fletcher, received the contract on November 20, 1973.⁵ By putting forward the lowest bid, Thiokol became the prime contractor to design and build the world's largest solid boosters ever conceived for space flight.⁶ It would be a little more than twelve years until NASA and the world would see the true price to be paid for these solid rockets.

In its evaluation report the board stated, "Thiokol could do a more economical job than any of the other proposers.... [T]he cost per flight to be expected from a Thiokol built motor would be the lowest."⁷ There was no acknowledgement that the Thiokol design was ranked last. Dr. Max Faget, designer of the Mercury and Gemini spacecraft, with over thirty-five years in the space business, one of the fathers of Shuttle technology, stated it best when he said, "The biggest mistake we have ever made was putting solid boosters on the Shuttle." That mistake has never been corrected, as Shuttles continue to fly with these boosters which are also being looked at to provide power for the next generation of craft intended to travel back to the moon and on to Mars.

The Presidential Commission on the *Challenger* Accident would later find that "cost considerations overrode any other objections." It was penny-pinching at its finest as far as the contractor was concerned. Yet, when the choice was cheaper liquid boosters over a more costly solid rocket, NASA went with the solids, so cost could not have been the only consideration. In the end, in the unforgiving realm of space flight, if you don't intend to properly fund a project, it should not be built. President John Kennedy, when asking Congress to endorse his vision of a man on the moon stated, "If we are to go only half way ... then it is better that we not begin at all."

Thiokol engineers decided during their design process to rely heavily on the design then being used on the Titan boosters, which, as stated earlier, were much smaller than the SRBs.⁸ By doing this, they felt they could cut out a lot of the unforeseen problems that come up when a new system is being developed. What the engineers did not take into consideration and subsequently did not include in their design, was that the Titan insulation from one segment fits tightly against the next segment.⁹ This tight seal is in place to prevent the gases of ignition from burning

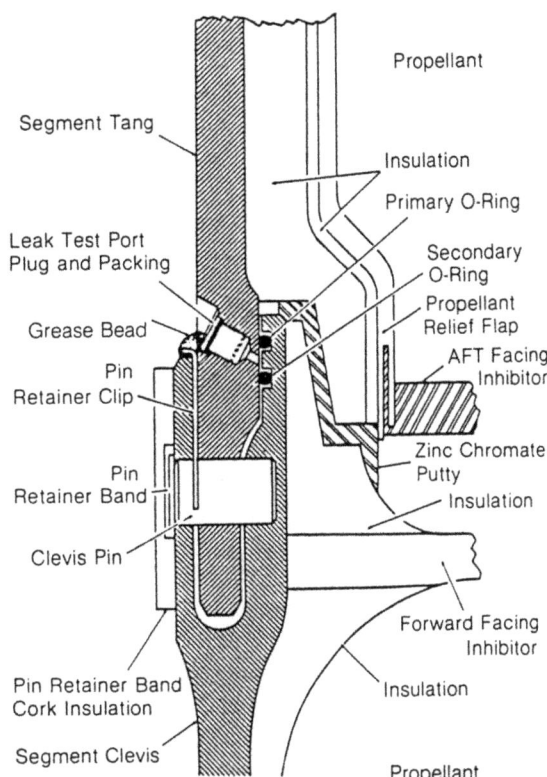

Solid Rocket Motor cross section shows positions of tang, clevis and O-rings. Putty lines the joint on the side toward the propellant.

Solid rocket motor cross section (graphics from *Challenger* report, page 57).

through.[10] To this tight fit was added an O-ring which could, but was not designed to, fully keep the hot gases from escaping. In Thiokol's design, the O-rings were expected to hold back the pressures of ignition using nothing more than insulating putty as an aid, expecting these to hold back the super hot gases with no tight fit.

After the first ten Shuttle flights, the putty that NASA was using to help seal the rubber, fluorocarbon, "elastomeric" O-rings on the field joints was changed. NASA had been using what could have been called "off the shelf" putty. This product was produced by the Fuller O'Brien Paint Company of San Francisco, California.[11] The putty had contained a base of asbestos which aided in its protective qualities in a superheated environment, but due to a ban on asbestos in some paint products by the Consumer Product Safety Commission, the company had stopped producing the asbestos-based putty. NASA then changed to using non-asbestos-based putty, which did not do the job as well. In fact, rocket motor engineers within NASA reported that the putty did not function as well as the old putty. Nevertheless, the space agency continued to use the new putty as problems with the O-ring seals become more common.

Things would go from bad to worse when Thiokol changed its design before production began. As the *Challenger* Commission reported, "Originally the joint seal design (O-rings) incorporated both a face seal and a bore seal. However, the motor that was eventually used had double bore rubber O-rings (built by Parker Hannifin Corp., Lexington, Kentucky). The original bore seal/face seal design was chosen because it was anticipated that it would provide better redundancy over a double bore ring seal since each is controlled by different manufacturing tolerances and each responds differently during joint assembly." Thiokol changed the design because it would be easier to produce, something that translated into greater profit.

Marshall engineers and managers who had oversight responsibility for the SRB designs soon found that they did not have enough solid-rocket background to fully oversee the project. They had to go outside the agency for expertise required for design review of Thiokol's work. The center soon found that the joint design, including the O-rings, was something that no one in the industry at the time had the expertise to fully understand. Therefore, when Thiokol went ahead with the design change to the less effective double-bore seal design, NASA gave its full approval, even though this was not the originally approved contract design.

Two other differences from the unmanned Titan O-rings were also evident in the SRB design. The first was the inclusion of leak-check ports in the new solid-rocket design on the sealing of the O-rings. The second was that, unlike the single-piece Titan O-ring, the SRBs' O-rings came in five pieces which had to be glued together. Both of these possible problem areas could have helped hot gases push past a faulty design.

Thiokol originally tested its design in the mid 1970s, which turned out to be less than satisfactory. By 1977, the company was using water pressure for what was then called a "hydro burst test."[12] During these burst tests Thiokol discovered a problem in the design position of the O-rings. These tests produced an effect which had not been anticipated. When high-pressure water at 1,500 pounds per square inch was used to simulate a booster-engine firing, it was found that the segments rotated.[13] This phenomenon referred to as "joint rotation"[14] showed that the metal tang, which fits inside the other booster segment's clevis joint, bent outward away from each other. Thiokol had expected that the joint would close when the pressures of launch pushed against it, but instead the joint opened. The joint would only open for a few milliseconds, but this would soon prove to be sufficient to cause grave problems with the solids.[15] This problem was discovered almost four years before any Shuttle was flown and nine years before the *Challenger* disaster. At this point there was plenty of time to fix the problem.

When this problem was reported to Marshall management, the results were downplayed by Thiokol's executives, but not by the engineers.[16] The managers reported that the joint rotation situation was not significant and that it would not pose a safety problem.[17] In fact, Thiokol was so blasé about the rotation that they decided not to conduct any further tests, and Marshall management did not object. If Marshall had demanded more tests then, the design might have been changed and a disaster might have been avoided.

Within Marshall engineering conclaves, doubt was being raised.[18] On October 21, 1977, engineer Leon Ray submitted a report entitled *Solid Rocket Motor Joint Leakage Study*. This report stated that Thiokol's "no change" in the design was "unacceptable."[19] Ray further stated that the "tang can move outboard and cause excessive joint clearance resulting in seal leakage."[20] This was the seal leakage of hot gases that Thiokol stated was not a safety problem. The company would continue to press this "no problem" view well into the launch schedule of the Shuttles. America's space agency could have demanded a redesign or stopped the contract. NASA did neither.

On January 9, 1978, John Miller, then chief of the Solid Rocket Motor Project Office at Marshall, sent a memorandum to his boss at Marshall, Glenn Eudy. In that memo Miller stated that the joint seals needed the use of shims to maintain enough O-ring pressure and that this was "mandatory to prevent hot gas leaks and resulting catastrophic failure."[21] NASA's SRB management was thus made fully aware of the O-ring problem more than three years before the first Shuttle flight, and eight years before *Challenger* was launched on its last mission.[22] Why didn't Glenn Eudy act on this life-threatening design flaw?

Static-motor tests were conducted at Thiokol's test station in July 1978. These tests again showed that the tang/clevis joint movement was greater than the company had expected, that the joint opened and did not properly seal, just as in the hydro burst tests. These tests, as far as joint rotation was concerned, were still not accepted as showing a problem. In fact, Thiokol had decided that the data from their own test was not valid and therefore that they would not spend any money to correct the problem. They had convinced themselves the problem did not exist, even though it was an engineering fact.

When no response was forthcoming from the NASA chain of command on the use of shims, John Miller wrote a second memo on January 19, 1979. In this memo he stated, "We find the Thiokol position regarding design adequacy of the clevis joint to be completely unacceptable."[23] Miller was concerned about many of the design problems, but one area sticks out above all others. He stated that when the joint rotated it caused the O-rings to "become completely disengaged from its sealing surface on the tang."[24] This disengagement was a fatal flaw.

A copy of this memorandum was sent to Marshall Solid Rocket Booster manager George Hardy, but according to testimony no copies were ever sent to Thiokol. Was NASA pushing itself to keep the Space Shuttle on their ever-expanding timeline despite known dangers? By this time, NASA had passed its originally planned launch date, with Congress and the press both asking questions about launch dates and readiness. The space agency was not about to announce a redesign this late in the program. NASA had no money for a redesign and was trying to do far too much with far too little, and in that respect Congress owns a very real part of the blame for the Shuttle's destruction. At the very least, NASA should have been redesigning as fast as possible at the same time they were launching Shuttles with great care.

On March 23, 1979, the phase-one certificate report was released. In it were reports of leak-check failures and problems with the joint-case assembly.[25] There were many more problems than NASA expected on the project. But the certificate made no mention of any potential O-ring failures due to joint rotation. O-ring blow-by was simply not a problem that NASA or Thiokol was willing to look at, even though the data was now becoming overwhelming.[26]

In 1980, NASA established a Space Shuttle Verification/Certification Committee. The committee's job was to review all aspects of the Shuttle program before a launch. In one section of the committee's report it stated that "redundancy of the O-rings was a verification concern," but the report concluded that "the primary purpose of the second O-ring is to test the primary and that redundancy is not a requirement."[27] It would appear from that report that NASA did not even know its own design requirements. When confronted with this information in testimony before the president's commission, George Hardy stated, "I don't know where they would have gotten that infor-

SRB CRITICAL ITEMS LIST

Sheet 1 of 2

Subsystem: SOLID ROCKET BOOSTER	Criticality Category: 1 Reaction Time: Immediate 10 Sec.
Code: 10-01-01	Page: A-6A
Item Name: Case, P/N (See Retention Rationale) (Joint Assys, Factory P/N 1U50147 Field: -1U50747-	Revision:
No. Required: 1 (11 segments, 3 Field joints, 7 plant joints)	Date: December 17, 1982
FMEA Page No. A-4 of MSFC-RPT-724	Analyst: Garber
Critical Phases: Boost	Approved:

Failure Mode & Causes: Leakage at case assembly joints due to redundant O-ring seal failures or primary seal and leak check port O-ring failure.

NOTE: Leakage of the primary O-ring seal is classified as a single failure point due to possibility of loss of sealing at the secondary O-ring because of joint rotation after motor pressurization.

Failure Effect Summary: Actual Loss - Loss of mission, vehicle, and crew due to metal erosion, burnthrough, and probable case burst resulting in fire and deflagration.

RATIONALE FOR RETENTION

Case, P/N 1U50129, 1U50131, 1U50130, 1U50185, 1U50147, 1U50715, 1U50716, 1U50717

A. DESIGN

The SRM case joint design is common in the lightweight and regular weight cases having identical dimensions. The joint concept is basically the same as the single O-ring joint successfully employed on the Titan III solid rocket motor. The SRM joint uses centering clips which are installed in the gap between the tang O.D. and the outside clevis leg to compensate for the loss of concentricity due to gathering and to reduce the total clevis gap which has been provided for ease of assembly. On the Shuttle SRM, the secondary O-ring was designed to provide redundancy and to permit a leak check, ensuring proper installation of the O-rings. Full redundancy exists at the moment of initial pressurization. However, test data shows that a phenomenon called joint rotation occurs as the pressure rises, opening up the O-ring extrusion gap and permitting the energized O-ring to protrude into the gap. This condition has been shown by test to be well within that required for safe primary O-ring sealing. This gap may, however, in some cases, increase sufficiently to cause the unenergized secondary O-ring to lose compression, raising question as to its ability to energize and seal if called upon to do so by primary seal failure. Since, under this latter condition only the single O-ring is sealing, a rationale for retention is provided for the simplex mode where only one O-ring is acting.

The surface finish requirement for the O-ring grooves is 63 and the finish of the O-ring contacting portion of the tang, which slices across the O-ring during joint assembly, is 32. The joint design provides an OD for the O-ring installation, which facilitates retention during joint assembly. The tang has a large shallow angle chamfer on the tip to prevent the cutting of the O-ring at assembly. The design drawing specifies application of O-ring lubricant prior to the installation. The factory assembled joints have NBR rubber material vulcanized across the internal joint faying surfaces as a part of the case internal insulation subsystem.

A small MS port leading to the annular cavity between the redundant seals permits a leak check of the seals immediately after joining segments. The MS plug, installed after leak test, has a retaining groove and compression face for its O-ring seal. A means to test the seal of the installed MS plug has not been established.

The O-rings for the case joints are mold formed and ground to close tolerance and the O-rings for the test port are mold formed to net dimensions. Both O-rings are made for high temperature, low compression set fluorocarbon elastomer. The design permits five scarf joints for the case joint seal rings. The O-ring joint strength must equal or exceed 40% of the parent material strength.

B. TESTING

To date, eight static firings and five flights have resulted in 180 (54 field and 126 factory) joints tested with no evidence of leakage. The Titan III program using a similar joint concept has tested a total of 1076 joints successfully.

---

mation because that was the design requirement for the joint."[28] Even with a full review of the system, NASA fell short of its requirements to test and confirm the Shuttle hardware. No one in upper management, it would seem, was watching the O-ring store or understood the product.

In May 1980, the Verification/Certificate Committee recommended that full-scale tests be conducted to "verify the field joint integrity." It was further suggested that the tests be conducted at "propellant temperatures ranging from 40 [to] 90 degrees Fahrenheit."[29] It would be reported to

SRB CRITICAL ITEMS LIST

Sheet 2 of 2

Subsystem: SOLID ROCKET BOOSTER   **Criticality Category:** 1   **Reaction Time:** Immediate to Sec.

Item Code: 10-01-01   **Page:** A-68

Item Name: *Case, P/N (See Retention Rationale) (Joint Assys. Factory P/N 1U50347 Field: 1U50717)   **Revision:**

RATIONALE FOR RETENTION (CONT'D)

A laboratory test program demonstrated the ability of the O-ring to operate successfully when extruded into gaps well over those encountered in this O-ring application. Uniform gaps of 1/8-inch and over (TWR-13486) successfully withstood pressures of 1600 psi. The Hydroburst Program (TWR-11664) and the Structural Test Program (STA-1) for the standard weight case (TWR-12051) and the Lightweight Case Joint Certification Test (TWR-12829) all have shown that the O-ring can withstand a minimum of four pressurizations before damage to the ring can permit any leakage.

Further demonstration of the capability of joint sealing is found in the hydro-proof testing of new and refurbished case segments. Over 540 joints have been exposed to liquid pressurizations at levels exceeding motor MEOP with no leakage experienced past the primary O-ring. The only occasions where leakage was experienced was during refurbishment of STS-1 where two stiffener segments were severely damaged during cavity collapse at water impact.

A more detailed description of SRM joint testing history is contained in TWR-13520, Revision A.

C. INSPECTION

The tang -A- diameter and clevis -C- diameter are measured and recorded. The depth, width and surface finish of the O-rings grooves are verified. The surface finish of the tang is also verified. Characteristics are inspected on each O-ring to assure conformance to the standards to include:

- Surface conditions
- Mold flashing
- Scarf joint mismatch or separation
- Cross section
- Circumference
- Durometer

Each assembled joint seal is tested per STW7-2747 via pressurizing the annular cavity between seals to 50 5 psi and monitoring for 10 minutes. A pressure decay of 1 psig or greater is not acceptable. Following seal verification by QC, the leak test port plug is installed with QC verifying installation and torquing.

D. FAILURE HISTORY

No failures have been experienced in the static firing of three qualification motors, five development motors and ten flight motors.

Opposite and above: SRB Critical Items List, December 17, 1982 (document from Challenger report, pages 241/42 (two pages).

NASA that Thiokol had tested their motor designs to as low as 21 degrees, but this was not correct. In fact, these tests were never conducted. This information came in testimony before Congress and was reported by Rep. James Sensenbrenner (R–WI) of the House Science and Technology Committee. In its final report, dated September 1980,[30] the committee stated that "tests already conducted satisfied the intent of the commission recommendations." Once again NASA reversed itself, leading itself and the country down the wrong path. No further tests were conducted to insure the integrity of the O-ring seal. Seven months later NASA would launch its first Space Shuttle using a faulty, but certified, solid rocket motor system.

On 15 September 1980, the Propulsion Committee of the Verification/Certification Group officially certified the Thiokol-built solid rockets as being satisfactory for manned flight.[31] It was hardly a glowing endorsement of critical flight hardware, but NASA was on the ropes and, due to outside pressure, needed to get the Shuttle into launch mode as soon as possible. NASA would on November 24, 1980, list the SRBs as being "Criticality 1R," which meant the space agency believed that the secondary O-ring would seal under any flight condition.[32] This was contrary to three years of testing and data submitted by NASA engineers and booster managers who knew better. NASA had even written a critical-items list that same year which stated that "redundancy of the secondary field joint seal cannot be verified" and that "it is not known if the secondary O-ring would successfully reseal if the primary O-ring should fail." The list raised the possibility of "total element failure of which could cause loss of life or vehicle."[33] NASA was flying with faulty hardware it did not fully understand, and was not trying to fix.

Countdown to Disaster

On November 12, 1981, NASA launched *Columbia* on STS-2, its second Shuttle test flight. After the mission, cut from 5 days to 2 days 6 hours due to a faulty fuel cell, a postflight inspection showed the first erosion of the primary O-ring in flight. It was the worst ever found to date on a field joint.[34] This anomaly was not reported during the flight-readiness review for the next flight, STS-3.[35] Nor was this problem reported to Marshall Space Flight Center, or given a tracking number along with other flight anomalies.[36] A problem with a life-threatening possibility should have at least been given some type of tracking, with a solution or at least a temporary fix worked on with as much speed as possible. When the toilet failed to work properly on the Shuttle it was given a tracking number.

Press coverage was mixed. From the *Los Angeles Times* of November 15, 1981: "It could be that this second flight will prove to be the only aberration in a long string of successful space missions." From *The Wall Street Journal* of November 16, 1981: "*Columbia* survived, but the operator's confidence wavered." It would seem that the press felt that perfection in the test program was to be expected soon, but they were unaware of — or failed to report — the critical problem with the SRBs, and debris hits seemed to be of little interest. "36 tiles had been lost and 19 others had been damaged."[37]

During May 1982, Thiokol conducted high-pressure O-ring tests which again showed that when the solid rocket boosters ignited, the secondary O-ring would not properly function. It was after these tests that Marshall finally concluded that the secondary O-rings were not viable.[38] On December 17, 1982, Marshall changed the O-ring system area from Criticality 1R to Criticality 1.[39] Its failure report concluded: "Failure Effect Summary: Actual loss — loss of mission, vehicle and crew due to metal erosion, burn through, and probable case burst resulting in fire and deflagration." However, the report also included a "rationale for retention": "Full redundancy exists at the moment of initial pressurization. However, test data shows that a phenomenon called joint rotation occurs as the pressure rises, opening up the O-ring extrusion gap and permitting the energized ring to protrude into the gap. This condition has been shown by test to be well within that required for safe primary O-ring sealing. This gap may, however, in some cases, increase sufficiently to cause the unenergized secondary O-ring to lose compression, raising question as to its ability to energize and seal if called upon to do so by primary seal failure." Both NASA and Morton Thiokol used this document as justification to continue flying with only one functional O-ring. Clearly they ignored the test results, ignored the change from Criticality 1R to 1, and ignored their own tests.

It would appear that some engineers and managers, such as Deputy Director of Science and Engineering Directorate George Hardy and Lawrence Mulloy, Solid Rocket Booster Project supervisor at Kennedy, felt that the secondary seal would still be redundant despite the tests. The presidential commission on *Challenger* reported that these men felt failure would occur only in "situations of worst case tolerances." It can be argued that a launch at perhaps as low as 8 degrees constitutes worst case tolerances, at 24 degrees below freezing and 43 degrees colder than any previous launch temperature. These O-rings were rubber not steel, and frozen solid rubber at that!

In testimony before the *Challenger* Commission,[40] given on May 2, 1986, Glynn Lunney, Manager of the Shuttle program at the time the criticality change was initiated, as well as when the waiver was signed, allowing the flights to continue, answered questions about the change.

> GLYNN LUNNEY: Well, the approval of the waiver in March of '83, ... I was involved in that. I was operating on the assumption that there really would be redundancy most of the time except when the secondary O-ring had a set of dimensional tolerances add up, and in that extreme case there would not be a secondary seal. So I was dealing with what I thought was a case where there were two seals unless the dimensional tolerances were such that there might only be one seal in certain cases.
>
> CHAIRMAN ROGERS: Now, to me, if you will excuse the expression, that sounds almost contradictory, what you just said. What you first said was you came to the conclusion that you could only rely on the primary seal and therefore you removed the R.
>
> GLYNN LUNNEY: Yes, sir.
>
> CHAIRMAN ROGERS: And now you're saying, if I understand it, that experience showed that there was redundancy after all.
>
> GLYNN LUNNEY: No, I don't know of any experience showing that. What I'm saying is that the removal of the R is an indicator that under all circumstances we did not have redundancy. There were a certain number of cases under which we would not have redundancy, of the secondary O-ring. Recognizing that, even though there were a lot of cases where we expected we would have redundancy we changed the criticality designation.
>
> CHAIRMAN ROGERS: It was saying to everybody else you can't necessarily rely on the primary seal, and if the primary seal fails, as you've said here, there may be loss of vehicle, mission and crew.
>
> GLYNN LUNNEY: I would adjust that to only say you cannot rely on the secondary O-ring but we would expect the primary O-ring to always be there.

Associate Administrator for Space Flight L. Michael Weeks stated, "The Solid Rocket Booster was probably one of the least worrisome things we had in the program." After Mr. Weeks was replaced by former astronaut Richard Truly, the solids became a worrisome thing.

After only five flights, which included the failure and erosion of the right solid rocket booster on the STS-2 flight, Morton Thiokol ran the following ad copy in newspapers around the country on November 12, 1982: "Experience pays off. Following a completely successful development and qualification program for the National Aeronautics and Space Administration, Morton Thiokol twin rocket motors have performed flawlessly on all Space Shuttle launches. And, with our 25 years of large solid rocket experience, we're batting 100% on quality, performance, cost effectiveness and on-schedule delivery for every mission. We are proud of our contribution to America's Space Program."

Representative James Scheuer (D–NY) asked Admiral Richard Truly, who had just been appointed as associated administrator for space flight after the *Challenger* accident, about his STS-2 flight. He asked if Truly had been told of the O-ring erosion on his flight and others by NASA management.

> RICHARD TRULY: No, sir.
>
> JAMES SCHEUER: How do you feel in retrospect? And not being told about this known ... life threatening condition?
>
> RICHARD TRULY: I think our investigation in this report clearly shows that the failure of communications of that problem or the proper recognition of it was a major failure and led to the cause of the accident. I'm more concerned not that I as an astronaut did not know but that the total program ... didn't work [on] that problem and others. In answer to your question we should have known. It should have been worked [on] as a major technical problem.

It is easy to see that if the astronauts had been made aware of the critical situation, they would have demanded a fix to the flawed SRB O-ring design. To prevent an interruption in an ever-increasing flight program, NASA signed waiver after waiver which allowed flights to continue despite major flaws in critical hardware. The Shuttle would be flying without a backup system, without any reliability, a clear violation of flight rules in place since the earliest days of the one-man Mercury program. The design was clearly faulty, but when it was recognized as such within NASA, the agency continued to fly. It became an acceptable risk, which gained a life of its own. Management at both Morton Thiokol and NASA continued to accept the risks as not desirable, but acceptable. Neither called for a redesign, nor any added safety features such as launching above 53 degrees or altering the flight schedule to fix the problem.

In September 1983, Lockheed Space Operations was awarded the contract for the NASA/USAF Space Shuttle processing. As part of that contract Pan American Airlines was given responsibility for establishing what was called "Airline Turnaround Procedures." Not only was NASA once again cutting costs by putting out a single processing contract to a company which had not built the system, but also, it would seem that after only eight flights NASA was attempting to get into the spaceliner business. This was test flight, pure and simple, not a spaceliner!

More Flight Damage — STS 41-B

STS 41-B was launched on February 3, 1984, and the flight caused great concern when it was found that the left solid rocket booster field joint, as well as its right nozzle, received a great deal of damage during the flight, damage to primary O-rings.[41] The temperature at launch was 57 degrees and this was the worst damage yet to an O-ring in flight. Because of this finding, Thiokol engineers filed a problem report. Thiokol management went to Marshall and presented a series of charts to the Solid Rocket Booster Engineering Office. During that presentation Thiokol attempted to assure NASA that their tests showed that if the primary O-ring failed, the secondary O-ring would seal.[42] Clearly something was amiss in that assessment since both NASA and Thiokol knew the second O-ring would not perform as required or stated. Both teams knew that tests showed it would not seal. Further, in that presentation Thiokol did not address the erosion problem at all, and once again NASA did not press the issue. The space agency had lost yet another chance to solve the problem.

It should also be noted that NASA was continuing to accept the fact that each mission would bring damage to the tiles as well. Many would be replaced, with many more being repaired for the next flight. It would only get worse for this so-called reusable system.

The press gave little or no coverage to problems on the solid rockets or tiles, if indeed they knew there were any. For the most part photographic launches and smooth landings were what members of the press had in mind. From the Associated Press newswire of February 3, 1984: "Perfect Shuttle liftoff no. 10. Then the crew gets down to work." There was no word of the erosion which occurred on both solid rockets that helped lift the Shuttle into space. By February 12, 1984, the Knight-Ridder News Service was reporting, "The Space Shuttle *Challenger* astronauts came full circle for the first time this Saturday, making a dream of a touchdown." As far as the press was concerned, this was just one more successful flight for NASA and not much more needed to be said. It was by this time getting very hard to find much coverage on the TV news about flights other than the few nightly sound bites. Live coverage of Shuttle launches was becoming a thing of the past, except on CNN.

In a memo dated February 29, 1984, sent to George Hardy, Keith Coates described his thoughts on the SRB problem this way, "Thiokol definition of their plans on resolution of the problem is very weak."[43] NASA was now clearly flying with problem hardware. Marshall Center then issued a report on the incident, which stated in part, "Remedial action — none required; problem occurred

Astronaut Bruce McCandless flies the Manned Maneuvering Unit (MMU) for the first time (NASA).

during the flight." This is a very strange comment. Was there somewhere other than during a flight when erosion was to be expected? And why did "problem occurred during the flight" not send up a red flag at NASA?

A further entry in the same report, dated March 8, 1984, stated, "Possibility exists for some O-ring erosion on future flights. (A clear violation of the design and flight requirements.) Analysis indicates max. erosion possible is .090 inches. Therefore, this is not a constraint to future launches." During a meeting, NASA and Thiokol management were briefed on the fact that the success of the Titan solid rockets did not mean that the Shuttle SRBs would also perform as well. It was noted that on the Titan system there was a segment-to-segment case insulation design, which prevented the hot gases from reaching the O-rings. The Shuttle had no such system, and the SRBs were three times larger than the Titans, so even more problems could be expected during their development and use.

When the next flight review was conducted on March 30, 1984, NASA had accepted the erosion problem as a "technical issue"[44] and recommended that STS 41-C be launched. NASA had accepted that some erosion due to hot-gas impingement could happen on that flight and others in the future. A mind-set was moving forward that a little erosion was not a big problem. Despite this seeming acceptance of problems, some members of NASA continued to see the O-ring problems for what they really were. After reviewing the results of the 41-C flight, Lawrence Mulloy, manager of the Solid Rocket Booster office, forwarded a letter on April 5, 1984, to Thiokol[45] asking for a formal review of the booster field joint, as well as the nozzle-joint sealing procedures. The letter directed Thiokol to "identify the cause of the erosion, determine whether it was acceptable, define necessary changes, and reevaluate the putty then in use."[46] But this was somewhat akin to asking the fox to guard the chicken coop. This should have been an outside independent study, overseen by Marshall engineers, not by Thiokol.

On May 4, Thiokol responded with a formal program called "Protection of SRM Primary Motor Seals."[47] In it the company outlined a plan to solve the problem and isolate the cause of the O-ring erosion. But this was only a proposal and in the meantime the Shuttle would continue to fly with the faulty design. NASA also found a problem with the putty they were using. The putty was found to be sensitive to humidity and temperature. And it was a problem with the putty which Thiokol felt added to the erosion on the STS-2 flight. The problem was clearly known.

United Space Boosters, Inc., a subsidiary of United Technologies Corporation, won the contract to assemble and refurbish Shuttle solid rocket boosters in August 1984. Once again NASA chose a company that had not designed or built a major Shuttle system to essentially become the operators of that system. Once again a low bid had undercut the safety of the Shuttle system.

Another area of concern was the amount of pressure applied during the leak checks. According to the *Challenger* report, "The nozzle joint leak check was changed from 50 psi to 100 psi before STS-9 launched in November 1983. After this change, the incidence of O-ring anomalies in the nozzle joint increased from 12 percent to 56 percent of all Shuttle flights. The nozzle pressure was increased to 200 psi for mission 51-D flown in April 1985, and all subsequent missions. Following the implementation of the 200 psi check on the nozzle, 88 percent of all flights experienced erosion or blow-by." Both NASA and Thiokol knew that the increased pressure with these tests helped cause the O-ring problem, but Thiokol continued to recommend to NASA that the increased pressure be continued. NASA did not override that decision. Thiokol felt it would prove the integrity of the seal, but it actually helped cause the seals to malfunction. No matter what the thinking was, it was clear that the higher the pressure test, the greater the O-ring problem became.[48]

During this period, NASA was fighting to keep the Shuttle program on track and doing battle with the Defense Department not only over control of the Shuttle, but over just how much would be taken out of the Shuttle payload bay and launched on expendable boosters. NASA had counted on the Defense Department to pay for many of the Shuttle flights, which would help defray the overall cost of the program to NASA. As reported by the Associated Press on September 5, 1984, "The Air Force is looking at alternatives to the Space Shuttle and a new study says any of the three rocket boosters under consideration could do the job for about $10 million. The National Aeronautics and Space Administration is concerned about the military's interest in alternative boosters because it expects the Air Force to book one-third of the Shuttle flights." In February of that year, Air Force Undersecretary Edward C. Aldridge, Jr., in testimony before the House Science and Technology Committee stated, "We need a hedge against technical and operational problems that the Shuttle may develop." Aldridge was looking for a secondary capability to launch at least one-fifth of the large spy satellites then being prepared by the Defense Department.

Both NASA and the Defense Department would soon take a very close look at the next Shuttle mission, 51-C. It would be the first dedicated top-secret military Shuttle mission, with a lot riding on the flight. It would also have more O-ring problems than any previous flight.

I can tell you that as long as I'm in this job, politics will continue to take a back seat to readiness.
— James C. Fletcher, NASA Administrator, January 11, 1988

O-Ring Anomalies and Failures

Flight	Date	Problem Area	Booster	Damage	Joint Temp.
STS-2	11/12/81	Aft field/prime	Right	Erosion	70 degrees
STS-6*	4/4/83	Nozzle/prime	Right	Heat / O-ring	67 degrees
		Nozzle/prime	Left	Heat / O-ring	
41-B	2/3/84	Nozzle/prime	Right	Erosion	57 degrees
		Field/prime	Left	Erosion	
41-C	4/6/84	Nozzle/prime	Right	Erosion	63 degrees
		Aft field/prime	Left	Heat / O-ring	
		Igniter/prime	Right	Blow-by	
41-D	8/30/84	Field/prime	Right	Erosion	70 degrees
		Nozzle/prime	Left	Erosion	
		Igniter/prime	Right	Blow-by	
DM-6**	10/25/84	Inner gasket/primary		Erode / blow-by	52 degrees
51-C	1/24/85	Field/prime	Right	Erode / blow-by	53 degrees
		Field/secondary	Right	Heat and soot	
		Nozzle/prime	Right	Blow-by	
		Field/prime	Left	Erode / blow-by	
		Nozzle/prime	Left	Blow-by	
51-D	4/12/85	Nozzle/prime	Right	Erosion	67 degrees
		Igniter/prime	Right	Blow-by	
		Nozzle/prime	Left	Erosion	
		Igniter/prime	Left	Blow-by	
51-B	4/29/85	Nozzle/prime	Right	Erosion	75 degrees
		Nozzle/prime	Left	Erode / blow-by	
		Nozzle/secondary	Left	Erosion	
DM-7**	5/9/85	Nozzle/primary		Erosion	61 degrees
51-G	6/17/85	Nozzle/prime	Right	Erode / blow-by	70 degrees
		Nozzle/prime	Left	Erode / blow-by	
		Igniter/prime	Left	Blow-by	
51-F	7/29/85	Nozzle/prime	Right	Heat / O-ring	81 degrees

*First use of a new, lighter-weight external tank and lighter-weight solid rocket.
**DM = Demonstration Motor Test, not a flight motor

Flight	Date	Problem Area	Booster	Damage	Joint Temp.
51-I	8/27/85	Nozzle/prime	Left	Erode 2 places	76 degrees
61-A	10/30/85	Nozzle/prime	Right	Erosion	75 degrees
		Field/prime	Left	Blow-by	
		Field/prime	Left	Blow-by	
61-B	11/26/85	Nozzle/prime	Right	Erosion	76 degrees
		Nozzle/prime	Left	Erode / blow-by	
61-C	1/12/86	Nozzle/prime	Right	Erosion	58 degrees
		Field/prime	Left	Erosion	
		Nozzle/prime	Left	Blow-by	
51-L	1/28/86	Field/prime and secondary	Right	Burn thru (orbiter destroyed)	28 degrees

Section II

Countdown to Disaster

Chapter Three

51-C: Too Cold for the Military

> *Extreme weather conditions in the area are projected to cause icing on the External Tank that could be hazardous.... The launch is scrubbed for the day.*
> —NASA spokesman, January 22, 1985

One year, nearly to the day, before *Challenger* was launched in bitter cold weather, *Discovery* sat on launch pad 39A, waiting to place into orbit a top-secret air force satellite. It was Tuesday evening, January 22, 1985, and temperatures had fallen below freezing for the third night in a row. With ice on the pad and a cold front in the Florida area, NASA scrubbed the launch for the next day. The lowest temperature predicted for that night was 30 degrees, which was warmer than the night before *Challenger*'s launch. Yet this time, NASA managers did not launch. The question must then be asked: What was so different or important about this flight that was unimportant about *Challenger*?

Space for the Air Force

The Pentagon, and the air force specifically, was never gung ho on the Space Shuttle. Air force officials thought that focusing entirely on the Shuttle would leave the air force unable to place its national defense payloads into orbit if a problem with the Shuttle developed. They were proven right. This was, after all, a new, experimental technology and problems were bound to come up.

As early as 1956, the U.S. Air Force was developing ideas to place men in space, even before any unmanned satellite had orbited. With the X-15 hypersonic rocket plane as a research tool, NASA and the air force were learning much about how winged craft would perform at great speeds and very high altitudes. Built by North American Aviation, the X-15 first flew on September 17, 1959, with test pilot Scott Crossfield at the controls. By the time it had finished its last test flight on October 24, 1968, it had completed 199 flights (the 200th was canceled due to bad weather), and flown higher and faster than any winged craft had ever flown. Not until the Shuttle had flown would those records be broken. It should be remembered that even after 199 flights the X-15 was still a research aircraft and never declared operational. The Space Shuttle would be called operational after only four flights. For the most complicated space/aircraft ever built to be called operational so soon was not only deceptive on NASA's part, but did not make sense.

The air force had developed the X-20 Dyna-Soar (Dynamic Soaring)[1] in the early sixties as a way of pushing past the X-15 program and placing military men into orbit, as part of a purely military space program. The X-20 would have been a 35-foot-long, shuttle-type space/aircraft, with a 20-foot wingspan, and weighing only 4.5 tons. Looking like a lifting body, the X-20 would have been launched by a Titan III rocket to a velocity of 25,000 feet per second, then glided to an altitude of at least 60 miles. After its reconnaissance mission was completed, the X-20 would fly to an unpowered (deadstick) landing on a conventional runway.[2] It has been suggested that the air force actually got around to launching a wooden and plastic, full-scale mock-up in 1966, into a suborbital flight.[3]

Seven pilots were selected for the program[4]—Major James W. Wood, Major Henry C. Gordon, Major Russell L. Rogers, Captain Albert H. Crews, Captain William J. Knight and NASA pilots Milton O. Thompson and Neil Armstrong. Armstrong would later command the first lunar landing mission of the Apollo program on July 20, 1969, and place "one small step for man" on its surface. Thompson, Knight and Armstrong had also flown the X-15. In fiscal year 1962, the Department of Defense budgeted $100 million for the project. When President Kennedy came into office, the X-20 was again funded, this time for $921 million, which would keep the program alive through test flights up to 1963.

Another look at the project by the Defense Department and Congress showed that the NASA-developed Gemini program would more easily perform the reconnaissance duties—expected to be assigned to the X-20 (Space Fighter). Funding was cut to $130 million in 1963, but the first test flight was still scheduled for 1966. By then the air force was finding it difficult to develop any real mission for the X-20, and when NASA and the Defense Department began looking at orbiting a laboratory, it was all but over for the X-20.[5] Even when the air force attempted to sell the X-20 as a supply craft to the Manned Orbiting Laboratory, there was simply no longer any support for the Dynamic Soaring idea. With the development of the manned Mercury Program, funding for the X-20 ($400 million total projected[6]) was finally canceled in December 1963 by Defense Secretary Robert McNamara. The air force had lost its "Space Fighter," but not its desire to enter the Space Age.[7]

On the same day that the X-20 was cancelled, the Defense Department announced its new project as a follow-up program to the two-man Gemini space flights being planned in the mid–1960s. It was the air force's Manned Orbiting Laboratory (MOL) which the air force hoped would finally put them into space.[8] This program would use modified Gemini hardware (Gemini B or Blue Gemini)[9] already being developed to place air force personnel into orbit for stays on board a military laboratory.[10] The air force would spend $200 million on MOL development, creating a laboratory 10 feet in diameter and 18 feet long.[11] The lab was supposed to be launched from pad SLC-6 at Vandenberg Air Force Base, on a Titan 3C booster. Air force astronauts would then dock and perform experiments in low earth orbit. But the program was scrapped before it could ever get to a launch pad. In fact, the launch pad was never close to being completed, and no X-20s were ever built.[12] The Congress and the White House, with help from NASA, had used development of the MOL to shoot down the X-20; now unmanned satellites would do the same to the Manned Orbiting Laboratory in August of 1969.[13] With the loss of the MOL project, contract-holding McDonnell Douglas would need a new project and the company was later awarded the contract for Skylab. The $3 billion cost and lack of any real missions requiring military crews finally ended the MOL project. With MOL gone, so went Blue Gemini.[14]

More and more the air force would be using unmanned reconnaissance satellites. The Lockheed-built KH-9 "Big Bird" reconnaissance satellite could do the job cheaper and much longer than any MOL crew, which doomed the lab. The KH-9 would orbit around 150 miles up, and not only record film but deliver real-time television transmission to the intelligence community. Later models would be operational for up to two years and could reportedly read license plates on parked cars from orbit. These new spy birds could even track, in real time, single targeted individuals from space. The lab was simply no longer needed and well behind the technology curve. Even before the MOL project was canceled, however, there was a human price to pay when MOL test pilot Robert H. Lawrence was killed in an air crash at Edwards Air Force Base on December 8, 1967. Flight test has never been an easy or safe occupation, and every Shuttle flight is flight test.

However, all was not lost. Seven of the MOL-trained astronauts were selected and transferred to NASA. Selected August 14, 1969, were Major Karol J. Bobko, Lt. Cmdr. Robert Crippen, Major Charles Gordon Fullerton, Major Henry W. Hartsfield, Major Robert F. Overmyer, Major Donald H. Peterson and Lt. Cmdr. Richard H. Truly. They would constitute the seventh group of individuals selected by NASA for the astronaut corps. All would eventually fly into orbit on board a Shuttle, and Richard Truly would later become NASA chief. In time the idea of a small manned lab, now called Spacelab, would come up again and was built to be flown in the cargo bay of the Shuttle. Overmyer would command Spacelab 3 in April 1985; Fullerton would command Spacelab 2 in July 1985; and Hartsfield would command Spacelab D1 in October 1985. With MOL pilots flying the Shuttle with Spacelab in the cargo bay, perhaps the Manned Orbiting Laboratory did arrive in orbit after all.

Air Force Problems

With the Shuttle program now funded, the air force believed they would finally be able to put military crews into space. They were not happy with the costs however, and knew that of the four Shuttles built only two could be counted on to carry the full 65,000 pound payloads the Pentagon needed to place in orbit. The Shuttle system was the only show in town.[15] With only two of the Shuttles available, there was little room for error or accident. Any problems with the program would keep the Pentagon and its top-secret payloads planted firmly on the ground. Since 1973, the air force had been designing its classified payloads, such as the new KH-13 spy satellite, for deployment from the Shuttle's cargo bay. They did not yet have any other boosters capable of placing these new birds into orbit. The air force was also designing their satellites with a refueling capability to be accomplished by space-walking military astronauts, which no unmanned system could accomplish. But despite this planning, the Department of Defense would not tell NASA management about the types of missions they were planning to put on the Shuttle. With this lack of real data, planning for military payloads became a guessing game. NASA was forced to hire the Aerospace Corporation, out of El Segundo, California, to put together fictional missions in order to aid NASA with military-payload mission planning. The military had no confidence in NASA's ability to keep their secrets, and little confidence in the ability of the Space Shuttle to deliver their payloads into space. For their part, NASA did not trust the air force when it came to launch commitments or financial support. It would not be a happy or financially successful arranged marriage. Before the *Challenger* disaster, however, they would still work as a team.

Air force officials had more on their minds than security and a lack of confidence in the Shuttle system. On April 4, 1983, NASA launched *Challenger* on her first flight, designated as STS-6. During that mission the four-man crew deployed NASA's first Tracking and Data Relay Satellite (TDRS-A) from the cargo bay. The TDRS had used as its booster a new air force-contracted Inertial Upper Stage (IUS), which was to be used to push TDRS into a geostationary orbit some 22,500 miles into space. However, the IUS failed to boost the satellite into the planned orbit and left the TDRS tumbling out of control for three hours. It was later discovered that the second stage had failed and that an IUS thruster had been stuck closed. The booster, when it finally separated, may have even bumped the TDRS. This was the same booster rocket the air force was depending on to place its classified payloads into similar orbits as the TDRS. Air force managers were concerned — photographing the IUS as it failed some 20,000 miles in space, using a classified camera on top of a mountain in Socorro, New Mexico.

The fact that NASA was able to use smaller thrusters on the TDRS itself to eventually raise its orbit did not impress the air force. Bottom line for the air force was that their IUS had failed.

This lack of confidence in the Boeing-developed Inertial Upper Stage (IUS), forced the Pentagon, in June 1983, to postpone an upcoming military Shuttle mission. Later, the air force would state that its primary payload was also not ready, so the mission slated for November 1983 was simply canceled. But was the payload not ready or was the air force simply unwilling to risk its payload on a questionable IUS?

The pressure for a successful military Shuttle mission was mounting at NASA as 1984 came around, but problems would still plague the missions. In February of that year, the Pentagon announced the cancellation of its next attempt at a military Shuttle mission, which had been scheduled for July 14, 1984. At the same time it was reported that the Department of Defense had reversed itself when it came to using the Shuttle exclusively to launch its payloads. The Pentagon would seek funds for the development of a heavy-lift expendable booster and would use it to orbit at least one-fifth of its new generation of large classified satellites then being built and tested. The military was losing what little confidence it had, not only in the IUS, but in the Shuttle system itself. NASA managers were now on the ropes. Costs were going to rise dramatically and the space agency was not developing ways to work around the problem other than launching more Shuttles. Launch pressure would become a critical safety issue.

Finally, in September 1984, the air force and NASA felt they were ready, and mission 51-C was announced. It would be the first all-military mission with a top-secret payload to be deployed from the Shuttle's cargo bay and boosted into geostationary orbit using the troubled IUS. Target launch date was December 8, 1984. This mission was critical to the air force because this satellite would replace an aging one then on station, but in many ways it would be much more critical to NASA. The agency had to gain the confidence of the Defense Department and show that the Space Shuttle could actually deliver what it had promised, a reliable, human-operated satellite delivery system.

The air force was also creating a complete Shuttle system of its own, and one that would be completely independent of NASA. In Colorado Springs, Colorado, the Air Force Consolidated Space Operations Center was being constructed, at a cost of $1.2 billion. From this new 640-acre facility, the air force would direct its Shuttle missions much like NASA now does from Mission Control in Houston. Slated to be up and running by December 1985, this air force Shuttle system on the surface did not seem to make much sense when one considered the general air force lack of confidence in the Shuttle. But air force officials knew that only one manned program would get any kind of funding for years to come and that was the Shuttle. The air force would have to make the best of what they felt was a bad program. If they wanted to ride, they would have to ride the Shuttle.

At the same time, a new Shuttle launch complex was being constructed at Vandenberg Air Force Base on the coast of California.[16] SLC-6 (Space Launch Complex Six, nicknamed "Slick Six") was originally begun in the 1960s as the launch pad for the Manned Orbiting Laboratory program. But when that program was cancelled in 1969, the site was abandoned "in place," as the government likes to say. In 1973, Air Force Secretary John J. McLucus ordered that funds be used for the conversion of the site from its unfinished MOL configuration to a fully operational Shuttle launch site. In truth the site was way too small to support a Shuttle launch.

Construction and redesign began on the site for a planned Shuttle launch in October 1978. At the same time the air force also decided to fund the development of the IUS (Inertial Upper Stage) for use by the Shuttle system to boost its payloads into geostationary orbits. According to NASA Administrator Fletcher, the air force had also committed itself to purchasing two additional Space Shuttles, but without a signature on a contract, NASA did not really have a reliable commitment.

By February 1985, construction had almost been completed and the Vandenberg site was being fit tested using the flight test Shuttle, *Enterprise*. The *Enterprise* was mated to an empty, flight-ready external tank and two inert solid rockets. All went well, and it appeared that the air force was well on its way to having a fully operational Shuttle site of its own, at a cost now reaching $2.8 billion. In April 1985, before a Congressional hearing, Rep. George E. Brown, Jr. (D–CA) voiced his concern over the major role the military was playing in the Shuttle program. "Space is becoming primarily a military program. As a policy issue, that concerns me." He was not the only one voicing concern. Many in NASA viewed the military as a threat to the agency's ability to perform in a civilian mode. Military secrecy was beginning to bother many of the old NASA hands. Secrecy was not what they had signed on for. They were explorers and engineers, not military men.

As part of that polar-orbiting air force program, NASA won approval from the Chilean government to extend the existing runway at Easter Island in the South Pacific. That runway would be extended to 11,000 feet to allow an emergency landing if needed by a Shuttle launched from Vandenberg. It would never be used or needed.

By mid–1985, the air force had announced a target date of March 20, 1986, for its first launch from Vandenberg, using the Shuttle *Discovery*. It was planned that in the future, *Discovery* would be operated only out of SLC-6 and that it would become a fully dedicated air force Shuttle. *Discovery* was at that time scheduled to be permanently assigned to Vandenberg by September 1986. The space agency was not happy, but it was one of the prices NASA would have to pay for air force support (read "funding"). But before the air force could open its west-coast facility, it needed a successful operational Shuttle flight from Kennedy. That flight would be designated 51-C, then being developed for a Shuttle mission from pad 39A, a scheduled January 22, 1985, launch of *Discovery*.

Mission — Top Secret

Navy Captain Thomas K. Mattingly would serve as commander for the crew of 51-C. Mattingly was selected as an astronaut in 1966 and was to have flown on the ill-fated Apollo 13 moon mission. However, an exposure to German measles (which he never contracted) bumped him from the flight. He did fly on Apollo 16 as command module pilot in April 1972, in which he orbited the moon. In June 1982, he commanded STS-4, the fourth and final test flight of *Columbia*.

Air Force Lieutenant Colonel Loren J. Shriver would serve as pilot for this flight. NASA had selected Shriver in 1978. Prior to becoming an astronaut he had attended the Air Force Test Pilot School at Edwards Air Force Base, and subsequently served as a test pilot for the F-15.

Marine Corps Lieutenant Colonel James F. Buchli would serve as one of two NASA mission specialists. He was selected in 1978 to become an astronaut and had served as CAPCOM (capsule communicator) for the STS-2 Shuttle flight. Before becoming an astronaut he had served in several marine fighter/attack squadrons.

Air Force Major Ellison S. Onizuka was selected as the second mission specialist. He was also picked to fly for NASA in 1978. Previously he had attended the Air Force Test Pilot school at Edwards, and was then selected to stay on as an instructor. After the completion of this top-secret air force mission he would be assigned to 51-L and *Challenger*. It should be noted that all NASA astronauts have high government clearances and it would not be necessary to use only military astronauts on a military payload assignment. But the air force had waited a long time for this one and air force pride would only allow military men on this first top-secret manned space mission for the United States Air Force.

Rounding out the crew would be Air Force Major Gary E. Payton. He would become the Defense Department's first purely military astronaut to fly in space as a payload specialist. All other military personnel who had flown for NASA had been attached to the civilian agency and were not flying only for the military. (When Colonel Buzz Aldrin landed and walked on the moon with Neil Armstrong, he too was in the air force, but was attached to NASA in a civilian capacity, although his pay rate was on a military scale, give or take.) Payton had been a pilot since 1972,

The crew for the first top-secret Shuttle flight 51-C were Payload Specialist Gary E. Payton, Pilot Loren J. Shriver, Commander Thomas K. Mattingly II, Mission Specialist James F. Buchli, and Mission Specialist Ellison S. Onizuka (NASA).

and had been selected for the USAF Manned Space Flight Engineer Program, along with twenty-four others in 1980. The names of these Defense Department astronauts would be kept secret until they were assigned to a flight. It should be remembered that the other four crew members were at the very least well known, if not famous, and they were all military men as well. There was no real need for this secrecy either.

Originally, the Shuttle *Challenger* had been designated to carry the Defense Department payload on December 8, but problems with tiles and processing shifted the responsibility to *Discovery*. *Challenger*'s thermal tiles had been damaged during its previous mission in October 1984, and the air force did not want to wait for repairs. Replacement of its older on-orbit spy satellite, according to the air force, simply could not wait any longer. Now NASA would also have to put up with air force pressure to launch, but that was the business they were in.

On November 17, 1984, *Discovery* was towed to the Shuttle processing hangar at Kennedy to begin work on its classified mission. All the public would be told was that around January 21 or 22, *Discovery* would be launched on a top-secret mission. Originally, the Pentagon did not want to even announce that much. They would have preferred no notice at all. No launch date and no launch time were best for them. In private, Pentagon planners knew that all the secrecy about launch times would do no more than make it a little more difficult to track the new spy satellite, and only for a while. It was simply not real security and both the air force and NASA knew it.

While *Discovery*'s flight was being readied, NASA was planning a July 2, 1985, launch date for mission 51-L. This flight was to have carried two satellites, the TDRS-B and the Aussat, the latter a telecommunications satellite for the government of Australia. But the satellites were not ready, so the flight was postponed.

In its November 5 issue of 1984, *Aviation Week and Space Technology* reported, much to the displeasure of the air force, that the first secret military Shuttle mission would carry a satellite into geosynchronous orbit. It further reported that "no liftoff time would be disclosed to the public, to prevent the Soviet Union from computing the best way to monitor the new satellite." The magazine knew a great deal more, but decided not to publish any further details, much to their credit, until that information was released and had become general knowledge. It would not take long for that to happen. The press was all over that mission as far as the payload was concerned, but would report nothing about O-ring and debris-impact problems.

Meanwhile, NASA would have to wait for its release of the 51-C mission patch because it had Major Payton's name on it and the air force was not yet prepared to release the name of its astronaut. It simply did not seem to make any common sense, and some at NASA were privately saying as much. By December it was obvious that the Defense Department was getting a bit on edge about its first Shuttle mission. On December 18, Brigadier General Richard Abel announced that any news stories that even speculated on the top-secret payload of 51-C would be, "investigated by the Defense Department, as a breach of National Security." Did the air force really believe they could pull that off in a nation with a history of a free press?

The next day the point was lost to the winds, when both the Associated Press and the *Washington Post* (of Watergate fame) published articles which described the payload as a new generation Signal Intelligence Satellite (SIGINT). The satellite, it was reported, was designed to intercept radio, telephone and satellite transmissions, and would be placed to do its work over the Soviet Union. Air force officials were not pleased, to say the least, but it would not do a whole lot of good to make a big noise about it now that the secret was out. It made no sense to confirm the story by going after the press.

The satellite's parking orbit would be 3000 miles south of the Soviet Union at 22,300 miles above the equator. From that point it would be able to monitor communications from Asia, Africa, and Europe continually. This satellite was an improved version of older air force spy satellites that had been placed into orbit with expendable boosters. Four or five older ELINT satellites were in orbit, but fast running out of fuel and useful life. The Defense community badly needed this launch to go off successfully.

Despite the secrecy, the air force did inform the public that the launch would occur between 1:15 P.M. and 4:15 P.M. on the 23rd of January. Additionally, it was announced that only a small handful of special guests would be allowed inside Kennedy Space Center for the launch. There would be no general public viewing of the launch. The air force seemed to be tossing its power around, which was exactly what NASA workers feared and openly criticized. Also, contrary to expectation, there would be no press kits issued and no interviews with the astronauts before or after the mission. The only media relations the crew could be involved in were to smile and wave to the press corps. There would also be no air-to-ground voice commentary carried by NASA for the press after the second OMS (Orbiting Maneuvering System) burn, and the Shuttle was successfully in orbit. Only a regular statement on the status of the orbiter would be reported every eight hours.

After the successful flight of Mission 51-A in November, which had launched two satellites and recovered two others, all eyes were focused on how long it would take to prepare *Discovery* for its long-awaited top-secret military mission. *Discovery* had been in the Vehicle Assembly Building since December 21, to be mated to its solid rockets and external tank, when it was rolled out to pad 39A on January 5, 1985. The processing had picked up speed after the holidays and the Kennedy Space Center was in full gear for its upcoming flights. It was expected to be a very full and busy launch year.

On January 7, the crew conducted a full dress rehearsal and practice countdown, as the press was beginning to refer to the mission as "*Battlestar Discovery.*" Much was being made of its military payload, designed to keep an ear on America's potential enemies. The term "Star Wars," created by the Soviet propaganda press, was also bandied about, even though this flight would have nothing to do with SDI (Strategic Defense Initiative). It was a nice piece of propaganda for the Soviets. It gave the Soviets a chance to say "I told you so" about the Shuttle being a tool of the military, though it clearly was not, at least not as long as NASA and the American people had anything to say about it.

All seemed to be going well, not only by press reports, but behind the scenes as well, until a cold front passed through the area on January 21. This cold snap slowed work on processing the *Discovery* and conditions were beginning to look bad for the scheduled January 23 launch. As reported by *Countdown* magazine in its March 1985 issue, "On Tuesday, January 22, cold weather that moved in Monday [was] causing problems for [the next day's] launch. Freezing temperatures burst water valves and pipes in the launch pad fire extinguishing system during the night." "There was water spraying out there this morning," according to NASA spokesman, Jim Mizell. During the early morning, cold-related problems had set pre-launch schedules in the countdown about three hours behind, but by that evening most of the work was back on schedule. A launch was looking good, but always the weather could change things in a hurry.

As night came, Florida temperatures in the area around the Cape began to fall toward freezing for the third night in a row. Just before midnight, a NASA spokesman announced, "Extreme weather conditions in the area are projected to cause icing on the External Tank that could be hazardous." The launch was scrubbed for one day. The air force was taking no chances with this launch that could be avoided, and the weather was a problem it could avoid. There would be no similar avoidance for *Challenger* a year down the road, when the weather was much worse.

For the first time in U.S. manned-space-flight history, cold weather had postponed a scheduled launch from the Kennedy Launch Center. Low temperatures, which had been in the high teens and low twenties, were expected to rise to acceptable levels during the next two days. By the next scheduled launch day on January 24, it was forecast that temperatures would be in the high twenties overnight and mid fifties by launch time. NASA was concerned that if the launch had occurred in such low temperatures, ice might have been dislodged from the external tank and fallen onto the Shuttle's protective thermal tiles. NASA did not want to risk the damage and scrubbed. There were no stated concerns about the O-rings due to the cold weather. However, NASA did have tests which indicated problems could occur if the O-rings were frozen. These tests were ignored. There was also no concern that thermal insulation debris from the external tank could damage the tiles, even though Shuttles were taking debris hits on every flight.

With this identical problem, and the same and even lower temperatures for *Challenger* a year later, a scrub was not called. It can only be guessed what happened to the ice concerns for *Challenger*. And what about the frozen O-rings?

In a relaxed manner, to brush aside press questions about the delay, Air Force Public Affairs Spokesman Lieutenant Colonel Bob Nicholson would only say, "The only people we'd be frustrating is you." The air force was practicing a relaxed calm for the press, but in reality they were anything but relaxed.

On January 23, the lowest temperatures encountered were around 30 degrees. This was unexpectedly higher than what had been predicted, but humidity was also becoming a factor. And NASA would not be able to launch if cloud cover went over the 50 percent mark at 7000 feet around the Shuttle landing strip. In case of problems, the astronauts would need an unobstructed view of the Return-to-Launch-Site abort runway at Kennedy. Even with these warming temperatures, NASA officials stated that icing problems could have been encountered.

According to *Countdown* magazine, March 1985, "If temperatures at the Cape reached the upper 30s by 6 A.M. on launch day, fueling of the Shuttle would begin." Temperatures for launch afternoon were predicted to be near 60 degrees. All was go for launch. When *Challenger* was fueled for its last mission beginning at 1:25 A.M., the temperatures were near 20 degrees, with a wind chill factor perhaps as low as 10 degrees below zero. What were NASA managers thinking about

Discovery's launch on January 24, 1985 for STS 51-C (NASA).

at the time which allowed the fueling to continue? And when will they tell the public why they really launched? On January 24, 1985, temperatures well above freezing moved through the launch area, and these warmer temperatures brought with them the promise that icing problems were a thing of the past, at least for this mission.

Discovery left the pad at 2:50 P.M. EST, for a mission lasting 73 hours and 33 minutes, with air temperatures in the low 60s. O-ring temperature was 53 degrees.[17] It was perfect weather for

```
                        MESSAGE DISPL
  TO   LARRY WEAR                      TO   SANDY COLEMAN
 _om:  Larry Mulloy
Postmark:  Jan 31,85    7:39 AM
Status:    Certified  Urgent
Subject:  51C O-RING EROSION RE:  51E FRR
-----------------------------------------------------------------
Message:
       FRR DISCUSSION SHOULD RECAP ALL INCIDENTS OF O-RING EROSION, WHETHER
       NOZZLE OR CASE JOINT AND ALL INCIDENTS WHERE THERE IS EVIDENCE OF FLOW
       PAST THE PRIMARY O-RING.  ALSO, THE RATIONALE USED FOR ACCEPTING THE
       CONDITION ON THE NOZZLE O-RING.  ALSO, THE MOST PROBABLE SCENARIO AND
       LIMITING MECHANISM FOR FLOW PAST THE PRIMARY ON THE 51C CASE JOINTS.
       IF MTI DOES NOT HAVE ALL THIS FOR TODAY I WOULD LIKE TO SEE THE LOGIC
       ON A CHART WITH BLANKS TBD.
```

Urgent message by Lawrence Mulloy to Larry Wear, January 31, 1985 (document from *Challenger* report, page 247).

a Shuttle launch. On January 25, the Associated Press reported, "Military Space Shuttle operates flawlessly as satellite launch nears." Or so it seemed. After the Solid Rockets were recovered, the left SRB was found to have had erosion (80 degrees around the joint) on one of the field joint O-rings.[18] The right SRB had erosion on both the primary and secondary O-rings (110 degrees around the joint).[19] When returned for evaluation, the grease was found to be black as coal, indicating that a great deal of burning had occurred.[20] Cold overnight temperatures had done a job even when it was 66 degrees plus at launch time.[21] It was another red-flag time and NASA ignored it.

After the flight the public was treated to the usual stories from the press. From the *New York Times* of January 28: "Silence Attends Shuttles Landing, Ending 3-day Trip. Flight Had No Problems." And from the Associated Press the next day, "The Space Shuttle *Discovery* returned from its secret military mission in excellent condition. There were no significant Orbiter systems problems during the flight." It would seem from reports at the time that two major erosions and four actual blow-bys of hot gases during a cold launch constituted minor problems. At least this was the way it was viewed by many at NASA who by their training and experience should have known better. This flight could well have ended in disaster.

This was the first operational flight where a secondary O-ring had showed any effects due to heat. It was also the coldest O-ring temperature for a launch to date.[22] But most at NASA were unwilling to look at this as a major problem. On February 8, 1985, Morton Thiokol reported to the space agency that "O-ring erosion caused a concern that the gas seal could be lost, but its decision was to accept the risk." Thiokol had concluded that the next Shuttle flights, "could exhibit the same behavior, nonetheless the condition is not desirable but is acceptable." NASA seemed to be willing to fly and risk the lives of the crews knowing a very serious problem existed in a major piece of flight hardware. Dr. Richard Feynman, a member of the *Challenger* commission and a leading scientist, would later write, "The O-rings of the Solid Rocket Boosters were not designed to erode. Erosion was a clue that something was wrong. Erosion was not something from which safety can be inferred."[23]

Each of the next four Shuttle flights showed joint-seal problems and damage to the O-rings.

Flight 51-D launched on April 12, 1985, experienced O-ring erosion and blow-by.[24] 51-B launched on April 29, 1985, also had O-ring erosion and blow-by, as did 51-G launched on June 17, 1985.[25] The 51-F flight launched on July 29, 1985, had a nozzle O-ring blow-by.[26] Problems with the joint seals on the solid rockets continued,[27] as did the ability for NASA to consider them acceptable. This flight-erosion experience had taught NASA and its SRB contractor Morton Thiokol nothing.

Air force spokesman Nicholson stated, "We placed some unusual constraints on NASA personnel and to a person they lived up to the letter and agreement of the instructions. I think it worked very well." Was NASA still "living up to the letter and agreement," and if so, why? What was so important that *Discovery* was not launched in very cold weather, and what was so important that *Challenger* had to launch under identical or worse conditions one year later? Was the military payload considered too valuable to risk, but the *Challenger* and her crew simply not as valuable, or was there other pressure? Perhaps these questions need to be answered by NASA and the air force.

On February 27, 1985, the air force agreed to acquire fully one-third of all Space Shuttle flights, which would help give NASA the funding to continue Shuttle operations. For its part, NASA agreed to end its opposition in Congress to an air force plan to develop and launch ten expendable boosters for its heavy-lift program. The air force was also to get Shuttle launches on demand in the event that an unforeseen emergency occurred and a flight was required at once, within a few weeks, or when the next bird was ready. By this time, the Pentagon had lost most confidence in the Shuttle system, but they continued their support. At least five Defense Department Shuttle missions had been cancelled or postponed, and defense planners were now designing satellites which could be deployed from the Shuttle or launched on a new generation of military expendable booster. The military was now covering all bets. History would show that it was a very good decision.

On March 20, 1985, the Pentagon announced that the Shuttle *Discovery*'s commander, Thomas K. Mattingly, would be promoted to commodore. On May 1, 1985, Major Gary Payton was promoted to lieutenant colonel and Ellison Onizuka was also promoted to lieutenant colonel. These men had done a very good job, but it was becoming too risky to launch military payloads on Space Shuttles, if not civilian ones.

From a barefoot boy running around in the coffee fields, to an astronaut. That's quite a feat.
— U.S. Air Force Major Ellison Shoji Onizuka

Vandenberg Air Force Base, SLC-6

1969	A launch facility, built for the Manned Orbiting Laboratory program (MOL) is mothballed. It consists of a mobile service tower, a concrete launch pad, a single flame duct and a small launch-control center.
1971	An intergovernmental group named the Shuttle Launch and Recovery Board is formed to review possible launch sites for Shuttles.
1974	An air force task force is established to evaluate three possible launch sites at Vandenberg. SLC-6 is chosen as a cheaper alternative to building a brand new launch site.
April 6, 1978	GAO reports that SLC-6 will cost $1.2 billion to develop, and recommends not building the facility.
May 24, 1978	The House Appropriations Committee recommends $109.8 million to begin work at Vandenberg. NASA and the air force report that they have planned 144 launches from SLC-6 through 1986.
Sept. 1979	Construction begins at SLC-6 when existing mobile service tower is moved 150 feet back. Service platforms are added.

Shuttle *Enterprise* conducts fit checks at Vandenberg AFB, SLC-6, March 1985 (United States Air Force).

January 1980 Excavation of the facility is completed and major construction begins.
January 1981 Shuttle Assembly Building (SAB) is added to the facility construction plan.
May 1982 New flame ducts are completed along with the modification of the old duct, to direct Shuttle exhaust gases away from the launch pad.
June 13, 1983 The air force announces that the first Shuttle flight will be delayed until October 15, 1985.
March 1, 1984 The air force announces that they expect to launch up to ten Shuttle missions each year from Vandenberg.
May 15, 1984 Vandenberg Shuttle site is dedicated by Air Force Undersecretary Edward Aldridge, Jr. This dedication takes place as the Defense Department decides whether they will release the names of the military astronauts.
Aug. 15, 1984 SLC-6 is completed and the air force now reports that it will launch no more than four flights per year from the facility.
Oct. 21, 1984 First Shuttle-launch-ready external tank arrives at Vandenberg.
Jan. 9, 1985 The air force announces the delay of its first flight until January 29, 1986. The second launch is scheduled for September 29, 1986.
Feb. 1, 1985 A five-man crew is assigned to the first Shuttle mission from Vandenberg.
Feb. 15, 1985 Space Shuttle *Enterprise* is mated to an external tank and two inert solid rocket boosters as verification and fit checks begin on SLC-6 with a Shuttle on the pad. It is the only time an orbiter will sit on the launch pad.
Aug. 19, 1985 The air force reschedules its first launch for March 20, 1986, due to payload problems.

Sept. 11, 1985 Undersecretary Aldridge and Air Force Major John Brett Watterson are named to the first mission to be launched from Vandenberg.
Oct. 15, 1985 The air force officially opens the new $2.8 billion West Coast Shuttle facility.
Nov. 5, 1985 The air force schedules the third flight for March 18, 1987.
March 1, 1986 The air force announces that its first SLC-6 mission, designated 62A, will be postponed for one year.

During May 1986 before the House Appropriations Subcommittee Hearings, the NASA administrator was questioned on plans for the Vandenberg Shuttle launch facility.

REP. EDWARD BOLAND: "What about the rumor that the Air Force is going to mothball Vandenberg that cost like $2.8 billion. What does that do to the fourth orbiter requirement?"

JAMES FLETCHER: "Mr. Chairman, as you know that's a rumor and we have a pretty firm commitment from the Defense Department that that will not happen."

June 17, 1986 A congressional report is released on SLC-6. The report states that the facility has a "poor design of the main engine hydrogen gas escape ducts." Trapped gases held in the ducts after an engine shuts down could cause an explosion. It was also reported that launch-pad support structures were far too close to the launch pad. Sound bouncing off the structures could destroy a Shuttle as it was lifting off.
July 1, 1986 The air force recommends to President Reagan that he place SLC-6 into "Operational Caretaker Status" until 1991-92.
July 28, 1986 The air force places SLC-6 into Caretaker Status until 1991.
Dec. 5, 1986 A GAO report supports keeping SLC-6 and states that the launch site is needed for future Shuttle operations. It also reports that $3.2 billion has been spent through fiscal year 1986 for the site.
Jan. 15, 1987 SLC-6 is placed on Minimum Caretaker Status, which means a reduction in funding from operational status.
April 28, 1988 The air force announces that SLC-6 will be mothballed, which means that the facility will not be maintained, only checked occasionally for erosion and rusting. The cost is supposed to be held at from $7 to $8 million yearly.
June 14, 1988 The air force closes its Space Control Center in Colorado and disbands its secret team of 32 military astronauts.
Sept. 28, 1990 The air force announces that Launch Complex 6 will again be converted to a launch facility for unmanned Titan IV boosters.
March 16, 1995 Western Commercial Space Center signs a 25-year lease with the air force to use 100 acres of land and the payload processing center at SLC-6.

After more than 30 years of development and construction Launch Complex 6 at Vandenberg Air Force Base, NASA and the air force had yet to launch a single rocket from the facility.

Chapter Four

1985 Was Not a Good Year

All of us in the business recognize that it's still high technology and there's some risk associated with it.... Sooner or later you were going to have a failure.
— Deke Slayton, Astronaut, January 28, 1986

Between 1980 and 1985, the Department of Energy produced three separate studies commissioned to show the failure rate which could be expected to occur using the Shuttle system. These studies, conducted on behalf of NASA and the U.S. Air Force, focused on major failures which would end with the loss of the orbiter and its crew. Two of the three studies focused on the possibility of a solid rocket booster failure. In the "Sierra Study" it was predicted that a failure would occur in one of every seventy missions. A second study from the Sandia National Laboratories concluded that a loss would occur in one of 210 flights. Finally, a third report concluded that the program could expect to lose a vehicle and crew in one of 1000 flights using the Shuttle system.[1] This study predicted that with a program expected to fly from 400 to 500 missions there was a 50/50 chance of losing a crew on a Shuttle mission. As it turned out, these studies were overly optimistic, but not nearly as optimistic as the publicly announced failure rate published by the National Aeronautics and Space Administration. With these reports and others in NASA's hands, the space agency announced a failure rate of 1 in 100,000. Clearly, NASA management misled the American people.

Dr. Richard P. Feynman, again in a personal observation attached to the *Challenger* report wrote, "Since 1 part in 100,000 would imply that one could put a Shuttle up each day for 300 years expecting to lose only one, we could properly ask, what is the cause of management's fantastic faith in the machinery?" What indeed?

During this same period, as NASA public relations staff were busy assuring everyone of the safety of the Shuttle, the space agency was continuing to cut back on the number of quality control personnel assigned to the program. From 1975 through 1985, the number of quality assurance people dropped from 1,689 to only 505, which amounts to a 70 percent decrease. According to NASA's own records, many of the missions were being slowed down in their preparations for launch due to a shortage of quality assurance personnel at Kennedy as well as at the prime contractors. And yet, as fewer and fewer people kept an eye on safety, more missions were flying. At some point along the way it could easily be expected that safety was going to fall by the wayside and a disaster would occur. The question remains: has this problem been resolved? If not, when will the program and the American people be expected to again pay the price of that continued failure?

Keep the Flights Going at All Costs

The year 1985 was projected to be very busy. Payloads were beginning to back up as pressure to fly increased. NASA had a can-do image to project, and with budget cutters all around, success—keeping the launch schedule on time—was paramount. Image seemed to become more important than flight-crew safety, even as information mounted about the dangers inherent in using the flawed Solid Rockets, dangers now acknowledged within NASA.

For the American people, 1985 would be a year of wondrous images from space, as NASA astronauts performed one spectacular manned flight after another. Space-walking white-suited astronauts were seen performing a large number of difficult and challenging tasks with skill and grace, which included the rescue and repair of several satellites. It was becoming the norm to see such outer space spectaculars. The public now saw space operations as an everyday occurrence, almost akin to airline operations. Space travel was seen as mature, successful, regular and above all safe, but nothing could have been further from the truth.

Milton Socobab from the General Accounting Office (GAO) in congressional testimony given after *Challenger*'s accident stated, "Three of seven mandatory government inspections were not made on the motors critical O-rings.... Such defects," he said, "do have the potential for causing a mission failure." When Edward Dorsey, General Manager of Morton Thiokol, was asked about these inspections, his answers seemed deceptive. "A mistake by the human being that prepared the paperwork," he said. Was this an attempt to deny the facts that the inspections were never made, or simply to shift the blame?

51-D *Discovery*

NASA's much-delayed sixteenth Shuttle mission, designated STS 51-D, was launched from Cape Kennedy on April 12, 1985, twenty-four years to the day after the now defunct Soviet Union launched the first man into space. Along with the other six crew members, this flight launched Senator Jake Garn on his earth orbital boondoggle. Garn was chair of the Senate committee with oversight responsibilities for NASA, and for years had pressured the space agency with requests for a flight. On opening a NASA budget hearing on May 12, 1981, Garn stated, "My first question is, when do I get to go on the Space Shuttle?" He was also able to use his considerable political power to push Hughes Aircraft Company payload specialist Greg Jarvis off of this flight. Because of this political bump, Jarvis would eventually end up on the doomed *Challenger* flight.

The *Discovery* flight had been delayed by what NASA called technical problems, and had originally been designated mission 51-E, but that flight had been cancelled and combined into the 51-D mission. It was at that point that Jarvis was bumped from the mission. Payloads from 51-E and 51-D were then loaded onto *Discovery*, leaving *Challenger* on the ground. During the processing of *Discovery*, a 2,500-pound metal bucket which was used to move workers around near the Shuttle smashed into the payload-bay doors of *Discovery*, breaking the leg of a Lockheed Space Operations technician named Gary Sutherland. Sutherland would recover, but this damage would cause an already delayed flight to be pushed back yet another month.

In the meantime, NASA was fighting another Shuttle problem which the press had actually picked up on. On every Shuttle landing where the brakes had been applied, one or more of the units had suffered some type of damage. It was not a problem so far, but was a potential one if an emergency landing needed to be performed on a secondary and probably much shorter runway. NASA would only note that so far the brakes had never failed to work and had always safely stopped the 75-ton space plane. Perhaps the space agency had just been lucky.

On launch day, NASA would show just how much they wanted to get *Discovery* into space when it lifted off with only 55 seconds remaining in the launch window. The flight had been held

> **MORTON THIOKOL. INC** COMPANY PRIVATE
> **Wasatch Division**
>
> Interoffice Memo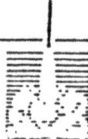
>
> 31 July 1985
> 2870:FY86:073
>
> TO: R. K. Lund
> Vice President, Engineering
>
> CC: B. C. Brinton, A. J. McDonald, L. H. Sayer, J. R. Kapp
>
> FROM: R. M. Boisjoly
> Applied Mechanics - Ext. 3525
>
> SUBJECT: SRM O-Ring Erosion/Potential Failure Criticality
>
> This letter is written to insure that management is fully aware of the seriousness of the current O-Ring erosion problem in the SRM joints from an engineering standpoint.
>
> The mistakenly accepted position on the joint problem was to fly without fear of failure and to run a series of design evaluations which would ultimately lead to a solution or at least a significant reduction of the erosion problem. This position is now drastically changed as a result of the SRM 16A nozzle joint erosion which eroded a secondary O-Ring with the primary O-Ring never sealing.
>
> If the same scenario should occur in a field joint (and it could), then it is a jump ball as to the success or failure of the joint because the secondary O-Ring cannot respond to the clevis opening rate and may not be capable of pressurization. The result would be a catastrophe of the highest order — loss of human life.
>
> An unofficial team (a memo defining the team and its purpose was never published) with leader was formed on 19 July 1985 and was tasked with solving the problem for both the short and long term. This unofficial team is essentially nonexistent at this time. In my opinion, the team must be officially given the responsibility and the authority to execute the work that needs to be done on a non-interference basis (full time assignment until completed).

up due to cloud cover and rainstorms in the area, as well as by a civilian cargo ship which had strayed into the solid-rocket recovery area. It would not be good press to have an expended SRB drop on top of a cargo ship in the Atlantic Ocean. As far as flight operations were concerned, one of two satellites deployed failed to fire its booster engine, which would have placed it into a proper orbit. Despite attempts by the crew to swat the ignition lever by using an improvised fly swatter, the satellite could not be saved. It would take another rescue mission later in the year during a very dangerous space walk to effect a proper fix. The rocket engine could well have ignited while the astronauts were working on it. NASA's luck seemed to be holding.

R. K. Lund									31 July 1985

It is my honest and very real fear that if we do not take immediate action to dedicate a team to solve the problem with the field joint having the number one priority, then we stand in jeopardy of losing a flight along with all the launch pad facilities.

Roger M. Boisjoly
R. M. Boisjoly

Concurred by:

Jack R. Kapp
J. R. Kapp, Manger
Applied Mechanics

COMPANY PRIVATE

Opposite and above: Roger Boisjoly's letter after STS 51-B on the critical O-ring erosion problem, July 31, 1985 (document from *Challenger* report, pages 249/50 [two pages]).

The landing occurred at Kennedy Space Center on April 19, and as the Shuttle rolled to a stop, its right main-landing-gear tire blew out. In addition, it was later found that 123 tiles were damaged, of which sixty would need to be replaced. Shuttles were continuing to take debris hits well above expectations. This was a clear indication of a major problem with the tiles, as every flight continued to show damage to the thermal system. Even more alarming at the time was the condition of the solid rockets, which had by then been recovered. Although launch had occurred at a mild 67 degrees, both left and right SRBs had erosion and blow-by of hot gases on the O-rings.[2] Clearly, there was another continuing problem showing up on nearly every flight, but flight waivers would continue to be issued by Lawrence Mulloy.

Due to the O-ring problem, now suspected (but not proven) to be exacerbated due to cold weather, Thiokol conducted temperature tests in early 1985. Thiokol's internal report stated that "at 100 degrees F, the O-ring would maintain constant contact and no hot gases would escape. At 75 degrees F the O-ring lost contact for 2.4 seconds. At 50 degrees F, the O-rings did not reestablish contact in ten minutes at which time the test was terminated."[3] The effect of cool weather on O-ring performance was by then a tested and scientific fact. It was then one year before *Challenger*'s final flight. Yet, in testimony before Congress and the *Challenger* Commission, NASA managers and Thiokol officials would both state that temperature was not a viable indicator of potential trouble, and the flights continued.

51-B Challenger

The second Spacelab mission being carried into space by *Challenger* lifted off on April 29, 1985, only ten days after *Discovery* landed on the Shuttle runway a few miles away. The mission appeared to be a complete success as it landed after a seven-day mission, but despite the orbital successes more O-ring problems had been found.[4]

On June 25, 1985, the solid rockets used for the 51-B flight were transported back to Morton Thiokol for unstacking and inspection. What Thiokol engineers discovered truly shocked members of the solid rocket team. Thiokol found that the primary O-ring had not sealed at all and had been eroded .171 inches.[5] The secondary O-ring did seal, but it had also been eroded.[6] As for the putty, it had been fully bypassed.[7] The putty fill had been totally useless and had been instantly evaporated by the hot gases. When describing the problem on the 51-B solids, Lawrence Mulloy stated, "This erosion of secondary O-ring was a new and significant event ... that we certainly did not understand." "It was evident that the primary ring never sealed at all, and we saw erosion all the way around the O-ring."[8] With the O-rings as a secondary seal only, Thiokol knew they were dealing with a major design flaw, enforced by the fact that the Shuttle had been launched at 75 degrees Fahrenheit, a very mild temperature indeed.

After the problem was reported to Rocket Manager Mulloy and the Marshall Space Center Problem Assessment Committee, a launch constraint was placed on the Shuttle system. But still Mulloy did not see it as a "threat to flight safety."[9] This piece of paper did not stop the progress of the flights, and, more critically, Thiokol officials stated that they were unaware of the constraint, which was issued by NASA in July 1985. Even if Thiokol was not informed of the launch constraint (and that is highly unlikely), it was NASA who gave the final go and no-go calls on all launches. It was clearly a NASA decision. The constraint would stay on for all flights from 51-F until just before *Challenger*'s 51-L flight, when an individual not yet identified removed it. After 51-L, a constraint would no longer be required. At this point NASA managers knew that the primary O-ring could fail because it had failed before in flight; they knew that cold affected the seals; they knew the secondary O-ring would not seal, and yet they did nothing but pass on some paperwork. No one stepped forward who could stop NASA from launching Shuttles.

51-G *Discovery*

NASA flew its eighteenth Shuttle mission on June 17, 1985, with *Discovery* and a crew of seven. It was 70 degrees Fahrenheit at the time of launch. NASA was now deep into its image-building program, sending individuals into space that represented companies and foreign governments. On this flight Prince Sultan Salman Al-Saud of Saudi Arabia would fly into space to watch Arabsat 1-B deploy from the cargo bay. It has been rumored for years that one of these guest astronauts became agitated while in orbit and had even tried to open the hatch into space. NASA has yet to comment on these rumors, but some have pointed to the prince[10] who is said to have been restrained during a portion of the flight.[11] It will probably be many years before we find out for sure who "lost it" in space.[12] NASA's guest-astronaut program would continue until the *Challenger* flight, when it was quietly put to an end, at least for a while. The *Los Angeles Times* reported: "The Space Shuttle *Discovery* blasted off perfectly Monday with a crew from three nations."

After the successful deployment of three communications satellites, Shuttle *Discovery* completed its seven-day mission, landing at Edwards Air Force Base in California due to the continuing brake problems. Again, as was becoming the standard, the solids suffered major problems. The right SRB had experienced two erosions and a separate blow-by of the O-rings and the left SRB experienced erosion and two blow-bys. Damage to the solids was now expected no matter what temperatures were present during launch. However, it was also noted that the colder the temperature, the more damage was to be expected. The brake problem was being worked, yet the SRB problem was not even being examined. As was also standard, the tiles took several solid hits from debris. Bt the Associated Press story of June 24 positively glowed: "Jesse W. Moore, Shuttle project director told reporters, 'This was one of the most successful missions of the Shuttle program.'"

Engineers at Thiokol's rocket team were becoming increasingly alarmed at the O-ring problems. Roger Boisjoly, a member of the structures section who was very familiar with the problems, warned management of the critical situation. He predicted, in a memo dated July 22, 1985, that "NASA might give the motor contract to a competitor or there might be a flight failure if Thiokol did not come up with a timely solution." Boisjoly wrote a second memo on July 31, which he entitled, "O-ring Erosion/Potential Failure Criticality." The memo was sent to Robert K. Lund, Thiokol's vice president of Engineering. In it Boisjoly flatly stated, "The mistakenly accepted position on the joint problem was to fly without fear of failure.... This problem is now changed as a result of the (51-B) nozzle joint erosion, which eroded a secondary O-ring with the primary O-ring never sealing. If the same scenario should occur in a field joint (and it could) ... the result would be a catastrophe of the highest order — loss of human life." This engineer was not about to mince words. He knew what the problems were and he wanted them solved. He would go as high as he could in Thiokol's management to attempt to solve the problem, and he would fail.

Boisjoly wanted Thiokol to set up a team as soon as possible to solve the O-ring problem. He knew, and he stated forcefully, that it was just a matter of time before one of the Shuttles would be destroyed in flight and the cause would be the SRBs produced by his company. He ended the memo, "It is my honest and very real fear that if we do not take immediate action to dedicate a team to solve the problem, with the field joint having the number one priority, then we stand in jeopardy of losing a flight along with all the launch pad facilities." This document proved that Thiokol upper management was informed of the situation and failed to correct the problem. It was the "smoking gun" that investigators would eventually find.

51-F *Challenger*

The next Shuttle mission — the nineteenth in the program and the fiftieth U.S. human space flight — was scheduled for launch on July 12, but the attempt was stopped three seconds before lift-off, after the main engines were ignited, due to a failed coolant valve in the number two engine. At that point all three engines were shut down.[13] The spacecraft was then sprayed with water on the pad to prevent any fuels from igniting and causing a pad fire. Forty minutes later the crew was removed from the orbiter, a bit shaken but unharmed. The *New York Times* reported

MORTON THIOKOL, INC.
Wasatch Division
Interoffice Memo

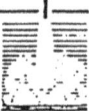

2871:FY86:141
22 August 1985

TO: S.R. Stein,
            Project Engineer

CC: J.R. Kapp, K.M. Sperry, B.G. Russell, R.V. Ebeling, H.H. McIntosh,
            R.M. Boisjoly, M. Salita D.M. Ketner

FROM: A.R. Thompson, Supervisor
            Structures Design

SUBJECT: SRM Flight Seal Recommendation

The O-ring seal problem has lately become acute. Solutions, both long and short term are being sought, in the mean time flights are continuing. It is my recommendation that a near term solution be incorporated for flights following STS-27 which is currently scheduled for 24 August 1985. The near term solution uses the maximum possible shim thickness and a .292 +.005/-.003 inch dia O-ring. The results of these two changes are shown in Table 1. A great deal of effort will be required to incorporate these changes. However, as shown in the Table the O-ring squeeze is nearly doubled for the example (STS-27A). A best effort should be made to include a max shim kit and the .292 dia O-ring as soon as is practical. Much of the initial blow-by during O-ring sealing is controlled by O-ring squeeze. Also more sacrificial O-ring material is available to protect the sealed portion of the O-ring. The added cross-sectional area of the .292 dia O-ring will help the resilience response by added pressure from the groove side wall.

Several long term solutions look good; but, several years are required to incorporate some of them. The simple short term measures should be taken to reduce flight risks.

A.R. Thompson

ART/jh

Interoffice memo reports on O-ring seal problems, August 22, 1985 (document from *Challenger* report, page 251).

that it was "a setback for the program of sophisticated and temperamental Shuttles, which have to be launched regularly if they are ever to be a commercial success." What the *Times* did not report was that the system was designed to stop a launch if any problems were detected, and that is exactly what it did. This was a success. The computers and the people who had programmed them had done their jobs well.

Four. 1985 Was Not a Good Year

Richard Cook, working as a budget analyst at NASA, wrote a memorandum on July 23, 1985, critical of the SRB performance and the many problems with O-ring seals. He wrote, "There is little question, however, that flight safety has been and is still being compromised by potential failure of the seals, and it is acknowledged that failure during launch would certainly be catastrophic." This memo failed to have any effect on the hard-pressed launch schedule.

After the first launch scrub, NASA felt they were ready again on July 29, but once again the main engines proved problematic. At 5 minutes and 45 seconds into the flight, engine number one shut down when an onboard sensor indicated that the engine was overheating.[14] As the crew increased power using the two remaining engines, a second sensor also indicated an overheating problem, but the crew, on orders from Mission Control to prevent a second engine from shutting down as well, shut the sensor off. NASA had more trust in the engines than in the sensors. Why? Did NASA know they were flying with faulty sensors? If the second engine had failed, the crew would have had to attempt a dangerous emergency landing, probably at Edwards Air Force Base, but only if they had enough energy on the Shuttle to accomplish the long glide to the emergency strip. As it turned out, the Shuttle was able to limp into a lower orbit and the crew conducted a maneuver called an Abort-to-Orbit (ATO). This was accomplished by increasing thrust from the two remaining main engines and programming a fuel dump using the on-orbit OMS (Orbital Maneuver System).

When the Shuttle *Challenger* returned from this mission, it was found that all three main engine sensors were defective. A reasonable individual might think that these unreliable sensors would never be used again, but just the reverse occurred. On succeeding flights the same unreliable sensors were still being used. Fewer and fewer eyes were watching Shuttle safety and reliability, and it was starting to show. Even though the Shuttle had launched in a warm 81 degrees, at least one of the SRBs showed that heat had touched an O-ring, but this was the best the SRBs had performed all year. It was, unless you accept the original design concept, which made it clear that not even heat was expected to touch the O-rings at any temperature during a launch.

On August 9, 1995, Brian Russell from Thiokol reported to Marshall Space Center that after motor ignition (rotation) transient there was "no reason to suspect that the primary O-ring would fail." But if burn-through had occurred prior to or during the rotation of the field joints (transient) it would make no difference if the problem would not occur after, because the damage would have already occurred. Therefore, Russell's report to Marshall did not seem to amount to much when it came to flight-safety issues. On August 19, Marshall and Thiokol program managers traveled to NASA Headquarters in Washington, D.C., to brief officials on the erosion problems with the O-rings on the field joints.[15] As reported in the *Challenger* report, "The briefing paper concluded that the O-ring seal was a critical matter, but it was safe to fly. The briefing was detailed, identifying all prior instances of field joint, nozzle joint and igniter O-ring erosion." (Thirty-six incidents on flight and test motors were known.) Despite a full briefing, NASA administrators and directors were determined to keep on flying, with no let-up in the schedule, already being pushed well beyond its capabilities. The reading was that it was safe to fly.[16] NASA was being asked to perform at levels which precluded safety considerations and they did so with full knowledge of SRB design flaws. All of NASA top management knew there was a major problem with the SRBs, but not one astronaut was informed of the situation.

The next day, Thiokol's vice president of engineering, Robert Lund, reported in a memo that "the result of a leak at any of the joints would be catastrophic." He announced that he had set up a task force to "investigate the Solid Rocket Motor Case and nozzle joints, both materials and configuration," and recommended both short-term and long-term solutions. What he did not do was recommend suspension of flights until the problem was corrected. That suspension would have cost Thiokol money and NASA bad publicity. As it turned out only a fatal disaster could suspend the flights.

> **MORTON THIOKOL, INC.**
> Wasatch Division
>
> Interoffice Memo
>
> 1 October 1985
> E150/RVE-86-47
>
> TO: A. J. McDonald, Director
> Solid Rocket Motor Project
>
> FROM: Manager, SRM Ignition System, Final Assembly, Special Projects and Ground Test
>
> CC: B. McDougall, B. Russell, J. McCluskey, D. Cooper, J. Kilminster, B. Brinton, T. O'Grady, B. MacBeth, J. Sutton, J. Elwell, I. Adams, F. Call, J. Lamere, P. Ross, D. Fullmer, E. Bailey, D. Smith, L. Bailey, B. Kuchek, Q. Eskelsen, P. Petty, J. McCall
>
> SUBJECT: Weekly Activity Report
> 1 October 1985
>
> EXECUTIVE SUMMARY
>
> HELP! The seal task force is constantly being delayed by every possible means. People are quoting policy and systems without work-around. MSFC is correct in stating that we do not know how to run a development program.
>
> GROUND TEST
>
> 1. The two (2) GTM center segments were received at T-24 last week. Optical measurements are being taken. Significant work has to be done to clean up the joints. It should be noted that when necessary SICBM takes priority.
>
> 2. The DM-6 test report less composite section was released last week.
>
> ELECTRICAL
>
> As a result of the latest engineering analysis of the V-1 case it appears that high stress risers to the case are created by the phenolic DFI housings and fairings. As it presently stands, these will probably have to be modified or removed and if removed will have to be replaced. This could have an impact on the launch schedule.

51-I *Discovery*

Discovery flew NASA's twentieth Space Shuttle mission on August 27, 1985. Launching from pad 39A, the temperature recorded was a balmy 76 degrees. Two earlier launch attempts were scrubbed by hardware and weather problems.[17] On August 24, poor weather at the launch site as well as overseas kept the Shuttle on the pad, and a faulty computer which had to be replaced using one from *Challenger* delayed the August 25 attempt.[18] Once on orbit, the five-man crew deployed three commercial communications satellites and repaired Syncom IV-3 which had been left in a useless orbit from a previous mission. After the solids were returned to Thiokol for examination, the left booster was found to have eroded in two places. This was not an unexpected event.

> A. J. McDonald, Director
> 1 October 1985
> E150/RVE-86-47
> Page 2
>
> FINAL ASSEMBLY
>
> One SRM 25 and two SRM 26 segments along with two SRM 24 exit cones were completed during this period. Only three segments are presently in work. Availability of igniter components, nozzles and systems tunnel tooling are the present constraining factors in the final assembly area.
>
> IGNITION SYSTEM
>
> 1. Engineering is currently rewriting igniter gask-o-seal coating requirements to allow minor flaws and scratches. Bare metal areas will be coated with a thin film of HD-2 grease. Approval is expected within the week.
>
> 2. Safe and Arm Device component deliveries is beginning to cause concern. There are five S&A's at KSC on the shelf. Procurement, Program Office representatives visited Consolidated Controls to discuss accelerating scheduled deliveries. CCC has promised 10 A&M's and 30 B-B's no later than 31 October 1985.
>
> O-RINGS AND PUTTY
>
> 1. The short stack finally went together after repeated attempts, but one of the o-rings was cut. Efforts to separate the joint were stopped because some do not think they will work. Engineering is designing tools to separate the pieces. The prints should be released tomorrow.
>
> 2. The inert segments are at T-24 and are undergoing inspection.
>
> 3. The hot flow test rig is in design, which is proving to be difficult. Engineering is planning release of these prints Wednesday or Thursday.
>
> 4. Various potential filler materials are on order such as carbon, graphite, quartz, and silica fiber braids; and different putties. They will all be tried in hot flow tests and full scale assembly tests.
>
> 5. The allegiance to the o-ring investigation task force is very limited to a group of engineers numbering 8-10. Our assigned people in manufacturing and quality have the desire, but are encumbered with other significant work. Others in manufacturing, quality, procurement who are not involved directly, but whose help we need, are generating plenty of resistance. We are creating more instructional paper than engineering data. We wish we could get action by verbal request but such is not the case. This is a red flag.
>
> R. V. Ebeling

Interoffice memo begins with "HELP!," October 1, 1985 (document from *Challenger* report, page 252).

During testimony before the *Challenger* Commission, after all of the O-ring problems had occurred in 1985, Lawrence Mulloy was asked why the Shuttle system was allowed to continue operations, despite a clear launch constraint which he had a primary hand in establishing.

CHAIRMAN ROGERS: Do you have ultimate responsibility for waiving the launch constraints?
LAWRENCE MULLOY: Yes, sir. I have ultimate responsibility for launch readiness of the Solid Booster.
CHAIRMAN ROGERS: So there was a launch constraint, and you waived it?
LAWRENCE MULLOY: Yes, sir.

It was clear that the individual who helped place the launch constraint on the Shuttle system for a very good safety reason was the same individual who waived that same constraint for no

ACTIVITY REPORT

The team generally has been experiencing trouble from the business as usual attitude from supporting organizations. Part of this is due to lack of understanding of how important this task team activity is and the rest is due to pure operating procedure inertia which prevents timely results to a specific request.

The team met with Joe Kilminster on 10/3/85 to discuss this problem. He wanted specific examples which he was given and he simply concluded that it was every team members responsibility to flag problems that occurred to organizational supervision and work to remove the road block by getting the required support to solve the problem. The problem was further explained to require almost full time nursing of each task to insure it is taken to completion by a support group. Joe simply agreed and said we should then nurse every task we have.

He plain doesn't understand that there are not enough people to do that kind of nursing of each task, but he doesn't seem to mind directing that the task never-the-less gets done. For example, the team just found out that when we submit a request to purchase an item, that it goes through approximately 6 to 8 people before a purchase order is written and the item actually ordered.

The vendors we are working with on seals and spacer rings have responded to our requests in a timely manner yet we (MTI) cannot get a purchase order to them in a timely manner. Our lab has been waiting for a function generator since 9-25-85. The paperwork authorizing the purchase was finished by engineering on 9-24-85 and placed into the system. We have yet to receive the requested item. This type of

reason at all. Clearly the booster problem had not been solved, only the paperwork had been completed. On September 10, Thiokol showed Marshall Space Center a presentation of erosion predictions, which included engineering concerns about joint deflection and the secondary O-ring's ability to seal if the primary O-ring failed. Nothing about temperatures was discussed at the presentation.

By early October, it was clear that management delays within Thiokol were becoming critical. Roger Boisjoly sent another memo to company management on October 4, 1985, in which he example is typical and results in lost resources that had been planned to do test work for us in a timely manner.

I for one resent working at full capacity all week long and then being required to support activity on the weekend that could have been accomplished during the week. I might add that even NASA perceives that the team is being blocked in its engineering efforts to accomplish its tasks. NASA is sending an engineering representative to stay with us starting Oct 14th. We feel that this is the direct result of their feeling that we (MTI) are not responding quickly enough on the seal problem.

I should add that several of the team members requested that we be given a specific manufacturing engineer, quality engineer, safety engineer and 4 to 6 technicians to allow us to do our tests on a non-interference basis with the rest of the system. This request was deemed not necessary when Joe decided that the nursing of the task approach was directed.

Finally, the basic problem boils down to the fact that ALL MTI problems have #1 priority and that upper management apparently feels that the SRM program is ours for sure and the customer be dammed.

Roger M Boisjoly 10/4/85

Opposite and above: Activity Report by Roger Boisjoly, October 4, 1985 (document from *Challenger* report, pages 254, 255).

warned about a lack of management support of the O-ring teams redesign efforts. He further stated, "Even NASA perceives that the team is being blocked (by Thiokol management) in its engineering efforts to accomplish its task. NASA is sending an engineering representative (from Marshall) to stay with us starting October 14th. We feel that this is the direct result of their feeling that we (Thiokol) are not responding quickly enough on the seal problem."[19] Earlier, on October 1, the manager of Thiokol's Solid Rocket Motor Ignition System, Robert V. Ebeling, had written a report outlining his view of the problem to his upper management.[20] He began his report with the word "HELP," and continued, "The seal task force is constantly being delayed by every possible means.... MFSC [Marshall Space Flight Center] is correct in stating that we do not know how to run a development program."[21] And finally, he ended by saying that those "whose help we need, are generating plenty of resistance. We are creating more instructional paper than engineering data.... This is a red flag."[22] Simply put, Thiokol did not want to spend the time and money needed to fix the problem of its own flawed design, and NASA was unwilling to stop or even slow down the number of Shuttle flights. NASA allowed a major contractor to continue to supply dangerous products to the U.S. manned space program. NASA, through Marshall Space Flight Center, had in essence told one of its prime contractors that they did not know what they were doing. Yet the program continued to fly Shuttles on flawed SRBs. That was a space agency failure which would eventually cause the deaths of seven astronauts.

61-A *Challenger*

On October 30, 1985, *Challenger* lifted off on what would be its final successful mission. Viewing the launch from the VIP viewing area, along with 91-year-old German rocket pioneer Hermann Oberth, were Christa McAuliffe and her backup for the teacher-in-space program, Barbara Morgan. They were on hand as part of their training for McAuliffe's upcoming flight. In *Challenger*'s payload bay was the West German Spacelab-D1 mission, onboard was an eight-person crew, the largest ever to be launched into space in one craft. During the ascent into orbit, a helium pressure regulator in the right OMS (Orbital Maneuver System) fuel-feed system failed, but its failure did not affect operations at the time. It was just one more in a long list of system failures becoming commonplace. Also, for some as yet unknown reason, the fire alarm went off six times onboard *Challenger* during the flight, but these all proved to be false alarms. It was almost as if *Challenger* was trying to send out her own warning. A more critical problem developed in one of the Spacelab's experiments when an oxygen leak of two pounds per hour was detected. NASA management decided that if the leak did not increase, it could be supported with onboard supplies of oxygen, and the mission was allowed to continue.

After a flight lasting a little more than seven days, *Challenger* landed on runway 17 at Edwards Air Force Base. When the solid rockets were returned for inspection and evaluation at Thiokol, it was again found that problems had developed during the two minutes of powered flight the SRBs were operating. The right solid rocket had experienced erosion and the left booster had blow-by of hot gases at two separate points.[23] Once again, even in temperatures of 75 degrees Fahrenheit, the solids had failed to perform as designed. Yet, when the problems being reviewed and worked during the Level I flight-readiness review for the next flight were discussed, there was no mention of problems with the O-rings or temperatures. The only action taken was by Lawrence Mulloy who once again signed a waiver of the launch constraint on the solids for the next flight. It was also noted that there were many debris impacts on the thermal system. This problem was also waived.

61-B *Atlantis*

Shuttle *Atlantis* lifted off launch pad 39A on a warm November 26, with a crew of seven on a mission to deploy three communications satellites for the United States, Australia and Mexico. The crew was also assigned the task of assembling structures during space walks to help in the design of the long-planned and little-advanced Space Station Freedom[24] (a name which would change to Alpha and then International Space Station).

During the work required to assemble the Shuttle and its booster, a crane operator had improperly handled one of the solid rocket sections, causing a supporting pin to snap. A NASA board of inquiry later found that workers were using faulty equipment and were not properly supervised. It further reported that workers performing the stacking were unmotivated and inexperienced on the job at the Space Center. The workers were being asked to perform tasks they were not trained to do. A separate air force study conducted at the same time found that NASA's Shuttle ground-support operations were marred "by poor record keeping, disorganized maintenance work and other flaws." Further, maintenance monitoring was reported to be so lax and ineffective that their "use for long-range planning was not possible." These were the individuals tasked with launching and caring for the most sophisticated spacecraft ever built. More than just the SRB problem was becoming clearly evident within the whole of the Shuttle program and perhaps all of NASA.

When 61-B's solids were inspected, they too showed erosion and blow-bys, and once again the solid boosters would be cleared as acceptable for use on the next flight.[25] It was at this point that Robert Ebeling of Thiokol stated that he was so concerned about the O-ring problem that he told fellow members of the Seal Team Task Force that he believed "Thiokol should not ship any more motors until the problem was fixed."[26] Despite this and other concerns, his recommendation did not go any further.

Closure on the Solids

closure n. 1. A closing or shutting up. 2. An end; conclusion.

Normally when one discusses closure on an item it is expected that the problem will be solved or worked through to conclusion. For the solid rocket problem, this was simply not the case. For Thiokol it was simply a paper chase with the hope that the problem would not become critical. After all, with 23 flights under their belts by that time, Thiokol seems to have viewed the erosion and blow-bys as an anomaly, but not fixable (read "costly and profit reducing"), not something that should be firmly addressed and corrected.

On December 6, Brian Russell of Thiokol wrote a letter to the Thiokol Solid Rocket Motor Project director, Allen McDonald. In that letter Russell requested "closure of the Solid Rocket Motor O-ring erosion critical problems." Four days later, McDonald wrote a letter to Lawrence Wear at NASA's Solid Rocket Motor Office asking for closure of the O-ring problem. In that letter McDonald detailed all of the O-ring problems to-date, as well as the launch constraint brought on by the July 1985 launch. (Thiokol management did know about the launch constraint as shown by that document.) McDonald noted that the O-ring problem could not be solved for some time. Thiokol was looking for a way out, not an engineering fix to a problem.

Later, the *Challenger* Commission questioned Brian Russell on the closure issue:

CHAIRMAN ROGERS: Mr. Russell, when you say "close the problem out," what do you mean by that? How do you close it out normally?

BRIAN RUSSELL: Normally, whether it takes engineering analysis or tests or some corrective action, a closeout to the problem would occur after an adequate corrective action had been taken to satisfy those on the problem review board that the problem had indeed been closed out ... to make sure that wouldn't happen again, and then to verify that corrective action, and at that point that problem would be ready to be closed out.

CHAIRMAN ROGERS: What do you understand launch constraint to mean?

BRIAN RUSSELL: My understanding of a launch constraint is that the launch cannot proceed without ... everyone's agreement that the problem is under control.

CHAIRMAN ROGERS: "Under control" meaning what? You just said a moment ago that you would expect some corrective action to be taken.

BRIAN RUSSELL: That is correct, and in this particular case on this 51-B nozzle O-ring-erosion problem there had been some corrective action taken....

CHAIRMAN ROGERS: But really my question is: did you gentlemen realize that it was a launch constraint?

BRIAN RUSSELL: I would like to answer for myself. I didn't realize that there was a formal launch constraint on this one, any different than some of the other erosion and blow-by that we had seen in the past.

CHAIRMAN ROGERS: What was being done to fix it?

BRIAN RUSSELL: Well, we had a task force created of full-time people at Thiokol, of which I was a member ... and we had done some engineering tests. We were trying to develop concepts.

CHAIRMAN ROGERS: Can I interrupt? So, you're trying to figure out how to fix it, right? And you're doing some things to try to help you figure out how to fix it. Now, why at that point would you close it out?

BRIAN RUSSELL: Because I was asked to do it.

CHAIRMAN ROGERS: I see. Well, that explains it.

MR. ROBERT RUMMEL: It explains it, but really doesn't make any sense. On one hand you close out items that you've been reviewing flight by flight, that have obviously critical implications, on the basis that after you close it out, you're going to continue to try to fix it. So I think what you're really saying is, you're closing it out because you don't want to be bothered.

NASA's Marshall Space Flight Center received a letter asking for closure on the O-ring issue on December 18, 1985, and an entry was written on Marshall problem reports, which indicated, "Contractor closure received." Later, on January 23, 1986, only five days before *Challenger* was launched, a second entry was placed on the same report which stated, "Problem is considered closed."[27] Mysteriously the handwritten entry had no name on it, so we may never know which manager or other responsible individual personally helped destroy *Challenger*. Nothing had been done to solve the problem. It was at this point that *Challenger*'s fate was sealed. The O-ring erosion problem had been "closed."[28]

The President's Commission on the *Challenger* Disaster later found that many of the individuals responsible for the Shuttle program had (or stated they had) no knowledge about cold-weather concerns when it came to the O-rings on the solid boosters, even after the major problems, that had occurred. These men, all entrusted with the lives of the astronauts, included Arnold Aldrich, manager of the Shuttle program; Jesse Moore, associate administrator for manned space flight; Richard Smith, director of Kennedy Space Center; and James Thomas, deputy director of Kennedy Launch and Landing Operations. The year 1985 was certainly not a good one, but 1986 would be even worse.

Flight Readiness Review — A Final Look

On January 15, 1986, the National Aeronautics and Space Administration conducted its Level I Flight Readiness Review for mission 51-L.[29] This was the final review of all aspects of flight preparation about any questions which may have come up in the mission. There was no mention of O-ring problems, nor any concerns raised regarding the probability of cold weather on the way.[30] The main topic of discussion for 51-L was the launch window which would place the *Challenger* in the best position to view and photograph Halley's Comet. When the Spartan satellite was added to the flight, the launch was moved to the afternoon, and this change would have made a transatlantic abort landing a night affair. But Casablanca, a backup site, was not equipped for a night landing so it was dropped as an abort site for an afternoon launch. Later, as mission 51-L was pushed back from December 1985 to January 1986,[31] the changed lighting conditions for viewing the comet led to rescheduling for a morning launch. With that change, the Transatlantic abort site and Casablanca would both be well lit and thus fully available. The *Challenger* Commission report would find that "no outstanding concerns were identified in the discussion of flight design." The final check was complete.

Four. 1985 Was Not a Good Year

As far as NASA was concerned, *Challenger* and her seven-person crew were ready to go. She would launch on January 25, 1986, and a press kit was issued which reflected that date.

> *We understand it's ready to go and we are looking forward to fly.*
> —Pilot, *Challenger* 51-L Mike Smith

Chapter Five

Pre-Flight: Dangers Not Seen

Space is the great war that tests us, and sometimes destroys us.
— Ray Bradbury, January 29, 1986

The Lucky Flight of *Columbia*'s 61-C

The launch of *Challenger* on mission 51-L had been scheduled for 3:43 P.M. EST, January 22, 1986, but it slipped to January 23, then again to January 24, due to delays in launching *Columbia* on Shuttle mission 61-C.[1] It was slipped due to problems related to bringing *Columbia* back from space. Weather was causing more than just icing problems. *Columbia*'s original launch date was December 18, 1985.[2] However, due to longer checkouts required to close out the orbiter's aft compartment, because of its nearly two-year hiatus, it could not be readied in time, so the flight was put off for the first time till the following day.

Columbia had undergone modifications at Rockwell's Palmdale plant in California, modifications which had taken eighteen months at a cost of $42 million. But *Columbia*'s launch was scrubbed on December 19, just fourteen seconds short of liftoff, due to an electrical problem in its right solid rocket booster.[3] (This was later shown to be a false reading.) It would have been a record tenth launch for the year. At that point *Columbia* was rescheduled for a January 4, 1986, launch, to avoid having NASA teams or flight crews working during a planned Christmas and New Year's shutdown.

It was rumored that worker morale at the Cape was bad, and they needed the time off. Later studies would prove this to be true. At the time, Robert Sieck, manager of Shuttle Operations at Kennedy, was quoted as saying, "We're absolutely going to take Christmas off. And we're going to minimize the team requirements between Christmas and New Years."[4] Was this a reaction to morale, or an admission that teams working on the Shuttle program were being pushed far beyond their abilities to respond to developing problems? It was crunch time at the Cape and workers were starting to show wear and tear. NASA still wanted the Shuttle to become commercially successful and the pressure was on to perform. But a workforce shortage caused by budget cutbacks was holding up work on many areas. There were not enough quality control people available to approve the work. With not enough people needed to do the job, were shortcuts were being taken to keep on schedule? It was later discovered that some quality-control paperwork was even being completed before the inspections were done.

NASA was also fighting a spare-parts problem which had resulted in the cannibalization of parts from one Shuttle to use on another. "You rob one vehicle to fly another. You may have to support four vehicles (with one item), so you have to go find the damn thing," said Vince Vanderburg, ground crew supervisor at Kennedy, in an interview after the *Challenger* disaster. Costs had become a problem and the budget was simply unable to support four operational Shuttles. Eventually one of those single items was going to fail and the whole program was going to be halted. This was a chronic problem which dated back to the original flights in 1981. There were not four operational Shuttles, but a Shuttle parts system which required NASA to hunt down unique parts so it could reconstruct the next Shuttle for its upcoming flight. One NASA official told Congress that they had pirated one spacecraft so much that he "could carry [what remained of it] in his suitcase." This was dangerous, yet at congressional hearings space agency managers would be asked what happened to funds to cover this very problem, and no one could answer that question, nor did Congress press very hard to discover the answer. There was plenty of blame to go around after the disaster, which was just around the corner. As of this writing, the problem of spare parts has yet to be solved.

In the meantime, on December 22, *Challenger* was rolled out to launch complex 39B at 4:40 A.M., to prepare it for its scheduled January 22 launch.[5] This would be from the new Shuttle launch complex, located two miles from the *Columbia* then sitting on pad 39A. It had been reported that in November, during the stacking in preparation for that flight, a crane had hit one of the SRBs. Workers reported that a sharp cracking sound could be heard. However, NASA reported that no major damage to the booster had occurred, and the stacking continued.[6]

Pad 39B had not been used since an Apollo spacecraft was launched to dock with a Soviet Soyuz spacecraft in 1975 for the ASTP mission. Funding on pad conversion had been a long time coming in an already stretched budget. This would be the first time in the Shuttle program that two operational Shuttles would be on the pads, getting ready for missions at the same time. Only one time before had these pads held rockets at the same time. That occurred in 1973, when a Saturn V was sitting on pad 39A with the space station Skylab on top, and a Saturn 1B sat on pad 39B with Skylab 2 ready to launch and link with the station in earth orbit. Originally, pad 39B had been built for the moon program as a Saturn V launch facility. It was used to launch Apollo 10 on its lunar-orbiting mission, as well as all three Saturn 1Bs flown to Skylab. A final Saturn 1B for the Apollo-Soyuz Test Project, was also launched from that pad.

On December 23, *Columbia*'s 61-C mission was again pushed back, this time for two days. NASA stated that it was to give the crew of seven more time for training.[7] But if the crew were ready to launch on December 18, why would they need more training? What was NASA not telling the public? Was the crew truly not ready to go when the *Columbia* was set to launch on its original launch date? And if so, which managers allowed a crew not ready for a mission to climb into a spacecraft for an attempted launch? At the time NASA was playing a very dangerous game of balancing crew simulator training time, with a flight schedule that pressed hard on the abilities of the flight crews to develop the skills needed to safely fly. All areas, including ground and flight crews, were being pressed to their limits, and their limits were clearly shown by the errors which occurred on and before *Columbia*'s next mission. In fact, the 61-C crew had less simulator time before their flight than any other prior crew. NASA's simulator schedule was becoming saturated, and missions were beginning to pile up. Adding to that was the fact that NASA was using computer programs which had been online since before the approach and landing tests of the late 1970s, and an updated system was badly needed. The simulator training was not up to the high standards the American people would have expected for their manned space program. The program was starting to show its age. As of 2003, the space agency was still using 1970s computer technology for America's Shuttles.

NASA, in the same December 23 announcement, also stated that the delay in *Columbia*'s 61-C mission would force a one-day delay of *Challenger*'s flight which was then rescheduled for January 23.[8] The agency was beginning to grasp for things beyond its reach, and it was starting to show. NASA's response would be to push even harder and in doing so to accept even more risks.

Thirty-one seconds before launch on January 6, a computer detected a low temperature reading on a liquid-hydrogen fuel line, and *Columbia*'s count was halted.[9] Despite efforts to roll back the count and make repairs, public affairs officer Jim Ball was forced to announce, "We have scrubbed for today." This was the second time that the 61-C mission had been stopped inside of one minute from launch. The scrub on January 6 proved to be very lucky for the crew. Due to an error that NASA later blamed on worker fatigue, 18,000 pounds of liquid oxygen (news reports at the time consistently reported 18,000 pounds, but NASA in its publications says 4,000 pounds) were inadvertently drained from the shuttle external tank (the main fuel tank).[10] *Columbia* did not have enough fuel to reach a safe orbit. It should be noted that the loss of 4000 pounds of fuel would have allowed the orbiter to reach a safe runway, while 18,000 pounds would not have. The crew would have had to risk a dangerous emergency landing across the Atlantic at Dakar, Senegal, with a fully loaded orbiter. Some experts felt that a landing would have been nearly impossible to accomplish on the 10,000-foot runway, that even though Dakar was the TAL (Transoceanic Abort Landing) site for this launch, the Shuttle might well have passed the point of usability for that site when its fuel ran out. With insufficient fuel to go around the earth in a long sub orbit or some kind of lopsided full orbit, the Shuttle would have had only one chance: Dakar. A crash landing would have been much more likely, and at the time Shuttle crews had no bail-out capability.

Water landings of any type were simply not survivable, something that NASA reluctantly admitted after years of water rescue training for Shuttle crews, despite press reports to the contrary. Aborts are usually based on one or two engines going out, but with no fuel all three engines would be gone at the same time, leaving only the much lower-thrust OMS (Orbital Maneuvering System) engines to provide thrust for some kind of a landing attempt. *Countdown* magazine in its May 1988 issue reported that, "Unknown at the time — at least publicly — the true cause of the problem rested with a fatigued launch controller who misread computer data. The launch controllers were on duty for the eleventh straight hour at the time, having worked three 12-hour night shifts to prepare for the flight. If *Columbia* launched, it would [have] run dry of liquid oxygen — resulting in disaster." ("We will never compromise safety standards for schedule," Jesse Moore had said.) It was time to go back to three shifts of eight hours per day. But there was no extra money in the budget for additional qualified personnel. The Shuttle program was being run on an ever-decreasing shoestring budget, which was still being reduced before the *Columbia* disaster in 2003.

After corrections had been completed, *Columbia* was again ready to fly on January 7, but bad weather at the Shuttle landing site in Florida, and at both Transoceanic Abort Landing (TAL) sites, kept *Columbia* firmly bolted to its pad.[11] Launch was scrubbed with nine minutes till liftoff. Haze had obscured the mission's two TAL airstrips in Dakar, Senegal, and Moron, Spain, making it unsafe to land the Shuttle there in case of an emergency.[12] At Kennedy, the runway would be used in case of a return-to-launch site (RTLS) abort. But it too was covered by clouds (greater than fifty percent), making a landing at that site far too risky, as well. It was truly a rare occurrence to have weather problems at both TAL sites and Kennedy.

While all of this attention was directed at getting *Columbia* off the ground, the *Challenger* crew continued to train for their upcoming mission. On January 8, Commander Scobee and Pilot Smith conducted a practice countdown on board *Challenger* at pad 39B. It was the first opportunity to be on board for a training session on the pad. As clouds and rain kept *Columbia* on the ground, the rest of the 51-L crew, wearing yellow rain coats, practiced emergency escape procedures using the Shuttle slide-wire escape system. Although no one actually slid down the wires during the practice, it was a successful training session.

Controllers then began a weather waiting game, as conditions prevented a launch for two days until January 9. However, the storm had stalled over the Gulf of Mexico, and it was forecast

Five. Pre-Flight: Dangers Not Seen

to still be raining on the tenth at the Cape. By this time, NASA was starting to become concerned about meeting its schedule of fifteen launches planned for 1986 and it was only January. But *Columbia* was "go" on January 9, until a five-inch-long oxygen-line temperature sensor broke off due to a bad factory weld. The sensor flowed down the line and lodged in the oxygen pre-valve, above *Columbia*'s main engine Number Two. Had *Columbia* launched with the sensor jammed in the pre-valve, it might have caused what NASA refers to as "a destructive shutdown" of Shuttle engine Number Two. Regular people would call it an explosion. This "shutdown" would have destroyed the engine just as the Shuttle was entering orbit at MECO (Main Engine Cut Off). It does not take much effort to imagine what would have happened to *Columbia* and her crew if one of the main engines had exploded. Pieces of the destroyed Shuttle would have soon reentered the atmosphere and burned up along with the crew. At that point NASA might never have been able to discover what went wrong. Was this bad weld another indication of the "go" attitude of NASA moving all the way to the contractors? Perhaps. A future Congressional investigation would press hard for answers as to what went wrong on this pre–*Challenger* mission.

The press had no real idea about the problems and pressures pushing the space agency to launch Shuttles on an ever-increasing schedule. They never really tried to find out. The naive press looked only at the surface and never at the many complexities involved in launching a Shuttle. Only after the *Challenger* and *Columbia* disasters did the press ask the truly tough questions, and then only for a short period of time.

After a number of scrubs, Jesse Moore, the associate administrator for the Office of Space Flight, tried to put a little perspective on the situation. "I hope people don't think we're losing credibility by not launching. We want to find out about problems on the ground, not in orbit!" Were his concerns about credibility pushing him to launch when he should not have, or were other pressures more demanding than an uninformed press pushing even harder? The space agency had been increasing its Shuttle operations to the point where it was simply over-ambitious, and well beyond its ability to perform. Both people and equipment were being pressed to the limits.

January 10 proved to be the seventh time *Columbia* and its crew would not launch on a scheduled launch date, as wind and heavy rains came down on both Shuttles.[13] It was the most delayed mission of the program. It may have been these rains which helped cause the *Challenger*'s accident. Some of these windswept rains may have penetrated the booster seals on *Challenger*'s SRBs and become frozen while it waited for its cold early morning launch.[14] NASA did find on an earlier mission, STS-9, that water had gotten into field joints.[15] The conditions were so bad that a NASA weather plane, piloted by one of the astronauts, could not even take off to inspect the local conditions at Kennedy.

A Launch at Last!

Finally, on January 12, the much-delayed *Columbia* 61-C mission got underway at 6:55 A.M., with a spectacular predawn launch from pad 39A,[16] but *Challenger*'s delays had just begun. Getting *Columbia* down from orbit would prove almost as difficult as launching it. By January 17, *Columbia*'s landing had been halted for a second day in a row, by rain and clouds at the Kennedy Launch Center.[17] That weather system was also bringing high winds and high waves off the coast in the SRB recovery area. NASA wanted to land at Kennedy, because it would eliminate the need to transport the Shuttle back from a landing in California, which was a costly and time-consuming procedure. Also, NASA managers wanted the *Columbia* back as soon as possible to prepare it for its scheduled March 6 launch. (That flight would carry the Astro telescopes into orbit to observe Halley's comet.) This *Columbia* landing delay caused *Challenger*'s flight to again be pushed back, this time to January 25.[18]

On January 17, weather proved just as bad or worse for a landing, so Shuttle controllers decided to forgo the Kennedy landing attempt just 22 minutes before the *Columbia* would fire its engines for reentry, and instead bring the Shuttle back to earth in California on the next day at Edwards Air Force Base.[19] Once again the luck of *Columbia*'s crew would hold. As *Columbia* was in its last orbit, ground controllers noticed that the forward-reaction-control system in the nose of the orbiter was approaching low temperature levels. While looking at the temperatures, another problem was discovered.

Controllers were shocked to discover that if a thruster test was performed as scheduled during reentry the reaction-control system could be damaged or even explode.[20] The test was cancelled and *Columbia* reentered successfully, landing on January 18, at Edwards at 5:59 A.M. PST.[21] Why was the thruster problem not discovered before *Columbia* was even launched? The Shuttle landed with its crew of seven, including Congressman Bill Nelson who had bumped Greg Jarvis from the flight, landing Jarvis on the crew of *Challenger*. NASA officials were well aware of major problems throughout the Shuttle system, but had been brushing them aside, and keeping their fingers crossed. Their luck would soon run out.

After the *Challenger* disaster, Hoot Gibson spoke of his 61-C flight and remarked, "Someone was watching over us because in several of these incidents it could have been very serious if we had actually launched that day." This of course is NASA astronaut bravado for "we would have been killed." With all that had gone wrong on that mission it would seem that only luck saved that crew of *Columbia*, luck which would not hold for the crew of STS-107 in 2003.

Columbia was on the ground but not yet home, and as usual had taken plenty of debris hits. *Columbia* arrived at Cape Kennedy on top of its Boeing 747 carrier aircraft after being transported from California on January 23. It was in time to be viewed by the arriving *Challenger* crew of mission 51-L. *Challenger*'s flight had been rescheduled for January 25 at 4:21 P.M., but because of continued bad weather at the Transoceanic Abort Landing site (TAL) in Dakar, it was not launched.[22] Haze was forecast to be a problem at the site for several days. Flight controllers decided to utilize Casablanca's airport in Morocco, which was not equipped for night landings, as an alternate TAL site. With this change, *Challenger* would be required to launch in the cold of morning to insure enough light at Casablanca in case its runway was needed for an emergency. January 25 was the launch date found in the press kit for 51-L. Pilot Mike Smith spoke to the gathered press and told them, "We understand it's ready to go and we are looking forward to go fly."

One more day was lost when launch processing personnel were unable to complete their work in support of the new morning liftoff time. Was this also a result of fatigue? Processing the Shuttle fleet was Lockheed Aircraft Company. They had won the contract for Shuttle processing due to having submitted the lowest bid on servicing. The contractors who had designed and built the Shuttle systems and who would know them better than any other company were no longer involved in preparing the orbiters for flight. Due to being unfamiliar with the systems, Lockheed personnel actually damaged some of them. At times they were seen on live TV using improper servicing tools.

When a prediction of unacceptable weather at Kennedy was made for January 26,[23] the launch was again rescheduled, this time for the next day at 9:37 A.M. EST. Managers were becoming concerned that continued delays would impact the other 13 flights that were scheduled for 1986. The weather on January 26 turned out to be well within acceptable parameters, which would have allowed a launch, but a frontal system was moving toward the Cape with heavy rain and thunderstorms. There was even a possibility of light snow in the area. Shuttle managers took one last look at the forecast at 10 P.M. that evening and decided that there was little chance for a liftoff on the 26th. This time controllers would err on the side of caution.

The *Challenger* crew left for the pad on January 27,[24] in cool, partly cloudy weather at 6:45 A.M. with expectations that the weather would not be a problem for launch. By 7:21 A.M. the crew had entered the cabin of *Challenger*. As Christa McAuliffe entered the crew cabin, one of the close-out team handed an apple to the teacher. January 27 was going well for a launch that day at 9:37

A.M., but when an equipment-hatch-closing fixture could not be removed from the Shuttle's hatch by 9:30, the launch had to be delayed again.[25] The screws had to be drilled out.[26] The closeout team had used an unapproved tool on the hatch. Technicians had to borrow another handle from Shuttle *Discovery* to get the hatch open. It was starting to look like a comedy act as the minutes slipped by. In fact, the closeout team did not even have the required emergency closeout kit. Because of this, the team had to send someone to go find a drill.

Public affairs officer Hugh Harris came on the air to announce that the launch was "no go" due to crosswinds at the return-to-launch site runway. He also reported that snow had fallen for eight minutes in the Jacksonville, Florida, area, and the jet stream had brought in extremely cold polar air into the area. The launch was then scheduled for 12:06 P.M. if the winds were then within limits. The count would be taken as far forward as it could, and then held to make sure the weather was going to cooperate.

By the time the orbiter hatch had been secured, and the fixture sawed off and closed out, crosswinds were increasing at the Kennedy Shuttle Landing Facility. John Young, chief of the astronaut office, was checking the winds at the landing site at Kennedy, flying the Shuttle training aircraft and continuing to find unacceptable conditions. The count was allowed to proceed until nine minutes from liftoff and holding. At that point the winds in the area continued to pick up speed with no let-up in sight. These unacceptable conditions once again caused a delay of twenty-four hours until January 28, at 9:38 A.M. EST.[27]

Hugh Harris, a NASA public affairs officer at the Kennedy Launch Center, announced the third postponement in three days: "We have just had an announcement from Launch Director Gene Thomas to the crew and ... launch team that we are going to scrub for today and that the people will be coming out to the pad to let the crew out of the orbiter shortly as we recycle the vehicle back to a safe condition."

Wind gusts of more than thirty miles per hour, and sustained winds at twenty, were pushing across the Shuttle runway at Kennedy, which would be needed in the event of a return-to-launch-site abort. In order to land safely a Shuttle cannot encounter crosswinds at that runway of greater than 17 miles per hour. The crew was also becoming fatigued since they had been inside the Shuttle on their backs for almost four hours. Hugh Harris reported, "The forecast for tomorrow is looking at about 28 degrees F. in the launch area at launch time, which would be 9:38 A.M." *Challenger*'s dark passage would have to wait one more day.

That evening, as program managers were struggling with the decision to launch, because of the cold weather and its possible effects on the SRBs, Christa McAuliffe called her parents to tell them that no matter what, tomorrow they were going to launch.

When you find something you really like to do and you're willing to risk the consequences of that, you probably ought to do it.
— Commander Francis Dick Scobee, *Challenger*

SECTION III

CHALLENGER'S FINAL MISSION

CHAPTER SIX

A Cold Morning Like No Other

We will never compromise safety standards for schedule.
— NASA associate administrator Jesse Moore, January 26, 1986

It was cold, very cold, colder than any previous launch day in human space flight history. After the launch was scrubbed for the day, NASA officials at Kennedy, Marshall and the contractor of the Solid Rockets, Morton Thiokol, struggled to find a way to launch *Challenger*, as engineers told managers it was simply too cold to launch. This management flight-schedule pressure was about to cost the lives of seven highly skilled and dedicated American astronauts.

Pre-launch and launch coverage of *Challenger*'s final flight are taken primarily from the *Challenger* report[1] and onboard audio recordings recovered from *Challenger*'s wreckage:

January 27, 1986

12:36 P.M. Launch is scrubbed for the day, due to high crosswinds at the launch site.

1:00 P.M. All appropriate managers and other personnel at Kennedy Space Center, are polled as to the feasibility for a launch within the 24-hour cycle. No constraints on the solid rockets are discussed for a 9:38 A.M., January 28 launch.

About 1:00 P.M. Lawrence Wear, SRM Project Office manager, Marshall Space Flight Center, asks Boyd Brinton, Space Boosters Project manager, Morton Thiokol, if Thiokol had any concerns about predicted low temperatures the next day, and what Thiokol had said concerning cold-temperature effects on the O-rings following the 51-C flight, the coldest launch to date.

About 2:00 P.M. A mission management team meeting is held to discuss temperatures and weather conditions predicted at the Cape for launch the next morning.[2] Present are Jesse Moore, Arnold Aldrich (manager of the Space Transportation Systems Program JSC), Lawrence Mulloy (manager of the SRB Project, Marshall SCC), and Dr. William Lucas (director of Marshall Space Flight Center). Ice on the launch pad was the main topic for discussion, as well as possible winds at the Kennedy Shuttle landing strip.

About 2:30 P.M. At Thiokol in Utah, Roger Boisjoly, a member of the Booster Seal Team, is informed of cold temperatures at the Cape expected during the launch period. Also informed is Robert Ebeling, manager of the Ignition System and Final Assembly SRB Project. Both feel that a launch in cold temperatures would be a problem for the O-rings based on tested cold effects on other O-rings which have flown, as well as years of ground tests.[3]

About 4:00 P.M. At Kennedy Space Center, Allan McDonald, director of the SRM Project for Morton Thiokol, receives a phone call from Robert Ebeling who expresses concerns about the performance of SRB field joints at low temperatures.[4] He makes it clear that the solids are being asked to perform outside of their design abilities, but cannot prove they will fail. Only a flight can do that.

About 5:15 P.M. Allan McDonald calls Cecil Houston, Marshall SFC manager at Kennedy Space Flight Center, to inform him that Morton Thiokol engineers have grave concerns about O-ring temperatures.[5] Thiokol has no engineering data at those temperatures and it will be "a roll of the dice to launch." (This should have been a red flag to Marshall because Thiokol had earlier informed NASA managers at Marshall they did have such data, when in fact none existed. This call alone should have been enough to scrub the launch for the day with a critical investigation started.)

Around 5:45 P.M. During a teleconference involving Kennedy, Marshall and Thiokol, Thiokol engineers recommend a launch delay, based on known cold-temperature effects on O-rings.[6] Participating were Stanley Reinartz, manager of the Shuttle Project Office, Marshall SFC, and Judson Lovingood, deputy manager of the Shuttle Project, and others.

Around 6:30 P.M. Another teleconference[7] ensues at Kennedy and Marshall after Lovingood calls Reinartz and tells him they should not launch, if Thiokol continues to believe there are cold problems. Lovingood also suggests informing Arnold Aldrich, manager of the National Transportation System.

Around 8:45 P.M. A teleconference is held with Morton Thiokol's Jerald Mason, senior vice president of Wasatch Operations; Calvin Wiggins, vice president and general manager of the Space Division, Wasatch; Joe Kilminster, vice president of Space Booster Programs, Wasatch; Robert Lund, vice president, Engineering; Roger Boisjoly, Seal Task Force; and Arnold Thompson, supervisor, Rocket Motor Cases. Taking part at Kennedy are Stanley Reinartz and Lawrence Mulloy, SRB Project manager. At Marshall Space Flight Center are George Hardy, deputy director, Science and Engineering; Judson Lovingood; and Ben Powers, Engineering Structures and Propulsion. Using charts, a history of the O-ring erosion and blow-bys is shown.[8] Data showed that O-rings react slower with lower temperatures and that the worst blow-by occurred on STS 51-C with an O-ring temperature of 53 degrees. This is in direct contrast to later testimony that cold temperature data was unknown. George Hardy states that he is appalled by Thiokol's recommendation not to launch but will not override Thiokol. This was reported by both McDonald and Boisjoly.[9]

Around 10:30 P.M. Thiokol caucuses with the issue of cold temperature effects on O-rings and erosion discussed.[10] Thompson and Boisjoly do not want to launch, along with Lund. All three men feel that a disaster could happen due to expected cold weather. Boisjoly informs Thiokol management that they would be flying outside of known engineering data. No known data would support a "go" for launch.

Final management review is held with Mason, Lund, Kilminster and Wiggins. Lund is told by Mason to take off his engineering hat and put on his management hat. Lund is then pressured to change his mind and decides to launch.

Final Decision: "O-rings have a margin to erode by a factor of three from previous worst case, also if the primary fails the secondary is in position and will seal." The data and O-ring history however, did not support this decision, and Thiokol testing had already established that the secondary O-ring would *not* seal. Thiokol management had just made a decision to launch based on data which showed that some type of failure was possible on the flight hardware. They also had zero data at the low temperatures *Challenger* was being asked to fly in. Thiokol management knew that the O-rings would not perform as designed and could fail.

Around 10:30 P.M. McDonald continues to argue for a delay, based on erosion and blow-by data. He is convinced that the cold weather is a factor in the possible failure of the SRBs.

Around 11:00 P.M. Teleconferencing continues. Thiokol reverses its decision; reassessed data

> **MTI Assessment of Temperature Concern on SRM-25 (51L) Launch**
>
> o Calculations show that SRM-25 O-rings will be 20° colder than SRM-15 O-rings
> o Temperature data not conclusive on predicting primary O-ring blow-by
> o Engineering assessment is that:
>   o Colder O-rings will have increased effective durometer ("harder")
>   o "Harder" O-rings will take longer to "seat"
>   o More gas may pass primary O-ring before the primary seal seats (relative to SRM-15)
>   o Demonstrated sealing threshold is 3 times greater than 0.038" erosion experienced on SRM-15
>   o If the primary seal does not seat, the secondary seal will seat
>   o Pressure will get to secondary seal before the metal parts rotate
>   o O-ring pressure leak check places secondary seal in outboard position which minimizes sealing time
> o MTI recommends STS-51L launch proceed on 28 January 1986
>   o SRM-25 will not be significantly different from SRM-15
>
> *[signature]*
> Joe C. Kilminster, Vice President
> Space Booster Programs
>
> **Morton Thiokol, Inc.**
> Wasatch Division

Telefax sent by Thiokol to Marshal Space Flight Center and Kennedy Launch Center (document from *Challenger* report, page 97).

is called inconclusive. Thiokol recommends, despite no firm data, to launch in cold weather.[11] (Later investigation will show that George Hardy, who was appalled when the engineers at Thiokol recommended "no go," now wanted their recommendation to launch in writing.[12] It was an unprecedented situation. He wanted evidence that they had okayed the launch so that NASA would have some form of deniability.) The deadly recommendation is faxed to Kennedy and Marshall Space Centers.[13]

Around 11:30 P.M. McDonald continues to argue for a delay, asking how NASA could rationalize launching below qualification temperatures. He states that he would not launch due to O-ring problems at low temperatures, as well as problems in recovering the SRBs in rough seas off the shore of Kennedy. McDonald states that if anything happens he would not want to have to explain to a board of inquiry. He is told that it is not his problem, but that his concerns will be passed on in an advisory capacity. Present are Mulloy, Reinarty, Houston and Jack Buchanan, manager of KSC Operations for Thiokol.

Around 11:45 P.M. Joe Kilminster of Thiokol faxes a recommendation to launch[14] at 11:45 A.M. EST the next day. Allan McDonald refuses to sign the final launch recommendation. Even with this later launch time recommendation, the space agency continues to prepare for a 9:38 A.M. launch in very cold temperatures. Key launch managers at Kennedy are not informed of O-ring concerns.[15]

Challenger, PAD 39 B — January 28, 1986

The weather at the Cape that morning was forecast to be clear and very cold. It was colder than any other day that a Shuttle launch had been attempted. A severe overnight frost was forecast for most of Florida, to include Cape Kennedy. Bad weather off the coast was also predicted, as were crosswinds. Booster recovery ships were heading back in to Kennedy.[16]

Due to cold temperatures at 10 P.M. the night before launch, pad workers pumped antifreeze through lines and kept water spigots on drip to keep the pad from freezing. Still, ice had accumulated in the launch pad area during the night, causing great concern for the members of the launch team.[17] As shown in photos taken that night by the ice team, large icicles were everywhere, some a foot in length or longer. The combination of freezing temperatures and stiff winds, according to the *Challenger* Commission report, caused large amounts of ice to form below the 250-foot level of the fixed service structure, including the access area to the crew's emergency-egress slide-wire baskets. Ice hung from the baskets, and it was possible that icing would have precluded their use in an emergency.

By 1:25 A.M., the three-hour fueling procedure had begun. During the 51-C flight, fueling did not begin until temperatures were allowed to warm up.[18] This was another red flag.

The ice inspection team was sent to the launch pad at 1:35 A.M.[19] and reported back to Launch Control at 2:00 A.M. Charles Stevenson, supervisor of the ice crew at Kennedy and team member B. K. Davis conducted the inspection. They reported the overall ice situation to Launch Control, but the fueling continued, just in case a launch was possible that day. The team found large quantities of ice on the service structure, launch platform and pad apron. The temperature continued to drop.

At 1:45 A.M. the fueling was ordered halted and the launch was delayed for two hours due to a computer that sensed an electrical problem in the system that linked fire detectors to Launch Control.[20] The wind-chill factor at the time was as low as 10 degrees below zero. The super-cold fuel being loaded also contributed to the overwhelming cold at the pad.

By 3:00 A.M., sheets of ice and hanging icicles had formed on the rails and other areas on the launch pad and could be easily seen on engineering cameras monitoring the pad. Mission Control in Houston as well as those at the Cape could easily see the ice.

The electrical problems were repaired by 3:55 A.M., but time had run out on the time allowed for the hold. With the electrical problem solved and the fueling continuing, the liftoff time was pushed back to 10:38 A.M.[21] to give some of the ice time to melt. Public Affairs Officer Hugh Harris reported, "We're down several hours, but that doesn't mean we can't catch up some of that time."

Due to concern about ice freezing in the pipes at the launch pad that were part of the fire extinguishing system, the pipes were again opened a little at 4:10 A.M. to allow water to drip out.

In conversation at Kennedy Space Center, Lawrence Mulloy, manager of the SRB Project, informed William Lucas, director of the Marshall Space Flight Center, of Thiokol's concerns about low temperature and its effect on the O-rings. Lucas is also shown a copy of Thiokol's fax which recommends a "go" for launch. Lucas has now been fully informed of cold-temperature concerns, and the possible effects on the O-rings. It was approximately 5 A.M., and the launch was still on. Lucas did nothing to stop or postpone it.

At 6:18 A.M., the crew received wake-up call in their quarters, but most had been up for a while.[22] As is NASA tradition, the crew of 51-L sat down for a traditional steak-and-eggs breakfast and photographs at 6:48 A.M., before suiting up for the flight.

At 7:00 A.M. the temperature was 24 degrees Fahrenheit.[23]

At the crew weather briefing held at 7:18 A.M.,[24] the temperatures and ice problems on the pad were discussed, but the crew was not told of any concerns about the effects of low temperatures on the Shuttle orbiter or the booster's O-rings. It was later reported that Mike

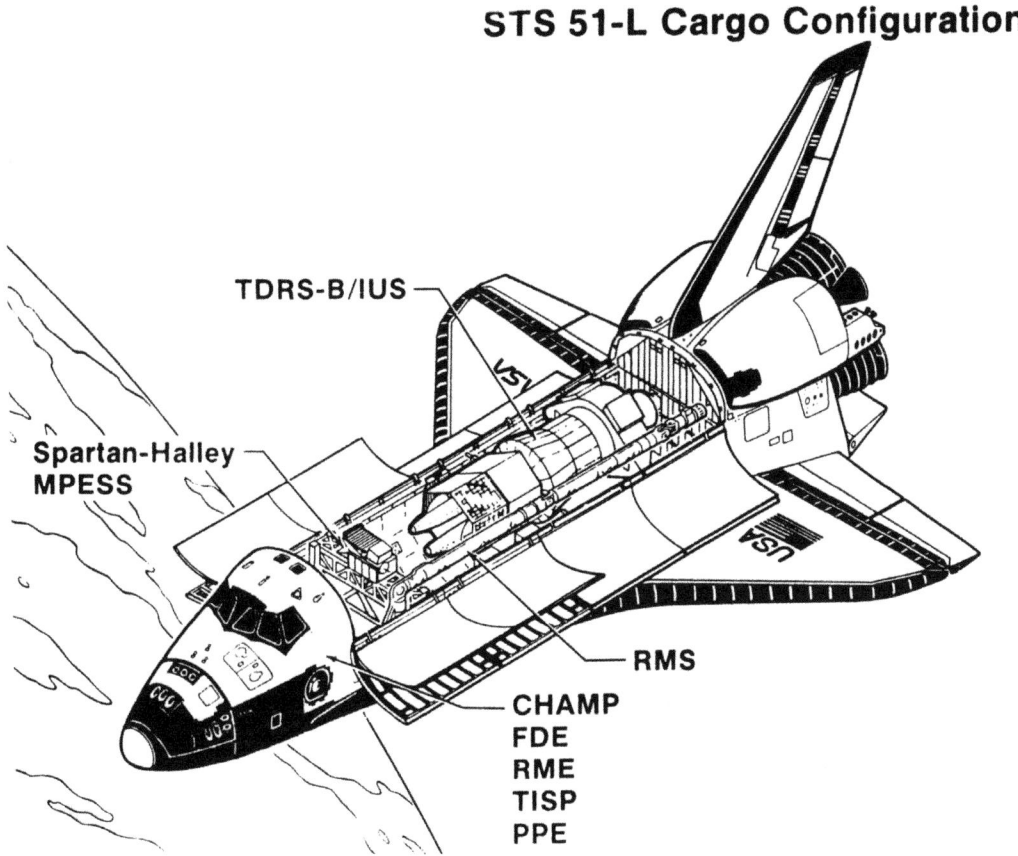

Cargo configuration of Space Shuttle *Challenger* for STS 51-L (graphics from NASA 51-L Press Kit).

Smith had informed friends that due to the cold weather there was no way NASA would launch that day.

At around 8:00 A.M., at Marshall Space Flight Center, Judson Lovingood, deputy manager of the Shuttle Projects Office informed Jack Lee, deputy director at Marshall, that Thiokol had at first said that a no-go launch condition existed, due to possible problems with temperature effects on the boosters. Lovingood then informed Lee that Thiokol had changed its position and was now supporting a go ahead for launch. At this point it should have been realized by Deputy Director Lee that the booster manufacturer (or at least some of the engineers) had serious doubts about a safe launch and it was time for him to take a very close look at the situation. He did nothing to stop or delay the launch.

The *Challenger* crew climbed into a Winnebago van at 7:48 A.M. and was driven to pad 39B. Chief Astronaut John Young escorted them, and for some reason did not have a smile on his face. Due to the cold weather the crew moved very quickly into the van as NASA Public Affairs reported, "Big smiles today, confidently getting into the van." At 8:03, the crew rode the launch pad elevator to the white room next to the Shuttle *Challenger*.[25]

> PAO-KENNEDY: This is Shuttle Launch Control. The countdown continuing very smoothly here at the Kennedy Space Center. While the crew members of 51-L get ready and get inboard the orbiter *Challenger*, the countdown clock continuing at T minus 1 hour 17 minutes and counting. This is Shuttle Launch Control.
>
> PAO-KENNEDY: The teacher observer Christa McAuliffe has been handed an apple by the closeout crew. They are going to be going very fast today to get out of the cold and into the cabin.

By 8:36 A.M., all crew members were in their assigned seats inside *Challenger*.[26] The cabin temperature was 61 degrees. Seated up front on the left side commander's chair was Dick Scobee. To his right sat Mike Smith. Behind Smith was Ellison Onizuka. To his left and behind the commander was Judith Resnik. On the mid-deck downstairs Ron McNair was seated next to the hatch. To his front right were Christa McAuliffe to the left and Greg Jarvis to the right. Some of the crew members' first comments after they were settled in were about the weather.[27] It was cold enough, even inside the cockpit, to fog up the crew's visors.

TEST CONDUCTOR: Lets hope we go today.
FRANCIS SCOBEE: We'd like to do that.
CONTROLLER: Brr!
FRANCIS SCOBEE: "Brr" is right.
TEST CONDUCTOR: PS1 OTC radio check. Good morning. I hope we launch today.
CHRISTA MCAULIFFE: OTC PS1 loud and clear. Good morning. I hope so too.
MIKE SMITH: Those clouds look like they're going away from us up here... Wow, Boy, the sun feels good this morning.
MANLEY CARTER (Astronaut trainee): You should have been here at 2 A.M. ice skating on the MLP [Mobile Launch Platform]?
FRANCIS SCOBEE: You guys up here working?
MANLEY CARTER: It's a lot of fun.
ELLISON ONIZUKA: Kind of cool this morning. My nose is freezing. Gunga Din doesn't know how to operate in cold weather.

Astronaut Manley L. "Sonny" Carter, Jr., the astronaut support person,[28] would later be assigned to his first flight on Shuttle *Discovery* scheduled to orbit the International Microgravity Laboratory in January 1992. However, before he could fly, Carter would die on April 5, 1991, in the crash of an Atlantic Southeast Airlines commuter jet.

At 8:44 A.M. the ice inspection team completed its second inspection and recommended a hold. The hold was called by the program manager to allow more of the ice to melt on the pad.[29] The team had found a temperature of 25 degrees on the left solid rocket booster and a temperature of about 8 degrees near the aft area of the right solid rocket booster. Because there were no "launch commit criteria" regarding surface booster temperatures that might cause a hold on the launch, the ice team did not report the temperatures to the launch controllers. The team did report finding sheet ice on the lower sections and skirt of the left SRB. The solid ice sheet went completely around the base of the booster.

At 9:00 A.M. the temperature was 28 degrees.[30]

By 9:08 A.M. Hugh Harris announced another delay so that controllers could determine if launch-pad water systems had caused a dangerous ice buildup. "One of the concerns is that these icicles, some of which are several feet long, could possibly break off during liftoff and damage the orbiter and its thermal protective systems." This came after a Mission Management Team meeting in which ice conditions were discussed, but there was again no discussion of possible O-ring problems.

At 10:00 A.M. the temperature was up to 32 degrees.[31]

JUDITH RESNIK: Is that snow?
FRANCIS SCOBEE: Yep, that's snow.
RONALD MCNAIR: You're kidding. You see snow on the window?
MIKE SMITH: Yep, it's clouded up out there and started to snow.
GREGORY JARVIS: Yeah. Fat chance.

Later, the weather and ice conditions were still the crew's primary topic of discussion as they waited for the go-ahead for launch.

ELLISON ONIZUKA: Ice team out here?
FRANCIS SCOBEE: Don't see anybody out on the top. Must be around the base.

JUDITH RESNIK: I don't know.
FRANCIS SCOBEE: Ice pickin'.
JUDITH RESNIK: [Laughs.]

At 11:15 A.M., the final ice inspection was completed during the final launch hold, and the team was given a "go" for launch. The ice inspection team had removed ice from water troughs and reported to the Mission Management Team that sheet ice was still on the left Solid Rocket Booster. The crew, who had been seated on their backs for nearly three hours, were starting to feel the effects of waiting for a launch to begin.

MIKE SMITH: We're all getting old.
JUDITH RESNIK: We gonna launch on time today?
FRANCIS SCOBEE: Just moments away.
ELLISON ONIZUKA: 10:38 launch?
JUDITH RESNIK: That's an indefinite.
ELLISON ONIZUKA: How about half an hour.
JUDITH RESNIK: It's half an hour from now, but it's still indefinite.
FRANCIS SCOBEE: We should've slept an extra hour this morning.
GREGORY JARVIS: They're probably making a fortune selling coffee and doughnuts out at the viewing areas.
FRANCIS SCOBEE: How about that. We should have gotten some.
JUDITH RESNIK: A few hot toddies.
FRANCIS SCOBEE: Yeah.

When Commander Scobee announced that the launch would be in nine minutes, Judith Resnik reported, "My butt's gonna like zero-G a lot better than these seats." Commander Scobee's response was, "Mine too." After many delays, the crew was more than ready for their flight into space, and the hold was about to end.

Challenger's Final Flight Timeline

Shuttle *Challenger*'s final flight timeline includes crew and public affairs comments reported live on NASA select TV, as well as reports by the Shuttle test conductor and controller for the flight, from T minus 9 minutes and counting. During the preplanned ten-minute hold, the entire launch center was polled, one console at a time, to make sure that everything in the system was ready for launch, and to insure that no one had any constraints or holds that would keep them from launching. This would include a final weather check at the launch pad and Shuttle runway, including a check with the Canaveral air force weather office. Areas polled included payload operations, the engineering director, range safety, Cape weather, operations manager, launch director, and finally the Shuttle itself. Overseas abort sites were also checked for any last-minute weather problems.

When the launch director satisfied himself that all areas had been taken care of, the controllers came out of the final hold and continued at the T minus 9 minute mark. The team was given a "go" and a final clear to launch. The full countdown is published in four large volumes and consists of 2000 pages of checks and procedures. If it had become necessary, it would have been possible to hold longer, depending on how much time was built into the launch window for that day. But for *Challenger* there would be no further holds.

Time Frame	Source	Comments or Events
T–9:15	PAO-Kennedy (Hugh Harris)	Fifteen seconds from resuming the countdown and looking at a launch of 51-L at 11:38.

Space Shuttle *Challenger* came out of its preprogrammed hold as the event timer was activated. This was the final segment of the countdown, after the last planned hold before launch, now rescheduled for 11:38 A.M. EST. Christa McAuliffe, Gregory Jarvis and Ronald McNair are not heard on the final flight recordings due to routine instructions for individuals to remain silent in the lower crew deck during the launch and reentry phases of the flight.

T–9 minutes	PAO-Kennedy	And we're at T–9 minutes and counting. Sequencer has been initiated. T–9 minutes and counting.
	Controller	
	PAO-Kennedy	The Ground Launch Sequencer program has been initiated.

The Ground Launch Sequencer is part of the launch-processing system and operates by relaying commands to the computers onboard the orbiter, which then reports back to Launch Control that the commands have been properly executed. The primary function of the Launch Sequencer is to check that all of the launch-commit criteria dealing with propellant loads, temperatures, pressures and other measurements are correct. It continues to monitor as many as a thousand measurements on the Shuttle stack during the final nine minutes of the count to make sure that they do not fall out of predetermined limits. Any measurements which do fall out of limits would cause a hold in the countdown.

	Test Conductor (Roberta Wyrick)	Start your APU [Auxiliary Power Units] bus recorders.
	Controller	LTC [Launch Test Conductor], will do.
	Test Conductor	And LTS, place all firing-room printer plotters online.
	Controller	LTS, coming.
	Test Conductor	And IFL, I'll need verification when you get motion on the aft recorders.
	Controller	IFL, copy.
	Controller	FPM 10 dash 15 was not performed.
T–8:30	PAO-Kennedy	T minus 8 minutes 30 seconds and counting. The flight instrument recorders are turned on. Mission Control has turned on the auxiliary data system. This package collects data from the aerodynamic information coming back as the orbiter flies through the atmosphere. Coming up on the eight minute point.

It would be data from these recorders that would show the *Challenger* struggling to stay on course due to problems with the right solid rocket booster. The computers however, would not relay enough data to ground controllers to indicate a problem with the launch in real time. Only later, when the information was reviewed, would these indicators become clear.

	Controller	Verify ops recorders one and two running.
	Test Conductor	Copy that.
T–8 minutes	PAO-Kennedy	T minus 8 minutes and counting.
	Test Conductor	Houston, send transmission of stored program commands.
	PAO-Kennedy	Orbiter test conductor Roberta Wyrick has requested that Houston send the stored program commands which is the final update on antenna management based on the liftoff time, and sets the system which makes the Orbiter compatible with downrange tracking stations.
	Test Conductor	Bring your ... to monitor, please.
	Pilot Smith	Okay, that's complete.
	Controller	Go for LH [Liquid Hydrogen] cap retract.
T–7 Min. 30s	PAO-Kennedy	T minus 7 minutes 30 seconds and the Ground Launch Sequencer has started retracting the orbiter crew access arm. This is the walkway used by the astronauts to climb in the vehicle, and that arm can be put back in place within about 15 to 20 seconds if an emergency should arise. Coming up on the seven minute point in the countdown.

T–7 Minutes	PAO-Kennedy	T minus 7 minutes and counting. The next major step will be when pilot Mike Smith is given a go to perform the Auxiliary Power Unit pre-start.

The orbiter crew access arm is the only way the crew can escape from a Shuttle while it is on the pad if an emergency situation occurs. When retracted, the crew is for the first time truly alone with their craft. If they look to the left out of the windows they will be able to see mankind's massive technology. Looking out the right window gives the crew a view of a primitive natural environment of swamps and waterways.

T–6:30	PAO-Kennedy	T minus 6 minutes 30 seconds and counting.
	Test Conductor	KRPS start your ... pressure recorders.
	Controller	Roger wilco.
	PAO-Kennedy	Coming up on the six minute point in our countdown.
	Test Conductor	PLT [Pilot], perform APU pre-start.
T–6 Minutes	Pilot Smith	OTC, PLT — it's in work.
	PAO-Kennedy	T minus 6 minutes, and orbiter test conductor has given Pilot Mike Smith a go to perform the Auxiliary Power Unit pre-start. Mike has reported back that it's in work. He will configure switches in the cockpit to put the Auxiliary Power Units in the ready to start configuration. [The APUs power the Orbiter's Hydraulic System, which controls the craft's aero surfaces during an emergency return to the Kennedy Launch Center, if such a return is required.]

If the APUs are used for more than seven minutes, there is not enough time to shut them down and service them and have full capability for another try at launch the next day. This would require a 48-hour delay before another launch could be attempted.

	Pilot Smith	OTC, PLT — APU pre-start is complete with three gray talk backs.
	Test Conductor	Copy that. Thank you.
	PAO-Kennedy	Mike Smith reported that pre-start is complete.

If there is a problem before the start of the APUs that can be fixed within the launch window, the count can be recycled to the T–9 minute mark or even earlier in the count.

T–5:30	PAO-Kennedy	T minus 5 Minutes 30 seconds and counting and Mission Control has transmitted the signal to start the onboard flight recorders. The two recorders will collect measurements of Shuttle system performance during flight to be played back after the mission. Coming up on the five-minute point. This is a major milestone where we go for Auxiliary Power Unit start.

The onboard flight recorders are similar to the famous black boxes on all commercial aircraft in that they will record how the Shuttle performed and can be reviewed by controllers if any problems occur, assisting in postflight investigations. Temperatures, pressures and physical stresses on the vehicle are recorded. *Challenger*'s onboard flight recorders were found on the floor of the ocean by underwater remotely powered vehicles weeks after the disaster. They had to be cleaned, dried and serviced before any useful information could be gleaned from them.

T–5 minutes	PAO-Kennedy	T minus 5 minutes.
	Controller	You are go for orbiter APU start.
	Test Conductor	PLT, OTC, perform APU start.
	Pilot Smith	In work.

At this point the three propellant-isolation-valve switches were thrown which allowed the hydrogen fuel to start flowing from the tanks toward the APUs.

| | PAO-Kennedy | And we've had [the] pilot ordered to perform the APU pre-start. LOX [Liquid Oxygen] replenish has been terminated and liquid oxygen drain-back has been initiated. Pilot Mike Smith now flipping the three switches in the cockpit to start each of the three Auxiliary Power Units. |

The liquid oxygen drain-back indicates that liquid oxygen is flowing through the main propulsion system and back to the large storage tanks. This is done to cool the system down slowly to minus 270 degrees so it will not be shocked by the massive flow of super-cold fluid pumped through its system during the launch. If this was not done, it is possible that cracks could occur at ignition, and destruction of the engine result.

	Pilot Smith	OTC, PLT. We have three good APUs.
	Test Conductor	Okay.
T–4:30	PAO-Kennedy	T minus 4 minutes 30 seconds and counting. The solid rocket booster and external [Tank] safe and arm devices have been armed. We've had a report back from Pilot Mike Smith that we have three good auxiliary Power Units. Main fuel-valve heaters on the three Shuttle main engines have been turned on in preparation for engine start.

Range-safety destruct devices are armed by a Ground Launch Sequencer command. A motor-driven switch called an arm-and-safe device accomplishes this. The system is then inhibited to prevent premature ignition until T–10 seconds when the system is fully activated. The system is then under the control of Air Force Range Safety personnel at the Canaveral Air Force Station.

| | Test Conductor | Flight crew, please close your visors. |

The crew will now experience oxygen flow in their suits as they close their visors in preparation for launch.

| | Commander Scobee | Roger. |
| T–4 minutes | PAO-Kennedy | T minus 4 minutes and counting. The flight crew has been reminded to close their airtight visors on their launch and entry helmets. And a final purge sequence of the main engines is underway. |

The purge is completed to insure that there is no surplus hydrogen or oxygen in the area at the time of ignition. If there were to be any surplus gases in the area, a small explosion could occur and could damage a portion of the Shuttle.

| T–3:45 | PAO-Kennedy | T minus 3 minutes 45 seconds and counting. The orbiter aero-surface test has started. The orbiter flight-control surfaces are now being moved through a preprogrammed pattern to verify that they are ready for launch. |

The Orbiter flight-control surfaces, elevon, speed break and rudders, would be needed to land in an emergency as well as later during a normal landing at the completion of the mission. They are moved using the APU system.

| T–3:30 | PAO-Kennedy | T minus 3 minutes 30 seconds and counting. Orbiter ground-support [equipment] power bus has been turned off, and the vehicle is now on internal power. [Fuel cells would still be receiving power for about one more minute.] |

Six. A Cold Morning Like No Other

T−3:15	PAO-Kennedy	T minus 3 minutes 15 seconds. Aero surface checks are complete and aero surfaces [are] in launch configuration. Gimbal [movement] checks of the Orbiter's main engines now underway. [After the three engines have gone through their gimbal checks they are moved into their flight readiness positions.]
T−3 minutes	PAO-Kennedy	T minus 3 minutes and counting. Gimbal checks now complete.
	Controller	Go for ET LO2 pressurization.
T−2:50	PAO-Kennedy	External tank liquid-oxygen pressurization has started, and purging of the Shuttle main engines is terminated.

The tank is pressurized after the valves are closed, the liquid oxygen is topped off. The tank will no longer be connected to the ground for supporting liquid oxygen replenishment.

T−2:44	PAO-Kennedy	T minus 2 minutes 44 seconds and counting. Retraction has started of the gaseous oxygen vent (valve) hood.
	Test Conductor	PLT, OTC: clear caution and warning memory. Verify no unexpected errors.
	Pilot Smith	OTC, PLT: In work.
	PAO-Kennedy	And the Ground Launch Sequencer will make a final check to make sure that the vent arm is fully retracted at T−37 seconds.
T−2:30	Telemetry	(Fuel cells stop receiving oxygen and hydrogen from the ground and are now using the required gases from onboard sources.)
	Pilot Smith	OTC, PLT: Caution and warning memory's cleared. No unexpected errors.
	Test Conductor	Copy.
T−2:20	PAO-Kennedy	T minus 2 minutes and 20 seconds, and ... the pilot, Mike Smith, has cleared the caution and memory system. No unexpected errors reported. Liquid oxygen allege pressure checks are underway. And the liquid oxygen tank approaching flight pressure.
	Telemetry	(Shuttle fuel-cell ground supplies are terminated; Shuttle now fully running on its own onboard reactants.)
	Controller	Go for ET [External Tank] LHT [Liquid Hydrogen Tank] pressurization.
T−2:05	Resnik	Would you give that back to me? Security blanket.[32]
T−2 minutes	PAO-Kennedy	T minus 2 minutes and counting. The liquid hydrogen replenish has been terminated, and liquid hydrogen pressurization to flight is underway. The vehicle is now isolated from all ground propellant and fluid loading equipment.
T−1:58	Commander Scobee	Welcome to space guys. Two minutes downstairs, you got a watch running down there?
T−1:47	Pilot Smith	Okay, there goes the LOX [liquid oxygen] arm.

This is the liquid-oxygen supply arm to the external tank. The Shuttle is now completely isolated and is ready for launch.

T−1:46	Commander Scobee	Goes the beanie cap [liquid oxygen vent cap].
T−1:45	GPC Computer	(The computer automatically verifies the readiness of the main engines for flight.)
T−1:44	Onizuka	Doesn't it go the other way?
	PAO-Kennedy	T minus 1 minute 44 seconds and counting. Coming up on the 90-second point in our countdown.

T−1:42	Onizuka	Now I see it; I see it.
T−1:39	Pilot Smith	God, I hope not, Ellison.
T−1:38	Onizuka	I couldn't see it moving; it was behind the center screen.
T−1:33	Resnik	Got your harnesses locked?
T−1:30	PAO-Kennedy	Ninety seconds and counting. The 51-L mission ready to go. The liquid-hydrogen tank now at flight pressure, and all three engines ready to go. Coming up on the one-minute point in our countdown.

The onboard computers have checked the three main engines for any anomalies and — having found none — continue the launch countdown. All systems still go for a launch.

T−1:29	Pilot Smith	What for?
T−1:28	Commander Scobee	I won't lock mine; I might have to reach something.
T−1:24	Pilot Smith	Ohhhkaaay.
T−1:04	Onizuka	Dick's thinking of somebody there.
T−1:03	Commander Scobee	Unhuh.
T−1 minute	Controller	T minus 1 minute.
	PAO-Kennedy	T minus 1 minute and counting. Sound-suppression water system now armed. [This controls the water deluge system on the Mobile Launch Platform or crawler.]

The hydrogen-burn igniters have been armed. These igniters will be fired at T−10 seconds to burn off any residual hydrogen gas. The hydrogen-burn igniters insure that any residual gases flowing through the engines do not accumulate causing a possible small explosion and pressure pulse at ignition.

T−59	Commander Scobee	One minute downstairs.
T−52	Resnik	Cabin pressure is probably going to give us an alarm.

This was a caution and warning alarm which is a routine occurrence during prelaunch. All is normal within the crew cabin at this point.

T−50	Commander Scobee	Okay.
T−47	Commander Scobee	Okay there.
T−45	PAO-Kennedy	T minus 45 seconds and counting. The solid rocket booster flight-instrumentation recorders have gone into the record mode. Coming up on the 30-second point in our countdown.

The solid rocket boosters produce 2.3 million pounds of thrust each at ignition. These boosters are made of sections connected to one another by tongue-and-groove joints sealed by O-rings and secured by 177 steel pins. The solid fuel inside contains a mixture consisting of 70 percent ammonium perchlorate, 16 percent powdered aluminum, 12 percent binder and a 2 percent epoxy curing agent. When they are ignited, the Shuttle stack is lifted off the pad for the start of the flight.

T−43	Pilot Smith	Alarm looks good. [Cabin pressure is acceptable.]
T−42	Commander Scobee	Okay.
T−40	Pilot Smith	Ullage pressures are up. [External tank ullage pressure.]
	Controller	You are go for auto sequence start.
T−34	Pilot Smith	Right engine helium tank is just a little bit low.
T−32	Commander Scobee	It was yesterday too.
T−31	Pilot Smith	Okay.
T−30	Commander Scobee	Thirty seconds down there.
	PAO-Kennedy	T minus 30 seconds and we've had a go for auto sequence start. The SRB Hydraulic Power Units have started.

Six. A Cold Morning Like No Other

The launch is now controlled by the Shuttle's five onboard, redundant set-launch-sequence computers (general purpose computers).

T–25	Pilot Smith	Remember the red button when you make a roll call.

This is a precautionary reminder for communications configuration prior to launch.

T–23	Cmdr Scobee	I won't do that. Thanks a lot.
T–21	PAO-Kennedy	T minus 21 seconds, and the solid rocket booster engine gimbal [movement check] now underway.
T–15	Cmdr Scobee	Fifteen.
T–15	PAO-Kennedy	T minus 15 seconds.
T–11	Telemetry	(Water suppression system begins to wash down the pad to protect it from the heat and sound vibrations caused by the launch.)
T–10	Controller	T minus 10 seconds. You are go for main engine start.
	Telemetry	(The hydrogen-burn igniters have been fired to burn off any residual hydrogen gas.)
	PAO-Kennedy	T minus 10, 9, 8, 7, 6…
T–6.566	GPC Computer	(Main Engine 3, Ignition Command) (Engine number 2021, flown on five previous flights).
T–6.446	GPC Computer	Main Engine 2, Ignition Command) (Engine number 2020, flown on five previous flights).
T–6.326	GPC Computer	Main Engine 1, Ignition Command) (Engine number 2023, flown on four previous flights).

After the three main engines are ignited they are run up to full programmed thrust. While this is occurring, the Shuttle stack is still bolted to the launch pad by way of the solid rockets. The thrust of the main engines bend the Shuttle assembly forward which the astronauts refer to as the "twang" motion. This energy is stored in the assembled pad structure until liftoff when it is transferred to the two SRBs. This structure load places even greater stresses on the solid rocket boosters than when the Shuttle stack flies through the period of maximum dynamic pressure. If a problem is going to occur with the solid rockets the odds are that now is the time it will occur.

T–6	Commander Scobee	There they go, guys.
T–6	Resnik	All right.
T–5	Commander Scobee	Three at a hundred.

At sea level, at 100 percent of rated thrust, each Shuttle main engine develops 375,000 pounds of thrust, for a combined total of 1,125,000 pounds of thrust.

T–5	PAO-Kennedy	We have main engine start, 4,3,2,1.
T 00	GPC Computer	(Solid rocket motor ignition.)

At ignition of the SRBs, the inside pressure of the motor cases goes from 0 to 1000 pounds per square inch.

	MS Resnik	Aaaall Riiight.
	PAO-Kennedy	And liftoff.

The *Challenger* report would conclude that "had [managers] known all the facts, it is highly unlikely that they would have decided to launch." But was that true, or simply a cover for a flawed system and a flawed decision to launch?

I want to de-mystify NASA and Space Flight.
— Payload Specialist Sharon Christa McAuliffe

CHAPTER SEVEN

The Final Ascent: The Flight of *Challenger*

We still are of the position that it's a bit of Russian Roulette.
— John Peller, Rockwell engineer, January 28, 1986

T–0 and Solid Rocket Booster Ignition

Space Shuttle *Challenger*, flying mission 51-L, was launched from pad 39B at the John F. Kennedy Space Flight Center, Florida, at 11:38:00 A.M., Eastern Standard Time, January 28, 1986. It was the first time a Shuttle had ever been launched from this newest facility, one of only two operational Shuttle launch pads. (The Shuttle launch facility SLC-6 at Vandenberg Air Force Base, California, would later be mothballed without ever being used.)

The temperature at the pad was estimated at launch by NASA to have been 36 degrees Fahrenheit. The coldest temperature for a launch before *Challenger* was 53 degrees. The temperature at the position of the failed O-ring on the right solid rocket booster was estimated to be 28 degrees Fahrenheit, plus or minus 5 degrees. Other temperature-sensing devices recorded near the right SRB registered as low as 8 degrees, but these were considered to be unreliable by NASA. Perhaps NASA did not want anyone to believe that they would risk a launch at those unforgiving temperatures. Temperatures were recorded at 24 degrees Fahrenheit on the pad the night before the launch.

The *Challenger* was carrying NASA's new Tracking and Data Relay Satellite (TDRS-B), and a flight crew of seven. The orbiter's launch weight was 268,829 pounds. The crew included Mission Commander Francis R. (Dick) Scobee who had served as pilot for *Challenger* on 41-C, Pilot Michael J. Smith, mission specialists Ellison S. Onizuka (who was previously on the classified 51-C mission), Judith A. Resnik (who flew on *Discovery* for the 41-D flight), and Ronald E. McNair (who had flown on *Challenger* for 41-B). Payload specialist Gregory B. Jarvis, flying for Hughes Aircraft Company, and teacher participant Sharon Christa McAuliffe filled out the crew. McAuliffe was chosen as the first private U.S. citizen to fly in space under a program credited to President Ronald Reagan.

The only commercial television network broadcasting the launch live was the Cable News Network (CNN). No other network felt that a live launch of a Shuttle was important enough to broadcast any longer, until *Challenger* was destroyed. (When ABC switched from its regular programming to live coverage of the Shuttle disaster, the network received 1200 complaints from viewers who demanded the network go back to their soap operas.)

There were some 500 reporters covering the launch of *Challenger* for mission 51-L. By the time the next Shuttle mission, STS-26 with *Discovery*, was launched 32 months later, there would be more than 4000.

T+0.008	E8 camera	(Hold down post number 2 PIC [Pyrotechnics Initiator Controller] firing.)
T+0.250	E9 camera	(First continuous vertical motion is recorded by cameras located on the launch pad.)
T+0.445	Camera	(First signs of black smoke can be seen near joint on right solid booster.)

At this point, as the SRB sections rotated against each other, 6000-degree gases from the right solid rocket booster had pushed past the cold primary and secondary O-rings into the joint. The gases have blasted outside to be caught on camera. The O-rings had failed to hold in the hot gases, sealing the fate of the *Challenger* and her seven-astronaut crew. The Shuttle had barely begun to move and already the crew was doomed. There was no turning back.

T+0.678	E60 camera	(Gray-black smoke is visible above the field joint on right Solid Rocket Booster.)[1]

The black color and dense composition of the smoke seemed to suggest that the grease, joint insulation and rubber O-rings in the joint seal of the right SRB were burned and eroded by the hot propellant gases. The vaporized material proved that there was not complete sealing within the joint at ignition.

The two pad cameras, which would have directly recorded the puffs of black smoke coming from the breached joint, were not in operation at the time of launch. Even if Mission Control had seen these puffs of smoke, there was nothing they could have done and it is doubtful the crew would have been told. Analysis of the film from other NASA cameras indicated that the initial smoke came from the 270–310 degree section of the aft field joint on the right[2] solid rocket booster. This was the area directly across from the external tank and the strut, which held the right solid rocket booster to that tank. The puffs of smoke were found to correspond to the frequency of the structural-load dynamics stored in the assembled structure until they were released during launch through the solid rockets. These vibrations appeared to have caused an opening/closing effect on the joints of the SRBs, allowing the blow-by to continue. Once the vibrations had subsided, the cameras appear to show that the puffs of smoke had halted.

T+0.836	E63 camera	(Eight puffs of smoke begin at this time, 0.836 through 2.500 sec.)[3]

Cameras at the pad showed puffs of increasingly blacker smoke. The smoke appeared to puff upward from the joint. While each puff was being left behind by the upward flight of the Shuttle, the next puff could be seen near the level of the joint.

T+01	Pilot Smith	Here we go.
	PAO-Kennedy	Liftoff of the twenty-fifth Space Shuttle mission. And it has cleared the tower.

A great deal of vibration, clanging and other noises were heard and felt by the crew as the twin SRBs ignited and the orbiter lifted from the pad. Flame poured out of the solids and the three main engines, as the seven astronauts were pressed into their seats.

T+2.733	CZR-1 camera	(Last positive evidence of smoke above right aft solid rocket booster/external tank attach ring.)[4]

Seven. The Final Ascent: The Flight of Challenger

At this point in the flight, the breached O-rings may have resealed themselves temporarily by the pressure of liftoff and by burning aluminum oxide from the rocket exhaust. If this is correct, it may have taken a second problem to once again force them open.

T+3.375	E60 camera	(Last smoke visible. After this point the smoke became indiscernible from the many gases mixing with rocket exhaust plumes on the pad.)[5]

Along with the very cold O-rings failing to seal properly, it is possible that water may have gotten into the field joints and become ice, adding another problem for the field joints to overcome. *Challenger* had been exposed to some seven inches of rain while on the pad, and water could have gotten in. When water becomes a gas, it expands a thousand times, adding a great deal of pressure. Later, tests did show that frozen water in the joints would at times prevent the O-rings from sealing property. (Tests from a destacked solid rocket from STS-9 showed that water could get into the field joints during a rainstorm.)[6] Photos taken before launch while the O-rings were being installed showed no abnormalities in the O-rings themselves.

When NASA reviewed launch films from previous missions, to determine if any of those flights showed the same black smoke found on *Challenger*'s flight, none was found. That review indicated that this O-ring blow-by was the worst yet for the program.

T+4.339	Telemetry	(Space Shuttle main engines 104 percent command: 104 percent of rated thrust for the main engines or 390,000 pounds of thrust.)
T+5.674	Telemetry	(Right solid rocket motor pressure 11.8 psi above nominal. This increased pressure was well within specifications and would not have caused an alarm if it had been noticed by Mission Control.)

A little after six seconds after liftoff, the spacecraft cleared the launch tower and began its roll program which pointed *Challenger* in the proper direction. The Shuttle's velocity was greater than 100 feet per second and increasing. Flight control was then transferred 1500 miles west to the Johnson Space Flight Center in Houston, Texas. This is NASA's Manned Space Flight Center. All American manned flights are controlled from there.

	PAO-Houston (Steve Nesbit)	Houston is now controlling.
T+07	Commander Scobee	Houston, *Challenger*, roll program. [This is a normal initiation of vehicle roll.]
T+7.724	Telemetry	(Roll maneuver initiated.)
	PAO-Houston	Roll program initiated.
	CAPCOM (Richard Covey)	Roger. Roll *Challenger*.
T+11	Pilot Smith	Go, you mother.
T+12–13	Camera	(Last signs of black smoke?)

For the next 45 seconds the Shuttle appeared to operate smoothly as it continued to push upward toward an expected SRB separation in one minute and twenty seconds. But the Shuttle stack would be passing through many layers of crosswinds, which were affecting its trajectory.

T+14	Resnik	LVLH [Panel switch configuration change: Local Vertical, Local Horizontal]
T+15	Resnik	Shit hot.

Opposite: Launch of *Challenger* showing black smoke between the solid rockets and external tank indicating the failure of the O-rings (NASA).

T+15	PAO-Houston	Good roll program confirmed. *Challenger* now heading downrange.
T+16	Commander Scobee	Ohhhkaaay.
T+19	Pilot Smith	Looks like we've got a lotta wind up here today. [Pilot Smith was taking note of the wind shear conditions, which were pushing the Shuttle in many directions. The spacecraft was fighting to stay on course.]
T+19.859	Telemetry	(Space Shuttle main engine 94 percent command.)
T+20	Commander Scobee	Yeah.
T+21.124	Telemetry	(Roll maneuver completed.)
T+22	Commander Scobee	It's a little hard to see out my window here.
T+24	Telemetry	(Main engines complete throttling down to 94 percent.)
T+27	PAO-Houston	Engines beginning throttling down now at 94 percent.
T+28	Pilot Smith	There's 10,000 feet and mach point five [half the speed of sound].
T+30	Crew member	[Garbled]
T+32	PAO-Houston	Normal throttles for most of the flight, 104 percent. Will throttle down to 65 percent shortly.
T+35	Commander Scobee	Point nine [nine-tenths the speed of sound].
T+35.379	Telemetry	(Space Shuttle main engine given 65 percent command.)
T+36.990	Telemetry	(Roll and Yaw attitude controls begin response to wind, from 36.990 to 62.990 sec.)

From this time until approximately 64 seconds into the flight, *Challenger* reacted to several high-altitude wind-shear situations.[7] These wind-shear conditions had been encountered on previous flights without problems and were within Shuttle design limits.[8] However, during this launch wind-shear conditions could have been directly responsible for causing a bending, and an opening and closing effect on the already damaged O-rings. Without these crosswinds and the Shuttle's thrust vectoring (changing the direction of the engines' thrust) to compensate for them, it is possible that the Shuttle O-rings, although damaged, may have held together long enough for the Shuttle to successfully separate from the solid rocket boosters and continue on into orbit. However, the temporary seal was now broken, and a disaster was in the making as the Shuttle punched through the sound barrier.

T+40	Pilot Smith	There's mach one [the speed of sound].
T+41	Commander Scobee	Going through 19,000 [feet].
T+42	Telemetry	(Main engines complete throttling down to 65 percent as the Shuttle readies itself for passage through maximum aerodynamic pressure on the vehicle.)
T+43	Commander Scobee	Okay, we're throttling down. [normal thrust reduction for the flight].
T+45	PAO-Houston	Engines at 65 percent. Three engines running normally. Three good fuel cells. Three good APUs.
T+45	NASA video	(Three bright flashes appear downstream of the *Challenger*'s right wing. Each lasts less than 1/30th of a second and may have been reflections. These flashes had been seen on other flights.)
T+51.191	Telemetry	(Space Shuttle main engines given 104 percent command.)
T+52	PAO-Houston	Velocity 2,257 feet per second [1,400 mph], altitude 4.3 nautical miles [4.9 statute miles], downrange distance 3 nautical miles [3.4 statute miles].
T+57	Commander Scobee	Throttling up.
T+58	Pilot Smith	Throttle up.

At this point the Shuttle had passed through the area of greatest pressure on the outside of the vehicle (maximum dynamic pressure of 720 pounds per square foot).⁹ The main engines had been throttled up (increased) to 104 percent and the two solid rocket boosters were still operating at full thrust. The solid rockets increase thrust automatically based on the design of the fuel inside the thrust chamber and cannot be throttled or changed in any way by the astronauts or Mission Control. Once the solids have been ignited they are impossible to shut down, unless they are destroyed by a command from range control, and must be ridden until separation after two minutes of powered flight.

T+58.774	E207 camera	(First evidence of smoke appears along the side of the right solid rocket booster.)

This was the first evidence that the right SRB was once again about to fail. The SRBs were also in full-thrust mode and pushing hard against the walls of their combined segments.

T+58.788	E207 camera	(First evidence of flame appears on right solid rocket motor. A very small flame was detected on enhanced film. One frame later the flame was visible without image enhancement. It appeared to originate at about 305 degrees around the booster circumference at or near the aft field joint.)[10]
T+59.000	NASA estimate	(Max Q, 720 psi [Maximum aerodynamic pressure on Shuttle].)
T+59	Commander Scobee	Roger.
T+59.262	E207 camera	(Continuous well-defined plume of fire on right solid rocket motor is recorded at this point in the flight.)[11]
T+59.753	E204 camera	(Flame from right solid rocket motor in +Z direction [down], seen from south side of vehicle.)
T+60	Pilot Smith	Feel that mother go.
	Crew member	Wooooo Hooooo!
T+60.004	Telemetry	(Solid rocket motor pressure divergence, right vs. left. The right SRB indicated a lower pressure and a lower thrust.)

Data developed later by Marshall Space Flight Center showed that at around 60 seconds into the flight a booster thrust mismatch began to occur between the Shuttle's two solid rocket boosters. This mismatch of thrust grew to approximately 85,000 pounds by the time the Shuttle was destroyed. The Shuttle was able to correct the trajectory and the mismatch was within specifications. The solid rockets using their thrust-vector control systems corrected it. This system gimbaled up to a peak of two degrees, which redirected the rockets' thrust, pushing the spacecraft back to the nominal preprogrammed trajectory. The solid rockets were starting to fight the trajectory problem.

The solid rocket boosters are never precisely matched and some type of thrust deviation can always be expected. Therefore, even if the ground or the astronauts noted this difference at the time it would not have caused any great concern. Nevertheless the system was beginning to struggle. The flame plume was now impinging directly on the strut, which attached the external tank to the solid rocket booster, and it would soon fail.

T+60.238	E207 camera	(First visible evidence of flame plume deflection, intermittent.)
T+60.248	E203 camera	(First visible evidence of solid rocket booster flame plume attaching to external tank ring frame.)
T+60.988	E207 camera	(First visible evidence of continuous right solid rocket booster flame plume deflection.)

As time increased, the flame plume increased in size and was deflected to the rear by the aerodynamic slipstream and circumferentially by the upper ring which attached the booster to the external tank. The flame plume was also directed onto the side of the external tank, which was later confirmed by examination of the recovered sections of the external tank wreckage.

During the second day of the initial investigation, NASA was to learn — using the videos taken during the flight — that films showed what to the investigators appeared to be an SRB burn-through. This burn-through was at or near one of the field joints. It was to be the first significant discovery in the investigation, and it became the driving force behind NASA's priority to locate the wreckage of the right solid rocket booster.

T+61.724	Telemetry	(Maximum Shuttle roll rate in response to wind.)
T+62	Pilot Smith	35,000 going through one point five [1.5 times the speed of sound].
T+62.084	Telemetry	(Greatest thrust-vector control response to crosswinds.)[12]
T+62.404	Telemetry	(Greatest yaw rate response to crosswinds.)

Due to the now rapidly increasing forces on the Shuttle, caused by the leak continuing to expand on the right solid rocket, the onboard systems attempted to counter them with thrust vectoring in an effort to keep the Shuttle on its preplanned trajectory. So violent were these movements that the computer registered a spike in the data. The entire Shuttle stack was shuddering back and forth, putting tremendous stresses on the solid rocket field joints, and the crew must have been shaken back and forth in their seats. It was not an expected sensation.

Challenger's right wing elevon was also being moved by computer commands in an effort to counter the yaw rates caused by the leak and flame. *Challenger* was struggling hard to fight the aerodynamic forces pushing on her and stay on course. This orbiter was not about to give up easily. This would be a fight to the very end.

T+62.484	Telemetry	(Right SRB outboard elevon actuator hinge movement spike.)

Once again, due to trajectory changes, the computer ordered the rocket engines to point in another direction in an attempt to keep the Shuttle on its preplanned trajectory.

T+63.964	Telemetry	(Start of preplanned pitch rate maneuver.)
T+64.660	Telemetry	(Liquid hydrogen tank leak near 2058 ring frame.)

This is the first telemetry indication that the flame from the right solid rocket had punched a hole in the external tank.[13] However, even with a hole burned through the tank and fuel pouring out it would continue to operate for another eight seconds, flying faster than the speed of sound.

T+64.705	E204 camera	(Bright sustained glow on sides of the Shuttle external tank can be seen between the external tank and the black-thermal-tiled underside of the Shuttle.)[14]

The bright glow on the sides of the external tank was a visible indication that hydrogen fuel was leaking from that tank, and mixing with the burning gases of the solid rockets.[15] The flame from the right booster had punched a hole through the external tank's wall and the fuels were beginning to mix with flame from the booster leak. This fire continued to grow until the final destruction of the Shuttle.

T+64.937	Telemetry	(Start of Space Shuttle main engine gimbal-angle large-pitch variations. Main engines adjusting to change in direction of flight due to winds and side thrust due to hole in right Solid Rocket.)
T+65	Commander Scobee	Reading 486 on mine. [Routine vehicle airspeed check.]
T+65	PAO-Houston	Engines throttling up. Three [main] engines now at 104 percent.
T+65.164	Telemetry	(Beginning of transient motion due to changes in aerodynamic forces caused by the flame plume as air moves past the plume.)
T+66.484	Telemetry	(External tank's liquid hydrogen ullage pressure begins to deviate.)

Seven. The Final Ascent: The Flight of Challenger

T+66.764	Telemetry	(Start of external tank liquid hydrogen usage pressure deviations.)
T+67	Pilot Smith	Yep, that's what I've got too.
T+67	CAPCOM	*Challenger*, go at throttle up. [Last radio report heard by the crew.]
T+70	Commander Scobee	Roger, go at throttle up. [Last voice contact with the crew heard on the ground.]

At Mission Control in Houston, Texas, sat capsule communicator and fellow astronaut Richard Covey, a veteran of mission 51-I, which had flown five months earlier. To his left sat Flight Director Jay Greene who was directing the mission and passing along information to the crew through the capcom. At Greene's left stood the public affairs officer for this flight, Steve Nesbit, who was calling the events of the flight for public transmission. It was scheduled to be his last launch call before moving on to his next assignment. For the most part he would keep his eyes on the readouts, rarely looking at the television set at his side. Next to Nesbit sat the flight surgeon.

| T+72.204 | Telemetry | (Start of divergent yaw rates, right vs. left solid rocket booster.) |
| T+72.284 | Telemetry | (Start of divergent pitch rates, right vs. left solid rocket booster.) |

At this point the right solid rocket motor gyro data showed that the right SRB was moving in relation to the rest of the Shuttle stack. This data indicated that the top or nose of the SRB was moving toward the external tank while the tail end was moving away. By this time the right solid rocket was no longer attached to the external tank by the lower strut, and was flying free at the still powerfully thrusting tail end. The lower strut had disintegrated as the booster rotated through the upper strut still attaching the right booster to the external tank. Despite this twisting, the upper strut would stay connected until the tank exploded. Wreckage of the Shuttle's right wing would later show that the underside may have been struck by the rotating SRB as indicated by some damage on the black thermal tiles in that area.[16] It would then have struck the external tank. This may have been what Pilot Smith saw out of his window.

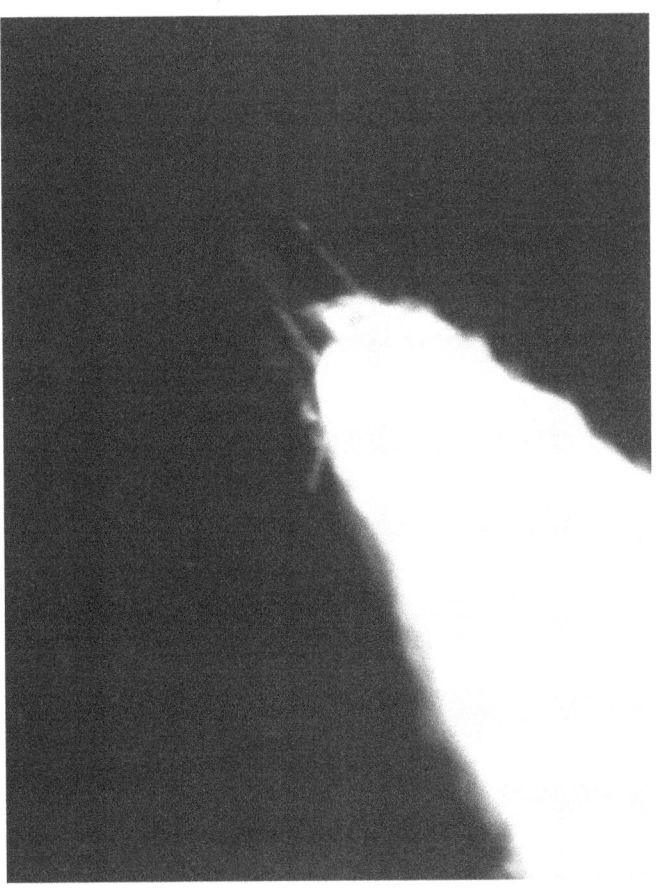

T+72 Seconds — External tank has been breached as fuel spills and burns (NASA).

The data indicated that a wide variety of flight systems onboard *Challenger* were reacting to events and attempting to correct the overwhelming events that were working to destroy it. The orbiter continued to correct the trajectory and kept *Challenger* on course right up until its destruction.[17]

T+72.478	Telemetry	(Solid rocket booster major high-rate actuator command.)
T+72.497	Telemetry	(Space Shuttle main engine roll gimbal rates 5 deg/sec.)
T+72.564	Telemetry	(Start of H_2 tank pressure decrease with two flow-control valves open.)

This was the first abnormal engine indication. As fuel pressure dropped, the control system automatically responded to the situation by opening the fuel flow-rate valve. The engines were now being starved for fuel but the computers were finding ways to keep the fuel flow going as the main engines continued to operate. Even as the Shuttle was beginning to break up, the main engines continued to push what remained of *Challenger*.

T+72.624	Telemetry	(Data reduction; last state vector downlinked from Shuttle.)
T+78.661	Telemetry	(Shuttle begins to yaw to the left.)
T+72.964	Telemetry	(Start of sharp Main Propulsion System LOX [liquid oxygen] inlet pressure drop.)
T+73	Pilot Smith	Oh, oh! [Last voice contact recorded from *Challenger*'s crew.]

This last transmission was acquired with great difficulty from the recorded data and was not heard as the flight was ongoing. The air-to-ground as well as intercom communication systems shut down at the time of the breakup. The sound of the breakup can be heard over this last taped crew comment.

It is impossible to know exactly what Pilot Mike Smith saw when he exclaimed, "Oh, oh!" We do know that from his right-hand seat he could have seen any dramatic movements made by the right SRB, and could have been the first to know that a disaster was imminent.

Video coverage showed that within one second of the breakup of the vehicle a bright orange flash was observed in the area of the Shuttle's nose. At the very least, all four flight-deck crew members would have been able to see that flash. It was later determined that the orange flash corresponded to

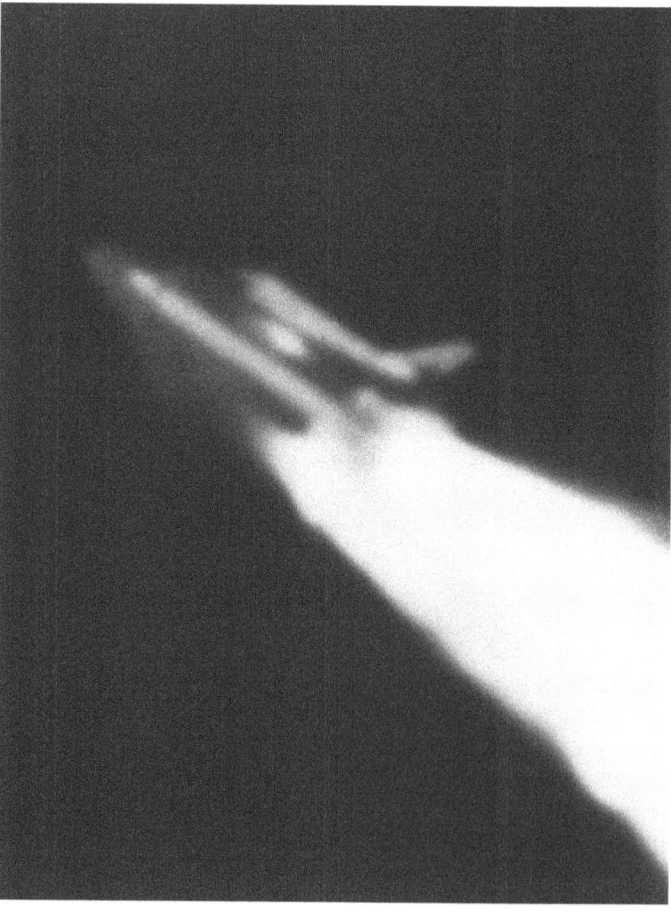

T+73.191 Seconds — Aft dome of external tank shows final structural failure (NASA).

Seven. The Final Ascent: The Flight of *Challenger*

the Shuttle's Reaction Control System fuels explosively burning, destroying the front section of the vehicle, and possibly killing one of the crew on the flight deck (possibly the commander who never used his emergency air supply).

There was virtually no visible hint of trouble on the flight, in the Shuttle or on the ground at Mission Control, until the explosion. The data — and the flight — just stopped.

T+73.010	Telemetry	(Data reduction, last full computer frame of Tracking and Data Relay Satellite (TDRS-B) data. The main payload of the *Challenger* was being destroyed as it sat inside the cargo bay.)
T+73.044	Telemetry	(Start of sharp Main Propulsion System LH_2 inlet pressure drop.)
T+73.124	E204 camera	(Circumferential white pattern on external tank aft dome; LH_2 tank failure.)[18]

At this point the external tank was beginning its break-up and final structural failure.[19] The entire aft dome which contained the hydrogen tank fell away, exposing a white vapor area trailing away from the aft area of the external tank. This tank failure added one G of force on the *Challenger* and her crew. As the aft end fell away, the tank collapsed, consumed in an orange-and-white, expanding ball of burning gases.

T+73.124	Telemetry	(Right solid rocket motor pressure 19 psi lower than left solid rocket motor. Lower rocket pressure equaled a loss of 85,000 pounds of thrust from the right solid rocket motor.)
T+73.137	E207 camera	(First hint of vapor at inter-tank area of the external tank. Liquid oxygen tank is failing.)

The release of the liquid hydrogen created an extra thrust of 2.8 million pounds, which pushed the remains of the hydrogen tank into the inter-tank structure.[20]

T+73.143	Telemetry	(All Shuttle engine systems start responding to loss of fuel and LOX inlet pressure.)
T+73.162	E207 camera	(Sudden cloud along the external tank between inter-tank and aft dome can be clearly seen.)
T+73.191	E204 camera	(Flash between Orbiter and LH_2 tank.)
T+73.213	E204 camera	(Flash near the right solid rocket booster forward attachment and brightening of flash between orbiter and external tank. Beginning of total vehicle breakup.)
T+73.282	E204 camera	(First indication of an intense white flash at the right solid rocket booster forward-attach point.)

This white flash was later found to be located in the area of impact of the right solid rocket booster when it struck the external tank.

T+73.327	E204 camera	(Greatly increased intensity of white flash.)
T+73.377	Telemetry	(Start of Reaction Control System jet-chamber pressure fluctuations in the nose of the Shuttle soon to explode.)
T+73.383	Telemetry	(All Shuttle engines approaching HPFT [High Pressure Fuel Turbo pump] discharge temperature redline limits.)
T+73.482	Telemetry	(Space Shuttle main engine-2 controller, last time word update.)
T+73.503	Telemetry	(Space Shuttle main engine-3 in shutdown mode due to HPFT discharge temperature redline.)
T+73.503	Telemetry	(Space Shuttle Main Engine-3 controller last time word update.)
T+73.523	NASA estimate	(Space Shuttle main engine-1 in shutdown mode due to HPFT discharge temperature redline.)

T+73.543	NASA estimate	(Space Shuttle main engine-1 last telemetered data point. Engines begin to automatically shut down.)
T+73.618	Telemetry	(Last validated orbiter telemetry measurement.)[21]
T+74.130	Telemetry	(Last radio frequency signal received from Space Shuttle *Challenger*: static, loss of downlink.)
T+74.587	E204 camera	(Bright flash in the vicinity of orbiter nose. Slow motion film shows that the fuels in the nose of the orbiter exploded downward away from the top front of the crew cabin. This secondary explosion did not destroy the crew cabin as it tumbled out of control, free from what had been the *Challenger*. It is possible, however, that the explosion did kill the Shuttle's commander seated right behind the RCS thrusters.)

At 73–74 seconds into the flight, *Challenger* suffered a catastrophic breakup due to the explosive burn of the fuel, which pushed the orbiter away from the destroyed external tank. The explosive burn itself did not destroy *Challenger*. There was no explosive shock wave because there was no overall explosion.[22] Extreme aerodynamic loads placed on it by the dramatic change in flight angle destroyed the spacecraft.[23] The orbiter then broke into several large sections, which could be seen flying out of the cloud of expanding gases.[24] The destruction was total and complete. Space Shuttle *Challenger* had lost its battle, and mission 51-L was over, but most of the crew were still alive.

The crew compartment separated at this time and continued an arched upward trajectory to about 65,000 feet.[25] The crew cabin could be clearly seen in the videos. *Challenger* was at an altitude of approximately 46,000 feet when it was destroyed while traveling at mach 1.92, almost two times the speed of sound. Downrange distance was almost seven nautical miles. The last view of *Challenger*, which could be seen on the live video, showed its front section nosing down as it was engulfed in burning fuel. Debris would continue to climb an additional eleven miles.

In an environmental impact statement, issued in 1977, it was calculated that if an external tank ever exploded, it would be equal to the explosive force of 6.3 million pounds of dynamite

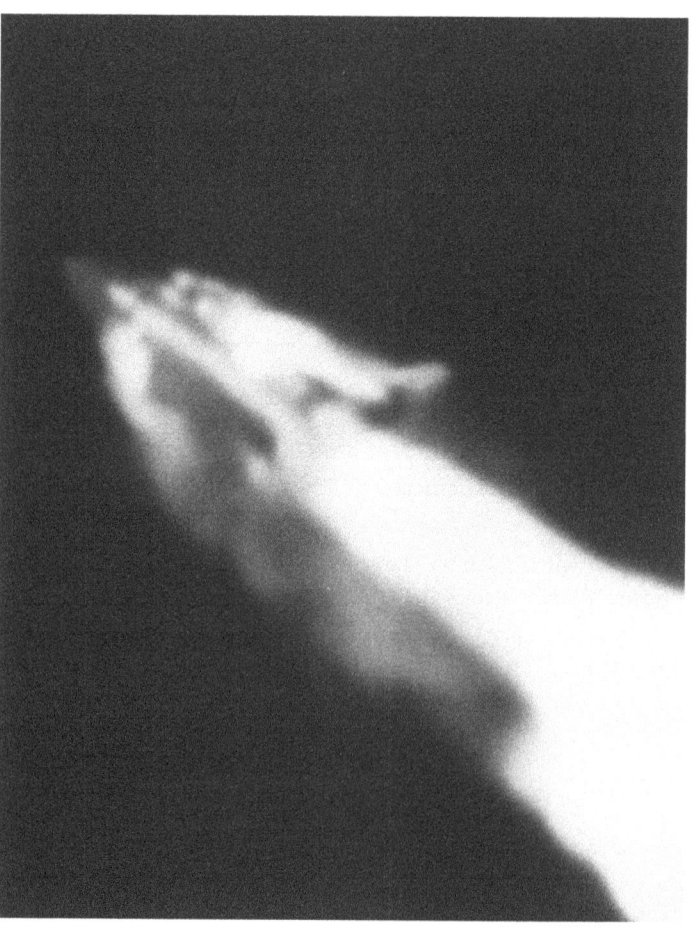

T+73.3 Seconds — Complete breakup of the external tank as *Challenger* flies free (NASA).

Shuttle debris falls from the explosive burn as both solid rockets fly away (NASA).

(a small atomic bomb). It was further stated that the solid rockets would also be destroyed and that there would be no chance for a crew to survive.[26]

Both of *Challenger*'s solid rocket boosters (SRBs) survived the breakup and explosively separated from the external tank, which had collapsed as its fuel and oxygen gases rapidly burned. The SRBs continued their powered flight, until they too were destroyed by a command sent by the air force range safety officer (RSO), Major Gerald F. Bieringer, stationed at Cape Canaveral Air Force Station. Bieringer had lost all data after the explosion and made a guess as to where the SRBs were and pushed the destruct button. If the Shuttle had survived the destruction of its external tank, the destruction of the SRBs could have endangered a Shuttle attempting a return-to-launch-site (RTLS) abort. Major Bieringer simply did not have enough information to make that critical decision. He did not see the explosion as it occurred nor did he know where the Shuttle was. There was no downlink!

The bright flash in the vicinity of the Shuttle's nose showed the destruction of the Reaction Control System and a hypergolic burn of its liquid propellant.[27] This is indicated by the reddish color which can be seen on the edge of the explosion just before the NASA video pulls back for a wider field of view.

Before the destruction of the Shuttle, there was no indication from any source that the vehicle was going off course. It appeared to be making a picture-perfect ascent. However, when the flight video is viewed at a faster-than-normal speed, its shimmy back and forth against the crosswinds can easily be seen. This view was later supported by telemetry data during the investigation.

T+76 Seconds — Complete breakup of *Challenger* as left solid rocket continues to fly out of the expanding fireball (NASA).

T+1:15   PAO-Houston   One minute 15 seconds. Velocity 2,900 feet per second [1,977 mph], altitude 9 nautical miles [10.35 statute miles], downrange distance 7 nautical miles [8.05 statute miles].

To explain why his commentary was still ongoing despite the fact that the Shuttle had been destroyed, Steve Nesbit, public affairs officer on the flight explained, "There was a monitor off to the left, but I would liken that to checking your rear view mirror while you concentrate on what is in front of you, and periodically you look to the side or to your rear-view mirror to verify what's going on. It was five or so seconds before I glanced over at the screen and saw what was remaining of the explosion." He had heard the flight surgeon ask "What was that?" and those words pulled him from the computer screen to the video monitor.

T+1:16	NASA video	(Video clearly shows the expanding cloud of burning fuels with both SRBs flying out of the cloud. Also visible are the three main engines still firing and flying off as a single unit, with the tail of the Shuttle still attached. The left wing of the *Challenger* as well as the forward fuselage containing the crew can also be seen. Enhanced still photos from the NASA video would show a group of umbilical lines trailing behind the falling forward fuselage.)
T+1:16.437	E207 camera	(Right solid rocket booster nose cap separation/chute deployment, still attached to the SRB.)

Air force radar operators were now able to track multiple objects[28] (but not the SRBs) falling toward the ocean and this information was forwarded to the flight dynamics officer.

Airline passengers and crews could see the explosive flash in the sky. First officer Ray Espinosa, onboard a Miami-bound Eastern Airlines flight from New York, stated, "I knew something was wrong when I saw the first flash."

Flight Director (Jay Greene)	Fido, trajectory?
Flight Dynamics Officer (Brian Perry)	Go ahead.
Flight Director	Trajectory Fido?
Flight Dynamics Officer	Flight, Fido. Filters got disagreeing sources; we're go.

The flight dynamics officer was indicating with the "we're go" comment that the orbiter should be on an RTLS abort, which indicated it would be flying back to the Cape for an emergency landing. He did not have a television screen to look at and did not know the extent of the disaster unfolding at the Cape.

T+1:38	NASA estimate	(*Challenger* crew compartment, with its seven-person crew, tops out of its upward trajectory and begins its 65,000-foot fall toward the Atlantic Ocean. At least three of the astronauts knew that death was only a few minutes away.)
T+1:50.250	E202 camera	(Right solid rocket booster destroyed via range safety system.)
T+1:50.252	E230 camera	(Left solid rocket booster, destroyed via range safety system.)

By 11:39:50 Eastern Time, many large pieces of Shuttle *Challenger*, and its external tank, were observed falling into the Atlantic Ocean east-northeast of the Kennedy Space Center.[29] NASA cameras positioned on top of the Vehicle Assembly Building clearly showed large pieces of the Shuttle falling into the ocean, while other tracking cameras followed smoking pieces of debris as they fell into the impact area.

After the destruction of the solid rocket boosters, they too fell into the Atlantic Ocean, east of the orbiter and external tank impact areas. The impact areas were estimated by NASA to be 18 miles downrange from Kennedy Launch Center.

Why were the two solid rocket boosters destroyed? The live video of the flight clearly shows both SRBs being pushed away from the external tank's explosive burn; however, both rockets then recovered from this change of trajectory and continued downrange, away from any land mass and back on a downrange trajectory. The space agency would later claim that they were destroyed because they may have been heading toward land or the coastline. What land? Africa, 3000 miles away?

When pressed at a news conference from the Johnson Space Center, Houston, held the day after the disaster, Ascent Flight Director Jay Greene stated NASA did not know why the SRBs had been destroyed. Yet, mission rules clearly state that the determination of controllability of the SRBs leading to a destruct call is the responsibility of the flight director alone. Further, he stated flatly that neither SRB was heading toward land. This view is supported not only by the video and still photography, but by the impact point of the SRBs on the Atlantic Ocean miles downrange from where the Shuttle wreckage was discovered.

Flight Director Jay Greene later said that the range safety personnel used the coastline as the area to gauge a destruct call. He was then asked if they were headed for the Florida coastline and again the answer was no. With no Shuttle to push into space, it makes sense to expect these two solid rockets to travel farther out over the ocean. The SRBs could only thrust for ten additional seconds of useful powered flight at the time of their destruction, therefore even if they were pointed directly at any land mass the boosters would never have gotten anywhere near the land. Were NASA and the air force getting rid of a possibly nasty problem? And, why destroy the solids if there was any possibility that the Shuttle had survived the explosion and was flying in the general area? It could have survived to make an emergency landing back at Kennedy Launch Center. The air force simply did not have enough information to make that decision, and it was not their decision to make. They made no contact with flight control before the solids were destroyed, contrary to written flight rules. They simply had no data, and no destruct authority.

	GNC	Flight, GNC. We've had negative contact, loss of downlink.
T+1:51	PAO-Houston	Flight controllers here looking very carefully at the situation. Obviously a major malfunction. We have no downlink.

At 11:40 A.M., the SRB recovery ships *Liberty Star* and *Freedom Star* were notified of the disaster by the launch recovery director (LRD), via the booster recovery director (BRD), and given the coordinates of the impact area. These two ships proceeded at a top speed of 15 knots toward the estimated impact point. Due to the cloud cover, the crew members of both ships were unable to see the accident when it happened.[30]

While en route, these ships recovered several pieces of debris. Why would these rescue ships waste valuable time recovering pieces of wreckage when the fate of the seven crew members was as yet undetermined?

T+2:05		At this point in a nominal launch profile, the solid rocket boosters would have separated from the external tank, leaving the Shuttle to continue into space with main engines thrusting into orbit.
	Flight Dynamics Officer	Flight, Fido.
	Flight Director	Go ahead.
	Flight Dynamics Officer	RSO [range safety officer] reports vehicle exploded.

There was a long pause at this point as the shock began to settle in.

	Flight Director	Copy.

It was at this point that the public affairs officer's voice began to show great stress as he struggled to find the words he needed to say to inform the public. Steve Nesbit would stay at his post and continue to report.

T+2:43	PAO-Houston	We have a report from the flight dynamics officer that the vehicle has exploded. Flight director confirms that. We are

		looking at, checking with the recovery forces to see what can be done at this point.
	Launch Safety Officer	Looks like about 50 minutes, five zero minutes before the helicopters are cleared in because of debris.
	Flight Director	Fifty minutes from what time, LSO?
	Launch Safety Officer	OK, from the time of the explosion.
	Flight Director	Copy.
T+3 minutes	PAO-Houston	Contingency procedures are in effect.
T+3 min 08s	PAO-Houston	We will report more as we have information available. Again, to repeat, we have a report relayed through the Flight Dynamics Officer that the vehicle has exploded. We are now looking at all the contingency operations and awaiting word from any recovery forces in the downrange field.

An H-3 helicopter with the call sign Jolly One was instructed to lift off for inclusion in ongoing rescue operations, but to hold short of the safety zone which NASA had established due to the debris falling in the area.[31]

	Flight Director	Don't reconfigure your console. Take hard copies of all your displays. Make sure you protect any data source you have.

Flight Director Greene, as a first attempt to understand what had just happened to the Shuttle, then polled the controllers. Not one of the Mission Controllers could give any indication of a problem. Many would later say that the data just stopped. Mission Controllers went about the task of protecting any information they had on their computers. They, like everyone else, wanted to know exactly what had just happened to their friends and coworkers. There would be no cover-up in Mission Control.

To protect the families of the crew members from the press, or others, they were escorted from the viewing area by NASA public-relations personnel to the astronaut crew quarters.

T+3:58	NASA estimate	(Two minutes and 45 seconds after the breakup, *Challenger*'s crew cabin shattered when it hit the Atlantic Ocean at an estimated speed of 204–207 miles[32] per hour. This impact deceleration force equaled about 200 times the force of gravity. All seven crew members were thought to be strapped into their flight seats on impact. The cause of death of the crew could not be determined due to the overwhelming forces of the impact.)

Dr. Joseph Kerwin, an astronaut and physician who flew onboard the first manned Skylab flight in 1973, was assigned to the investigation board for medical assessments. Dr. Kerwin would later report that he and a team of NASA scientists and engineers had determined that at least some of crew remained conscious long enough to activate their emergency Personal Egress Air Packs (PEAPs).[33]

An analysis of the gauges on two of the recovered air packs, including that of Pilot Mike Smith, concluded that from three-fourths to seven-eighths of the five-minute supply of air had been used, through normal breathing.[34] The air pack worn by Commander Dick Scobee was also recovered, but tests showed that it was not activated.[35] Is it possible that there was enough air in the cabin to not require the use of the air packs? If not, Commander Scobee was already dead before the cabin hit the ocean.

Dr. Kerwin also reported in his medical report (see Chapter Ten) that the crew possibly, but not certainly, lost consciousness in the seconds following orbiter breakup, due to in-flight loss of crew-module pressure. He stated that the loss of pressure would be deadly, despite the air tanks. "At 65,000 feet the crew would have had 6 to 15 seconds of useful consciousness, even with their breathing packs on. The number of seconds the crew may have retained consciousness would be a function of how rapidly the crew module lost cabin pressure, and that would depend on how large the hole was in the module." However, it is only a guess, based on circumstantial evidence, that the crew module did indeed lose pressure. This view is based on the usage of the breathing packs by only some of the crew. The crew module was far too damaged by impact on the ocean to give any indication of crew-module pressure-loss timing. The crew autopsy reports have yet to be released to the public.[36]

11:44 A.M. EST　　　　　　　　　　(GOES Satellite, from orbit over Florida, images the smoke and vapor cloud left by the *Challenger* explosion.)

Challenger's Search and Rescue Timeline

After the destruction of the Shuttle, Flight Director Jay Greene ordered all mission data, computers, flight recorders and launch facilities to be impounded, to secure any information on the disaster. This procedure helped insure that no information would be accidentally lost, destroyed or misplaced. Mission Control was not about to lose anything that could help in the investigation. All controllers, including astronaut Richard Covey, a personal friend of the crew serving as capsule communicator, stayed at their positions as search and rescue operations got underway. Very few conversations were going on as NASA cameramen continued to film inside of Mission Control. They had a job to do and they were going to do it no matter what, and the staff of Mission Control understood that. There was always a possibility that something at Mission Control had caused the problem, and these cameramen may have recorded it. Mission Controllers were too stunned to speak. Many stayed seated and stared straight ahead while others held their heads in their hands not willing or able to do much else. Control was no longer in their hands and there was nothing they could do.

T+5 minutes	PAO-Houston	This is Mission Control, Houston. We have no additional word at this time.
T+5:20	PAO-Houston	Reports from the flight dynamics officer indicate that the vehicle apparently exploded and that impact in the water [was] at a point approximately 28.64 degrees North (and) 80.28 degrees West. We are awaiting verification ... as to the location of the recovery forces in the field to see what may be possible at this point.
T+6:05	PAO-Houston	And we will keep you advised as further information becomes available. This is Mission Control.
T+8:42	NASA video	(Video clearly shows one rescue paramedic parachuting into the Shuttle *Challenger* impact area. Legs and arms can be seen moving as he fights the winds for control of his parachute, which can be seen to partially deflate. He is in trouble.)

After the NASA live video showed a long-range shot of the parachute for a few seconds, it cut away to show a view of the empty Launch Pad 39B. No other views were shown of the parachute. Why was this done? With a major disaster ongoing and with active rescue operations, it would be expected that NASA would attempt to show those operations. Witnesses at the Launch Center were also able to see the parachute coming into the impact area.[37] NBC News also reported that a hydrofoil and paramedics were sent into the impact area. One eyewitness to the unfolding disaster was *Countdown* magazine's Florida correspondent Carleton T. Bailie who was able to report a very clear and possibly unique view of the rescue parachute dropping into the recovery area. "I somehow managed to find the parachute in my telescope: it was still coming down. Some-

Seven. The Final Ascent: The Flight of Challenger

one shouted that it was a man. Through the telescope, it appeared that whatever was attached to the chute was as long as the chute was wide. I followed it for what seemed an eternity."[38] However, NASA chose to show an empty launch pad and views of Mission Control. Why did NASA discontinue the rescue video and what was the fate of these rescue personnel? That question has never been asked, and as such has never been answered. Where is that close-up video today?

T+9:10	PAO-Houston	This is Mission Control Houston. We are coordinating with recovery forces in the field. Range safety equipment, recovery vehicles intended for recovery of the SRB [are] in the general area. Those parachutes are believed to be paramedics going into that area.

A NASA Public Affairs contingency plan, dated March 30, 1984, developed in the event of a disaster stated, "In the event of an emergency involving crew injury or fatality, the fact will be apparent to radio listeners and T.V. viewers. Status of the crew will be the prime public consideration."

Associated Press writer, Howard Benedict reported on the wire that, "Paramedics leaped into the water in an effort to find any trace of survivors." He also reported, "One of them was seen floating down on parachute."

T+9:48	PAO-Houston	To repeat, we had ... an apparently normal ascent, with the data coming to all positions being normal, up through approximately the time of main-engine throttle back up to 104 percent. At approximately a minute or so into the flight there was an apparent explosion. The flight dynamics officer reported that tracking reported that the vehicle had exploded and impacted the water in an area approximately located at 28.64 degrees North, 80.28 degrees West. Recovery forces are proceeding to the area, including ships and a C-130 aircraft. Flight controllers [are] reviewing their data here in Mission Control. We will provide you with more information as it becomes available. This is Mission Control Houston.

The C-130 aircraft was carrying rescue divers trained to jump into just such a situation.

T+14 minutes	CNN News	(Cable News Network reports that medical personnel and divers had parachuted into the disaster area.)
T+26:15	NASA video	(Live video from onboard a NASA rescue helicopter shows a white round object floating in the disaster area. NASA makes no comment as the video is shown in silence.)

At a point thirty minutes after the launch, the Cable News Network (CNN) again reported that a C-130 in the search area had parachuted rescue medical personnel into the area to search for possible survivors, and showed a rerun of an individual parachuting into the area. Ten minutes later, CNN would announce that there were no survivors from *Challenger*'s final flight.

Two days later, Lieutenant Colonel Robert Nicholson, director of public affairs for the Eastern Missile and Space Center at Florida's Patrick Air Force Base would state, "The report [of paramedics jumping] was in error." Why would an air force public affairs officer be reporting on a NASA civilian disaster? And, if these were not paramedics, who did jump into the area? The video evidence is clear, irrefutable and unmistakable. The public affairs report did not state there were no parachutes, just that these men were not paramedics. Perhaps Nicholson was unaware of the USAF's 1730th Para-rescue Squadron whose only job was to drop into rescue situations, but that would seem to be highly unlikely. They use the same C-130 aircraft and parachutes seen by the many eyewitnesses. This was the same unit which was on standby for the next Shuttle flight designated STS-26.

At noon Eastern Time, the White House held a news conference conducted by White House Spokesman Larry Speaks. He reported that the president was informed of the accident by Vice President George Bush and National Security Advisor John Poindexter. Mr. Speaks further reported that the president viewed the reruns of the flight, and was saddened and concerned for the crew and their families.

Later, the White House reported that the State of the Union address, scheduled for that evening, would be postponed for one week due to the explosion. It was further reported that the President would address the nation that evening about the Shuttle tragedy.

PAO-Houston	This is Mission Control Houston. Repeating the information that we have at this time. We had an apparently [normal] liftoff this morning at 11:38 Eastern Time. The ascent phase appeared normal through approximately the completion of the roll program and throttle down and engine throttle back to 104%. At that point we had an apparent explosion.
	Subsequent to that, the tracking crews reported to the flight dynamics officer that the vehicle appeared to have exploded. And that we had an impact in the water downrange at a location approximately 28.64 degrees North, 80.28 degrees West. At the time the [contact] was lost with the vehicle, according to a poll by the Flight Director Jay Greene of the positions here in Mission Control there were no anomalous indications, no indications of problems with the engines or with the SRBs. There were also no other problems with any of the other systems at that moment, through the point at which we lost data. Again, this is preliminary information. It's all that we have at the moment. And we will keep you advised as other information becomes available. ... There are recovery forces in the general area ... including aircraft and ships. We saw what we believed to be paramedics parachuting into the impact area. And we have no additional word at this point. We will keep you advised as details become available to us. This is Mission Control Houston.

Within the next thirty minutes, aircraft and ship search-and-rescue support was requested by the Support Operations Center (SOC) using the search-and-rescue telephone network (SARTEL). The National Military Command Center (NMCC) was informed by SOC of the Shuttle disaster. The SOC requested that the Range Safety Office (RSO) provide an estimated time when the impact area would be clear, so that additional rescue forces could enter. The RSO made the determination based on tracking radar that it would be 55 minutes before any forces would be allowed in the area.[39] It was estimated by range-safety personnel tracking the many segments of the falling spacecraft that the pieces fell from at least 50,000 feet, as reported by Lieutenant Colonel Robert Nicholson, U.S. Defense Department spokesman.

The rescue helicopter code-named Jolly One reported sighting debris falling and landing in the Atlantic Ocean.[40] There was no indication if any of the debris could be identified. There was no word on the status of the crew. At the same time an HC-130 aircraft and two HH-3 helicopters departed from the Clearwater Coast Guard Station en route toward the impact area.

Within thirty-five minutes of *Challenger*'s destruction, the Joint Chiefs of Staff at the Pentagon formed a "Shuttle Response Cell."[41] This cell was formed before some of the pieces of the destroyed Shuttle had yet to finish their fall into the Atlantic. Because of the Shuttle's satellite-delivery capabilities, the military considered it to be a national military resource. Therefore a concern for its ability to continue operations may have been foremost on the minds of the Pentagon. Or was there something else? The Shuttle had been designed in part with large military payloads in mind. Was there such a payload onboard? If so, what was it? And why the rush? Was the Pentagon hiding something and was this why Defense Department spokesmen were being used on civilian Shuttle mission?

T+45 minutes PAO-Houston   This is Mission Control Houston. Repeating our earlier information for those who have not heard it previously.... Launch time was approximately at 10:38 Central Time. On launch, approximately a minute or so after tower clear, there was an apparent explosion of the orbiter. At the time data was lost, approximately a minute into the flight; that was shortly after throttle up to 104% of the three main engines. The flight director polled positions, flight controller positions, in the room later on this morning and was informed that there were no anomalous indications at the time. Tracking reported impact of the vehicle with the water. According to data, that was approximately 18 miles downrange at the time data stopped. Recovery forces being deployed to the field ... were unable [at first] to enter the specific area because of continuing falling debris and at about this time are [now] being admitted to the impact area. Contingency procedures are in effect and following ... all of the data available in Mission Control from the flight up to the point of the incident; data is being secured and will be carefully evaluated. We have no additional information at this time, and will keep you advised as other details become available. This is Mission Control Houston.

At 12:30 P.M., a second H-3 helicopter, with the call sign Jolly Two, reported to NASA that it was airborne and proceeding to the recovery area.[42]

T+70 minutes PAO-Houston   This is Mission Control Houston, at 11:48 A.M. Central Standard Time. Recovery teams are searching the impact area off the coast of Launch Pad 39B, where earlier this morning on ascent we had an incident. Approximately one minute after ascent an apparent explosion [took place] as the Space Shuttle had shortly before reached throttle back position. The search-and-rescue teams [were] delayed getting into ... the area because of debris continuing to fall from very high altitudes for as long almost as an hour after ascent. Those teams [are] in place now in the search area. We will provide additional information as it becomes available to us. This is Mission Control Houston.

If the rescue teams were not able to reach the search-and-rescue area as per the Public Affairs Office, then just who was the individual clearly seen parachuting into the area? Did something happen to that individual? If so, what was that person's fate? Did NASA later lie about the rescue workers to cover up the loss of additional personnel?

The search-and-rescue operations were being directed by the Launch Recovery Director (LRD), coordinated by the United States Coast Guard (USCG) and assisted by the Eastern Space and Missile Center (ESMC). The navy was requested by NASA to initiate preparation for an underwater salvage operation.[43] Within an hour of the *Challenger*'s destruction, NASA reported that pre-established sea- and air-rescue teams were operating within the impact areas, estimated to be twenty miles off the coast of Cape Canaveral. There were twelve aircraft involved in the search, craft from the air force, coast guard and the navy. Along with these were ten surface ships from the navy, NASA and the coast guard. This search would end up becoming the largest post–World War II search-and-rescue operation on record.[44] By mid afternoon, three U.S. Coast Guard cutters were en route to the impact area. By this time, the Explosive Ordnance Team members were placed on standby. The NASA launch recovery director requested that a major search-and-rescue operation begin.[45]

The booster recovery director (BRD) contacted Perry Off-shore Company to check on the availability of a Remotely Operating Vehicle (ROV) named *Sprint*. The director also contacted Sonat in Houston, Texas, to obtain a second ROV named *Scorpio*.

2:10 P.M. EST   PAO-Houston   This is Mission Control Houston at 1:10 p.m. Central Standard Time. There will be a press conference this afternoon held with Jesse Moore ... the associate administrator for the Office of Space Flight, NASA headquarters. That will be at 3:30 p.m. at the auditorium, Kennedy Space Center. And that's Eastern Time.

At 2:56 P.M. Eastern Time, CNN reported that recovery vehicles had spotted some floating debris, including what appeared to be debris with a parachute on it. Spokesman Nicholson of the Department of Defense reported that the debris could be part of the SRB. This debris with a damaged parachute on it would eventually be proven to have come from the right SRB. Since the left

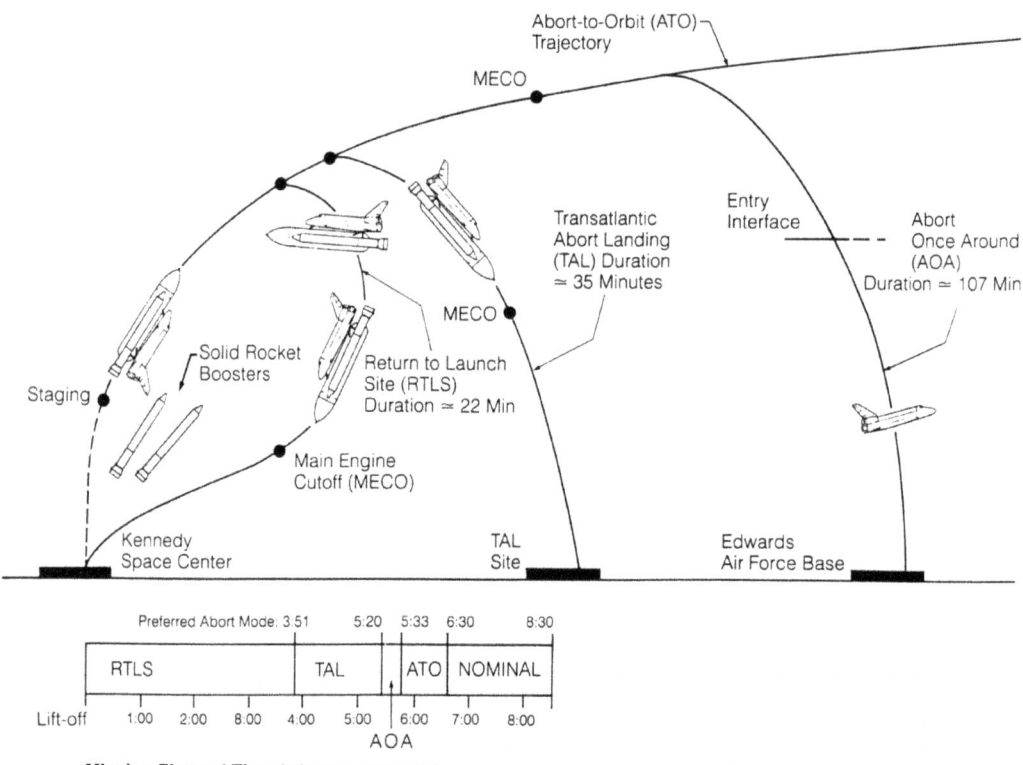

Space Shuttle abort regions (graphics from *Challenger* report, page 179).

SRB parachutes never deployed, this eliminates any possibility that the parachute seen live on national television was anything but an individual parachuting into the recovery area. No word of the orbiter's impact location was verified at the time. The swells in the recovery area were reported to be fifteen to eighteen feet with very low visibility. The helicopters were sent back to their base to wait for better visibility. No signs of life were ever reported.

4:10 P.M. EST	News conference with Jesse Moore, NASA Associate Administrator	It is with deep, heart-felt sorrow that I address you this afternoon. At 11:40 A.M. this morning, the space program experienced a national tragedy with the explosion of the Space Shuttle *Challenger*, approximately a minute and a half after launch from here at the Kennedy Space Center. I regret that I have to report that based on very preliminary searches of the ocean where the *Challenger* impacted this morning, these searches have not revealed any evidences that the crew of *Challenger* survived.[46]

As announced by Jesse Moore at his news conference, NASA formed a panel of experts to investigate the cause of the accident. This panel became known as the 51-L Interim Mishap Review Board.[47] Named to the board were William Lucas, director of the Johnson Manned Space Flight Center, Houston, Texas; Richard Smith, director of the Kennedy Space Center, Florida; Arnold Aldrich, manager of the National Space Transportation System at the Johnson Space Flight Center; James Harrington, director of Spacelab; and Walt Williams, a NASA consultant.

This was a temporary team developed until a formal investigation group could be convened by the head of NASA. However, before a formal panel could be formed, the task of investigating the accident would be removed from NASA by the president. One of the first actions of the panel was to view videos of the incident, which seemed to show the accident was caused by the explosion of the external tank. A list of action items was developed, and the panel prepared a fast presentation for a group arriving from Washington, D.C. The Washington group included Vice President George Bush, U.S. Senator John H. Glenn (who had flown on Friendship-7 in 1962 and would later fly a Shuttle mission), Senator Jake Garn (who flew on *Discovery* Mission 51-D in 1985), and NASA acting administrator Dr. William Graham.

That evening, at 5 P.M. EST, President Ronald Reagan addressed the nation from the Oval Office of the White House, on the loss of the *Challenger* and her crew.

> Ladies and gentlemen, I planned to speak to you tonight to report on the State of the Union. But the events of earlier today have led me to change those plans. Today is a day for mourning and remembering. Nancy and I are pained to the core over the tragedy of the Shuttle *Challenger*. We know we share this pain with all of the people of our country. This is truly a national loss.
>
> Nineteen years ago, almost to the day, we lost three astronauts in a terrible accident on the ground. But we've never lost an astronaut in flight. We've never had a tragedy like this. And perhaps we've forgotten the courage it took for the crew of the Shuttle. But they, the *Challenger* Seven, were aware of the dangers, and overcame them and did their jobs brilliantly.
>
> We mourn seven heroes: Michael Smith, Dick Scobee, Judith Resnik, Ronald McNair, Ellison Onizuka, Gregory Jarvis and Christa McAuliffe. We mourn their loss as a nation, together. To the families of the seven, we cannot bear, as you do, the full impact of this tragedy. But we feel the loss, and we're thinking about you so very much. Your loved ones were daring and brave, and they had that special grace, that special spirit that says: Give me a challenge, and I'll meet it with joy. They had a hunger to explore the universe and discover its truths. They wished to serve, and they did. They served all of us.
>
> We've grown used to wonders in this century. It's hard to dazzle us. But for twenty-five years, the United States space program has been doing just that. We've grown used to the idea of space, and perhaps we forget that we've only just begun. We're still pioneers. They, the members of the *Challenger* crew, were pioneers.
>
> And I want to say something to the schoolchildren of America who were watching the live coverage of the Shuttle's takeoff. I know it's hard to understand, but sometimes painful things like this happen.

It's all part of the process of exploration and discovery. It's all part of taking a chance and expanding man's horizons. The future doesn't belong to the faint-hearted. It belongs to the brave. The *Challenger* crew was pulling us into the future, and we'll continue to follow them.

I've always had great faith in, and respect for, our space program. And what happened today does nothing to diminish it. We don't hide our space program. We don't keep secrets and cover things up. We do it all up front and in public. That's the way freedom is, and we wouldn't change it for a minute. We'll continue our quest in space. There will be more Shuttle flights and more Shuttle crews and, yes, more volunteers, more civilians, more teachers in space. Nothing ends here. Our hopes and our journeys continue.

I want to add that I wish I could talk to every man and woman who works for NASA or who worked on this mission and tell them: Your dedication and professionalism have moved and impressed us for decades, and we know of your anguish. We share it.

There's a coincidence today. On this day 390 years ago, the great explorer Sir Francis Drake died aboard ship off the coast of Panama. In his lifetime, the great frontiers were the oceans, and a historian later said, "He lived by the sea, died on it and was buried in it." Well, today we can say of the *Challenger* crew their dedication was, like Drake's, complete.

The crew of the Space Shuttle honored us by the manner in which they lived their lives. We will never forget them nor the last time we saw them — this morning — as they prepared for their journey and waved goodbye and slipped the surly bonds of Earth to touch the face of God.

Thank you.

Arriving at the Kennedy Space Center at around 5:30 P.M., the Washington group met with members of the astronauts' families in private at the Operations and Checkout building.[48] Later, at around 7 P.M., they met with the members of the Kennedy Launch Team and the press.

At 8 P.M., Vice President Bush and his group were given a briefing by the review board in the LCC Operations Center. It was at this meeting that Richard Kohrs, deputy manager of the National Space Transportation System, was assigned to develop a plan to release information and material to the created teams involved in the Shuttle investigation. The review board ordered strict control on all items related to the disaster. NASA was not about to lose anything of material value when it came to this investigation. It became one of the first priorities to insure security of all items, including the launch pad itself. Three other members were added to the board, including astronaut Robert Overmyer, NASA Chief Counsel John O'Brien and NASA Chief Engineer Milt Silveira. Overmyer was later killed in the crash of an experimental aircraft in March 1996.

Later that evening, Vice President Bush and senators Glenn and Garn would fly back to Washington, leaving Acting NASA Administrator Graham behind to form teams which would investigate separate areas of the Shuttle accident.[49]

The mission was over. It was now time to recover the crew, and put the manned space program back on the road to space. No one on that cold and deadly day could have guessed that it would take thirty-two months to do the job.

It was just not our day.
— Robert Sieck, NASA launch director, Kennedy Space Center

CHAPTER EIGHT

Post Flight Operations

Yes, yes, it's true. Seven are gone; they're all gone.
— NASA public affairs officer, Kennedy Launch Center, January 28, 1986

As the nation mourned its seven heroes, the National Aeronautics and Space Administration quickly geared up for an unprecedented search operation.

Within the first twenty-four hours, better than 1200 square miles of the Atlantic Ocean's surface had been searched with no signs of human life being discovered.[1] However, several vessels in the prime recovery area retrieved 600 pounds of debris found floating on the ocean's surface.[2] These vessels included the Coast Guard cutter *Point Roberts*, which had rushed to the area to search for possible survivors and later brought wreckage to Canaveral Station. NASA's two SRB recovery ships, the *Liberty Star* and *Freedom Star*, offloaded the first debris with the largest piece reported to be approximately fifteen square feet, possibly from a wing.

To support the 51-L Interim Mishap Review Board, investigation teams were formed at the Johnson Space Center; the Kennedy Space Flight Center; and the Marshall Space Flight Center. These groups assembled and evaluated the recovered data on the systems and hardware related to the *Challenger* flight as it was received.[3]

In preparation for recovery operations, three areas were designated by the Mishap Board for storage and reconstruction of the destroyed Shuttle. The orbiter wreckage would be taken to the Logistics Building as well as a nearby area at the Kennedy Space Center, along with the remains of the external tank. The debris from the SRBs was stored at Hangar O and the EOD (Explosive Ordnance Disposal) range at Cape Canaveral Air Force Station.[4]

Along with NASA's probe, an additional level of investigation was added. This was developed when the National Transportation Safety Board (NTSB) was tasked with the structural evaluation of *Challenger* to help discover the cause of the explosion. The board would use the recovered wreckage of the Shuttle and its SRBs as well as the external tank for its evaluation, and report directly to the Mishap Board. They would attempt to reconstruct as much of *Challenger* as possible. Later, they would report to the *Challenger* commission when the Mishap Board was disbanded.

Challenger's Search and Recovery Operations

Wednesday, January 29, 1986

As the first full day of recovery operations got underway, a plan was developed by the review board to house any hazardous or explosive materials found in the search area. The Explosive

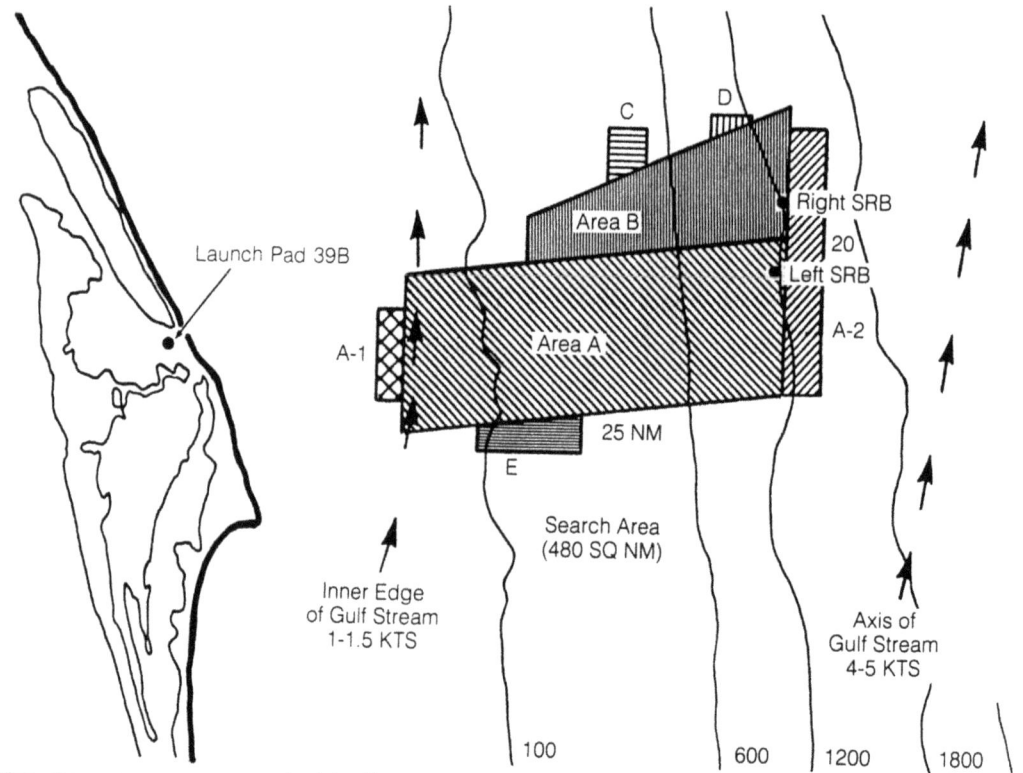

Map shows ocean areas searched for Shuttle wreckage in relation to Cape Canaveral and Launch Pad 39B. Wavy vertical lines indicate water depths.

Shuttle wreckage search areas in the Atlantic off of Kennedy Launch Center (graphics from *Challenger* report, page 67).

Ordnance Disposal Range at Cape Canaveral Air Force Station was the designated location where air force personnel were standing by to assist NASA. Explosive devices were located on the solid rockets as well as on the external tank, to prevent a straying rocket from hitting a populated area on the Florida coast. The range-safety destruct charges on the external tank were later photographed on the ocean floor, then recovered intact and found to have played no part in the disaster. These devices were of great concern during the early part of the investigation, as it was feared they might have been set off, causing the disaster. The fact that they were found intact is a testimony to their reliability and stability of design.

Lieutenant Joe Kyle of the coast guard, who was involved in the search and recovery, reported to the media that several small pieces of wreckage had been found ashore in the Cape Canaveral area during the night. He requested that individuals who found any debris which could have come from the Shuttle turn it in to the coast guard as soon as possible. NASA was not only concerned with recovery, but was aware that some orbiter debris could be dangerous. Pieces of the orbiter were beginning to wash ashore near *Challenger*'s launch site at pad 39B. At the same time, Lieutenant Commander James Simpson reported that recovery personnel at sea were finding "debris all over the place."

The U.S. Coast Guard cutter *Dallas* discovered two cone-shaped pieces of debris, which turned out to be the frustums from both SRBs.[5] They were recovered by the Coast Guard buoy tender *Sweetgum*. These cones would show conclusively that the right solid rocket booster had struck the external tank, and would be an important piece of evidence in the reconstruction of events. They would also show that their parachutes had not been properly deployed. That discovery would show conclusively that air force public affairs officers were incorrect when they had reported on the parachute sighting.

At a NASA news conference, Flight Director Jay Greene still appeared to be shock as he answered questions from the press. It was a very painful conference for this director, who had known most of the crew now thought to be lost. For Jay Greene, this flight was his ninth as ascent director for the U.S. space program. Again, responding to questions from the press, Greene stated that there was no indication whatsoever of any problems prior to the destruction of the Shuttle. "Everything was going normally, and then the data just stopped." Computer screens held the last data points with an "S" at the end to indicate static. This was the situation when controllers reported, "We have no downlink." "We train awfully hard for these flights, and we train under every scenario you can possibly imagine. There was nothing anyone could have done for this one. It just stopped."[6]

A NASA spokesman reported that an ice team had been sent out to the pad early that morning and all reports from that team were good twenty minutes before launch. Yet, photos taken before the flight would show icicles up to six feet long on the launch pad. Could NASA seriously have considered that a good condition? There were no comments as to what had happened to concerns about falling ice damaging the orbiter, which so concerned NASA on the top-secret Defense Department mission 51-C.

As a sign of the nation's loss, President Reagan ordered flags lowered to half-staff on all federal buildings throughout the country and all ships of the navy. At the same time, Representative Bill Nelson of Florida, who had flown on *Columbia* just seventeen days earlier, proposed naming Uranus's seven newly discovered moons for *Challenger*'s crew, but this proposal would prove unsuccessful. Later, the Soviet Union was successful in naming craters on the planet Venus for the two women onboard.[7]

In the evening, CNN reported that NASA officials were refusing to state why the solid rocket boosters were detonated, or which direction they were headed when they were destroyed. When pressed for an explanation, Robert Nicholson, the air force spokesman, would not comment on the specifics of the SRBs' destruction, nor would he state which land mass the solids were supposedly flying toward.

In Los Angeles, the city's Board of Supervisors ordered the Olympic Torch at the Los Angeles Coliseum, used for the 1936 and 1984 Olympic games, to be lit in honor of *Challenger*'s crew. It was but one of the many gestures of solidarity occurring throughout the nation.

By the end of the day, some 5,000 square miles of ocean surface off the coast of Florida had been searched by aircraft and ships for Shuttle debris. With all hopes of crew survival now gone, it would now be only a recovery operation.

Thursday, January 30, 1986

Early in the day, NASA announced that the following ships and aircraft were part of the recovery operation, with more available if requested by the space agency:[8]

USCG *Dauntless*, coast guard
USCG *Point Roberts*, coast guard
USCG *Sweetgum*, coast guard
USS *Sampson*, U.S. Navy
USS *Simpson*, U.S. Navy
USS *William Sims*, U.S. Navy
USS *Hubrey Fitch*, U.S. Navy
USS *Coelsch*, U.S. Navy
USS *Underwood*, U.S. Navy

	NASA *Liberty Star*—SRB recovery ship with *Sprint* RPV
	NASA *Freedom Star*—SRB recovery ship with *Scorpio* RPV
	USAF *Independence*, air force—SRB Recovery ship
Aircraft:	USCG C-130
	USAF C-130 (two)
	U.S. Navy P-3 Orion
Helicopters:	USCG H-32 (two)
	USCG H-52
	USAF H-3 (two)
	U.S. Navy (two)

The surface ships were using sonar to locate wreckage on the sea floor on this day. Divers would later go below the surface, if hits were made, to allow for positive identification of Shuttle wreckage. Searchers found there were many soundings on the bottom. There was a lot of old debris in the area because of the many rockets destroyed downrange from the Atlantic Missile Range, as well as crashed aircraft and lost ships. It made for a daunting task. It would also prove to be a gold mine of information for individuals interested in locating sunken wrecks. Lieutenant John Philbin from one of the search ships described the search area as "a calm, sunlit sea littered for miles with fragments. Little of it was identifiable."

Meanwhile, civilians near the launch site located some easily identifiable pieces of wreckage. One was a shredded glove found on the beach at the Cape which, reporters speculated, may have come from the *Challenger*, but was not positively identified by NASA. Another was "a piece of bone with a small section of human tissue" found on a beach near the small town of Melbourne,

Challenger nose section unloaded at Port Canaveral (NASA).

south of Cape Kennedy. It was still "attached to a shredded piece of light blue fabric."[9] There can be little doubt as to their origin. NASA would give no comment on these discoveries, but the remains were taken to Patrick Air Force Base for identification. If true, this would be the first recovery of remains from any of the *Challenger* crew members.

In news conferences however, NASA officials were speaking of their surprise at the number of large orbiter pieces that had survived the explosion, based on the live videos seen of the explosive destruction of the Shuttle. Later that evening, a ten-foot-square section of the right side of *Challenger*'s nose was brought back to the Cape by ship.[10] No one in the space agency at the time could have imagined that the entire crew cabin would survive the explosion, only to be destroyed upon impact on the ocean. With this evidence, it was clear that had the Shuttles been equipped with a crew-cabin parachute recovery system, at least some of the crew members might have survived the accident.

As reported by the Associated Press, Lieutenant Colonel Robert Nicholson, the director of public affairs for the Eastern Missile and Space Center, in answer to reporters' continuing questions about para-rescue personnel seen in the impact area, stated, "No paramedics jumped into the ocean in efforts to locate and assist any possible survivors following the explosion of Shuttle *Challenger*, despite some earlier reports to the contrary. In the heat of the moment there was commentary to that effect from Mission Control in Houston. The report was in error." In fact, someone did parachute into the area. The fact that at least one individual did could not be hidden from eyewitnesses at the Cape, nor from those who saw and recorded CNN coverage of the event as it happened, using home video recorders (as did the author), nor from the men and women at Mission Control. Nicholson was careful to say that no "paramedics" jumped into the ocean. He did not say that no person jumped into the ocean.

At the same time that Nicholson was putting forth the military's version, a NASA spokesman was continuing the story when he said that a parachute seen drifting to earth apparently came from a rocket booster. But what rocket booster could he have been referring to? This report continued, despite the fact that the parachutes used by the solid rocket boosters were in sets of three, due to the size and weight of the SRBs, and the chute seen on television and by eyewitnesses was alone and had a man hanging from it! Also, the two triangular shaped frustums with drogue chutes were both recovered, one destroyed and one intact and unused. That would account for all parachutes on *Challenger*'s SRBs. Some of the eyewitnesses even reported two individuals jumping into the impact area, as reported by NASA. This was the same type of parachute used by paramedics throughout the military. Also, the NASA spokesman seemed to have forgotten that the solids were destroyed at 1 minute and 50 seconds into the flight and the parachute was seen 8 minutes and 40 seconds after the launch. This is far too great a time discrepancy to allow the official story then being told by NASA to hold any truth. Also, cameras were showing the impact area from the time of launch through the period when the chute was seen. But who was wrong, NASA and the U.S. Air Force, or the millions of people who saw the parachute coming down, while it was being reported by NASA public affairs officer Steve Nesbit, 9 minutes and 10 seconds after *Challenger*'s liftoff? Did these represent one or more deaths to be added to the seven onboard *Challenger*? If so, what were their names, and why will NASA not come forward with the truth, even today? One truth is known; none of the crew members on board the Shuttle had parachutes. It was a failing which would be corrected before the next Shuttle was launched. The bottom line is clear: Live video does not lie.

Friday, January 31, 1986

The chair of the 51-L Interim Mishap Review Board, Jesse Moore, appointed two congressional committee members to the board as observers, to add a level of credibility to that board.[11] This was not what Moore or anyone else at the space agency wanted to do. The board also reported that wreckage impound areas had been found to be too small, and that NASA was looking for larger hangar areas. At Christa McAuliffe's school in Concord, New Hampshire, a letter from President Reagan was read to her students. In part he wrote:

> You could hardly do better than believe as she believed, that the lives and examples of every citizen alike—teacher, engineer, lawyer, policeman, mother, father—have the power to change the course of human events for the good.[12]

From NASA headquarters in Washington, D.C., Interim NASA Director William Graham requested and received a copy of two NASA video tapes of the launch which were taken from the north, showing a plume of flame emanating from the right SRB. Along with the video, a set of still photos was supplied, showing that plume. It had been little more than three days since the accident and the space agency already had its technical smoking gun.[13] The investigation team found one of the 39B launch pad cameras had taken photos of puffs of black smoke coming from the lower segment of the field joint between the right SRB and the external tank. This information would not be released to the public until February 13. The team wanted to be absolutely certain they had closed in on the problem area before going public with the information. But there was little doubt that the right SRB had a major problem during the flight as NASA officials began to question officials at Morton Thiokol on the history of the SRBs.

By this time the search for wreckage had expanded to 6,800 square miles of ocean surface, as fast-moving currents moved debris north up the coast. The area extended from twenty miles south of Daytona Beach and then northward sixty miles, to ninety miles off the Florida coast.[14]

The Coast Guard buoy tender *Sweetgum* returned an SRB frustum to the Cape late that night.[15] It was brought to the hazardous storage area because it contained four booster-separation motors normally used after the firing of explosive bolts, a procedure separating it from the external tank in a normal flight. Because the timing sequence which would have triggered the separation did not occur, the motors were still armed and ready to fire. That evening the navy's USS *Sims* was removed from the search, as other ships and aircraft continued to look for wreckage.

Saturday, February 1, 1986

By the first of February, the coast guard had sealed off a 200-mile section of the Florida coastline, keeping all private ships out of the area. The coast guard was concerned that civilians might go into the primary search area and recover debris and perhaps not return it to NASA for evaluation.

Meanwhile, the air search was expanded to 20,000 square miles, an area from Cape Canaveral north to Savannah, Georgia, and out approximately 90 miles to sea. Two coast guard Falcon jets, which had been added to the recovery forces, performed a search after a cone-shaped object was sighted 100 miles off of Savannah.[16] Currents were moving large sections of debris on the ocean surface and NASA needed to spot them as fast as possible. The surface area covered by ships was also expanded, to over 4,500 square miles.[17] Meanwhile, the NASA SRB recovery ship *Freedom Star* was being fitted with side-scan sonar, but it could not be deployed on this day due to strong Gulf Stream currents and winds in the search area. Later, the vessel would play a major role in locating large and important sections of the Shuttle.

At the Ordnance Disposal Range at Kennedy, a valuable piece of information was discovered when the nose cone of the right solid rocket booster was taken apart. Rockets used to push it away from the external tank at separation were not turned on and not fired. This was proof that neither the pilot nor the commander had initiated that action, according to NASA spokesman Charles Redman. This information cleared the crew of any responsibility for the SRBs separation, and showed that events had simply occurred too fast for them to react. NASA also confirmed that the nose section of one of the SRBs (the left one) had been recovered, complete with parachutes.

NASA's Mishap Review Board released a video taken during the flight, along with three photos depicting a rocket plume. NASA spokesman Hugh Harris stated that it was an unusual plume, shown to be located on the lower right SRB during *Challenger*'s flight.

Memorial services were held at Kennedy Space Center for the workers and their families on this day. Stands were positioned near the launch site so that guests could look out over the ocean where *Challenger* had flown and where recovery operations were ongoing. At the same time, the Soviet Union announced that it would name two craters on Venus for the two women killed on the flight. It was a gesture which illustrated the international effect the accident had on nations that have tested manned spacecraft. The memorial ceremony was timed to occur at the same time that the Shuttle had exploded, and a three-foot floral wreath was tossed into the ocean over the spot where the *Challenger* was destroyed, a few miles off of Cape Kennedy. After the wreath landed on the ocean near the impact site, a school of seven porpoises leapt from the water as if they understood the meaning of the ceremony. Richard G. Smith, director of Kennedy Space Center, later announced that a moment of silence would be observed each year at the Cape on January 28 at 11:39 A.M.

United Press International reported that it had conducted its own frame-by-frame video analysis, showing that the crew cabin remained intact after the breakup of *Challenger*. The news agency stated that their analysis was based on an enhanced close-up version of NASA's launch video, which clearly showed a cabin-shaped object flying out of the expanding gases. This was the first report of an intact crew cabin to be made public. NASA made no comment on the UPI report, as more and more wreckage was being recovered and returned to Kennedy Launch Center.

NASA requested and received from the air force two inflatable aircraft hangars[18] to be used to store wreckage, because the amounts that were being found were more than could be handled by the two areas already set aside. Space agency officials were now reporting that large sections of the Shuttle did survive the explosion and ocean impact and would probably be recoverable. The agency was also able to report that external fuel tank pieces were located, and that large pieces of *Challenger* may have been carried by ocean currents and moved up the eastern seaboard, based on what the coast guard was reporting.[19]

NASA, as would be expected, was being very cautious in identifying any item found which could turn out to be the personnel belongings of the crew out of respect for the crew and their families. At the same time it was again reported that a fragment of human bone had been found on a local beach, but the location was not identified. It was an indication of the destructive forces encountered when the crew cabin hit the surface of the ocean after its nine-mile fall.

NASA Acting Director Graham stated during a news conference that the solid rockets were "not susceptible to failure and had no credible failure modes that we could identify."[20] "The two booster rockets were considered so reliable they carried no sensors that might have detected a rupture in their casings. There was no credible belief that the boosters could fail."[21] At the very least this was a tremendously uninformed statement from the individual tasked with overseeing this nation's space program. It was also an extremely premature statement with no data whatsoever to back it up. Also, Graham had already seen photos which showed just the opposite of what he stated at the news conference.

Sunday, February 2, 1986

During a space agency news conference, enhanced close-ups of stills from the NASA video taken during the flight confirmed that *Challenger* had emerged from the explosive gases of the external tank, receded back inside, then reemerged as it was breaking up.[22] As the camera pulled away for a wider view, the Shuttle appeared to break in half, and the foreword-reaction-control area in the nose of the Shuttle could be seen to explode in a red cloud of gas.

Chairman William Rogers informed the press that the Presidential Commission would go into the investigation with no preconceived ideas. He further stated that not only would the explosion itself be reviewed, but "all decisions leading up to the launch would also come under scrutiny." NASA and Thiokol management would also be looked at very closely.

Recovery teams using photography on the ocean's floor identified a section of the Inertial

Upper Stage (IUS), which would have been used to boost the TDRS-B into geostationary orbit. The IUS had been spotted using the side-scan radar, which reported a likely target on the ocean floor.

By the fifth day after the mission failure, two coast guard Falcon jets were extending their search area to 33,000 square miles, from the coast to 150 miles offshore, from Cape Kennedy to Savannah, Georgia, in response to ocean currents which could move floating pieces of *Challenger* a long distance. The coast guard was still reporting strong Gulf Stream currents in the area.

The SRB recovery ships began searching for wreckage on the ocean floor based on air force tracking data of radar targets as they were monitored by range control after *Challenger* was destroyed. These were the multiple targets reported by NASA just after the destruction. Using the RPV *Scorpio*, the *Freedom Star* located two large objects on the ocean floor, as well as wreckage from a helicopter and a light aircraft. Three robot submersibles were being used in the search at this time, and were being called into use for identification of the most important possible wreckage sites. At the same time, a 6,300-square-mile area was being searched by six surface ships and seven aircraft from St. Augustine to Melbourne, up to 80 miles offshore.[23] Nearer the shore to three miles out, twelve USCG auxiliary ships, just added to the operation, began a search from Ponce de Leon to Jacksonville, Florida.[24]

The press reported that more human body parts were washing up onshore, with few details given, as NASA confirmed the finds but would not confirm that they were from the *Challenger* crew. This nonconfirmation was out of respect for the crew members' families who would be the first notified if positive identification was established. At the same time, a spokesman for the maker of the solid rocket booster that failed, Morton Thiokol, stated that its rockets were not to blame for the disaster. This was a very strange announcement considering the video evidence already shown to the public of a confirmed plume of fire coming from one of the solid rocket's sides.

By the end of the day, the Coast Guard cutter *Dallas* had located and delivered to Cape Canaveral, a thousand pounds of debris, with the largest piece measuring about six by eight feet.[25] By then, NASA had a very good idea where most of the larger sections of the Shuttle would be found, as they focused their search on locating the crew cabin and the solid rocket boosters.

Monday, February 3, 1986

Ronald Reagan established the Presidential Commission on the Space Shuttle *Challenger* Accident with an Executive Order Number 12546. That order gave the commission 120 days to find answers to the *Challenger* accident. This order effectively removed NASA from the lead in investigating the accident and placed William P. Rogers, the former Secretary of State, in full control from this point on.[26] The president said, "As we move away from that terrible day, we must devote our energies to finding out how it happened and how it can be prevented from happening again. It's time now to assemble a group of distinguished Americans to take a hard look at the accident, to make a calm and deliberate assessment of the facts and ways to avoid a repetition."[27] NASA's 51-L Interim Mishap Review Board, headed by Jesse Moore, held its final meeting on this day at Kennedy Space Center, handing over to the commission all of its data.

Eleven tons of *Challenger* wreckage had been recovered up to this date, but with less and less debris being found on the ocean's surface, the coast guard scaled back on the surface search. Lieutenant Commander James G. Simpson of the coast guard reported that two jets assigned to the coast guard had searched a surface area of over 33,000 square miles from the Kennedy area to Savannah, Georgia, using specially equipped surface-search radar.[28]

NASA spokesman Hugh Harris reported that 17 targets (including three large objects) on the ocean floor had been identified by radar-scan searches, and that they would be closely looked at by using remote underwater cameras. Radar printouts came from the Federal Aviation Administration, and these showed possible tracks of larger pieces of the *Challenger* as they fell to the sea. The submersibles would take video and still photographs to positively identify the targets as being from *Challenger*. NASA would not comment on whether or not any personal effects of the seven crew members had yet been recovered. NASA however, did acknowledge that if the *Challenger* had attempted an emergency detachment from its external tank and solid rockets, there would have been a very slim chance of its being successfully performed. Further, even if it had survived the explosion, the Shuttle would not have been able to return to the Kennedy runway and would have had to ditch in the ocean, which would have destroyed the *Challenger* and probably killed the crew. Impact on the ocean would have violently pushed the large payload it was carrying into the crew compartment. There were simply no viable, life-saving options available to the crew during that portion of the flight.

The *Los Angeles Times* reported that the SRBs did have sensors during the earlier flights, "but were removed to save weight after it was felt that the Solid Rocket Boosters were performing well. The boosters had suffered a 3 percent loss of power equal to approximately 100,000 lbs of thrust."[29]

According to an Associated Press report, "human parts were being examined, including "a portion of a foot attached to a blue fabric." Again, NASA would not comment about human remains. "One SRB was found about forty miles off the Kennedy Space Center. Don't know which one," Jim Mizeli, NASA spokesman, reported, as the sixth day of recovery operations came to an end. The SRB was nowhere near land.

Tuesday, February 4, 1986

The navy removed its four remaining ships from surface search duties because of the slowing of floating-debris sightings.[30] NASA reported that two objects spotted by sonar had turned out to be a light aircraft and a helicopter. However, NASA official Sarah Keegan stated that they were in error when the coast guard relayed a report of a light aircraft and helicopter being located on the ocean floor.[31,32] It was further reported that seven of 17 reported sonar targets were now identified as not being from the *Challenger*, after robotic submarines took closer looks.[33]

Hughes Aircraft Company announced that it had established a scholarship in memory of Hughes Aircraft engineer Greg Jarvis, killed on the flight.[34]

NBC reported, "Some of the human remains found so far and recovered have been identified as belonging to the seven astronauts." The crew-member-remains news blackout by NASA remained in place, as no comment was issued by the space agency.[35] It was a failing that would be corrected after the *Columbia* disaster.

Wednesday, February 5, 1986

The coast guard continued the surface search from St. Catherine's Island to as far north as Cape Fear off of North Carolina, using three aircraft and two coast guard cutters, due to the possibility that the Gulf Stream may have pushed debris that far up the coast. The search covered an area of 12,000 square miles.[36] The ocean search was now reduced to six surface ships and six aircraft, as high winds and rough seas prevented the use of the small robotic submarines.

A 17-foot section of undetonated explosive belonging to the external-tanks destruct package was found.[37] Back at Kennedy, team members from NASA and the National Transportation Safety Board were using identification numbers and blueprints to began creating an outline of the *Challenger* on a hangar floor, as pieces were being recovered. "They have a giant jigsaw puzzle over there," NASA official Jim Mizell reported. Major sections so far recovered included pieces of the fuselage, nose and wing flaps.

CBS News reported that part of the crew cabin had been located and that some crew equipment had floated to the surface.[38] Rumors circulated that Teacher-in-Space manuals had been located. NASA denied the report, while CBS also reported that some astronaut remains had also been recovered at the site and identified.

Thursday, February 6, 1986

Investigators from the President's Commission requested copies of NASA launch videos of Shuttle flights 41-D with *Discovery* and 61-B with *Atlantis*, as the Commission held its first meeting in Washington, D.C. Investigators believed that images of those flights could have shown problems similar to *Challenger*. The *Discovery* flight showed a flame at the tail end of the external tank. On the *Atlantis* flight there appeared to be a second plume of flame at SRB separation, on the side of the booster. However, these proved to be false leads.

NASA announced that the left SRB had been the one located by sonar in 150 feet of water forty miles off Cape Canaveral, as ships continued to search a 16,150-square-mile area.[39] Reports also indicated that nearly twelve tons of debris had been recovered to date in the continuing ten-day search operation, as ABC Evening News showed a film of an astronaut's damaged helmet that was found, floating, by the coast guard.[40] In all, three helmets worn by the *Challenger* crew would be located and returned to the launch center.

Friday, February 7, 1986

The coast guard announced that it would end its ocean-surface search by the end of the day. The operation had covered 150,000 square miles in eleven days. From this point on, all efforts would concentrate on identifying and recovering debris from the ocean floor, using radar data and side-scan sonar.[41]

Saturday, February 8, 1986

As 22 navy divers began working 18 miles northeast of Cape Canaveral,[42] high winds and choppy seas ended operations before they could begin a close-up search for the Tracking and Data Relay Satellite (TDRS) and its Inertial Upper Stage (IUS) booster.[43] NASA needed to know if any problems with *Challenger*'s largest payload to date contributed to the accident. The divers worked off of the USS *Preserver*, which had navigation and communication problems en route to the search area due to heavy rains accompanied by a great deal of lightning.[44] Also in the area was NASA's *Freedom Star*, which marked the area with flares.[45] Bad weather would be a constant problem, as Atlantic winter storms crisscrossed the area. Major search efforts at this point were focused on recovering the crew cabin and the right solid rocket. By the end of the day, the search had covered over 172,500 square miles of ocean surface.

Sunday, February 9, 1986

The U.S. Air Force SRB-recovery ship *Independence* used sonar to map an area of ocean floor forty miles off the coast of the Kennedy Launch Center during its search for the right SRB, as calmer seas allowed navy hard-hat divers to continue their search for the TDRS.

Monday, February 10, 1986

An Associated Press photo of a flight-crew helmet being held by a coast guardsman was published in newspapers around the country. NASA, once again, would not confirm that it had come from one of the crew members, but there was no doubt as to its origins.

In an interview published in *U.S. News & World Report*, former astronaut Frank Borman, who had commanded the first Apollo flight to orbit the moon in 1968, reflected on civilians in space, saying that "at this stage of development the Shuttle is still a very high-risk operation. Each of these missions should be treated as an experimental launch. There probably should not be any civilians aboard unless they have important scientific or astronomical experiments to perform."[46] In that same issue, *U.S. News & World Report* reported on a poll on continuing the manned space program. Public opinion seemed to be for a continued American presence in space.[47]

U.S. should keep spending level for Shuttle as is.	Yes 69%
America should continue with manned space probes.	Yes 75%

U.S. should keep civilians on space flights despite risks.	Yes 55%
America should switch entirely to unmanned probes.	Yes 17%

Tuesday, February 11, 1986

Rough seas off the Cape Canaveral coast hampered recovery operations,[48] as most of the ships did not leave port. These conditions were reminiscent of seas at the time of *Challenger*'s launch. Major Ron Rand, an air force official, reported that after a review by the Pentagon, "it's too early to address the impact [of the disaster] until we know how long the fleet will be grounded, what went wrong and what the fix will be." For a while however, the military would have to rely on its fleet of expendable rockets. They would consider using rockets launched by the European Space Agency. Even launches by communist China would eventually become part of the mix.

The USS *Preserver* was in the debris area conducting diving operations.[49]

Wednesday, February 12, 1986

The navy announced that its nuclear-powered research submarine NR-1 was en route from New London, Connecticut, to aid in the search of the ocean floor.[50] Later in the evening, 22 pieces of Shuttle wreckage were unloaded at Port Canaveral by USS *Preserver*, under cover of darkness.[51] Crews were spending as much time as they could on site in order to find the debris as soon as possible.

Thursday, February 13, 1986

Sections from the TDRS were recovered by navy divers and returned to Cape Canaveral. Inspection of this wreckage would show that the TDRS did not contribute to the accident.

Acting NASA Administrator Graham announced that the Teacher-in-Space Program would proceed and added that backup teacher Barbara Morgan would be given an opportunity to fly on the Shuttle.[52] Accepting the offer, Morgan said, "We have the opportunity to teach an entire generation a very important lesson. Those students need to see that challenges are meant to be met and overcome."[53] NASA would later reverse itself, ending the Civilian-in-Space program. Barbara Morgan would never teach any lessons from space, although she continued to support the Civilian-in-Space Program, and is an outspoken supporter of NASA. She would, however, train as a regular mission specialist in 2000 for upcoming Shuttle flights. In March 2002 it was announced that Morgan would fly a Shuttle mission sometime in 2004 to the International Space Station, but her flight would again be put on hold after the *Columbia* was destroyed. On August 8, 2007, astronaut/mission specialist Barbara Morgan flew on STS-118.

Friday, February 14, 1986

At a news conference in Washington, D.C., William Graham stated that he had not known that NASA documents dating from 1982 raised concerns about the safety of the O-ring seals until he was briefed four days after the explosion.[54] He added that he found NASA employees to be "extremely careful, extremely conscientious, extremely rigorous in technical and engineering and judgmental aspects."[55]

Saturday, February 15, 1986

The Presidential Commission released new enhanced photos showing puffs of black smoke coming from the side of the right solid rocket booster at ignition.[56] Recovery operations continued to focus on locating the right SRB on the bottom of the ocean 45 miles off the coast[57] northeast of the Cape, as more pieces of *Challenger*'s TDRS payload were being recovered from the ocean bottom by navy divers. NASA announced that the four-person Johnson *Sea Link II* submarine had been assigned to the search.[58]

Press reports indicated that a number of high NASA officials were expected to be barred from any active investigations due to their possible involvement in the accident as decision makers. Included were Richard Smith, director of Kennedy Space Center; William Lucas, director of Marshall Space Flight Center; Jesse Moore, associate administrator for space flight and chairman of the 51-L Interim Mishap Review Board; and Arnold Aldridge, manager of the National Space Transportation System, Johnson Space Center.[59]

Sunday, February 16, 1986

The Johnson *Sea Link II* remote submarine made images of wreckage on the ocean floor of what appeared to be the right SRB.[60] A NASA Spokesman reported, "They have photographed it (the right SRB) and we will compare that to prelaunch pictures."[61]

Monday, February 17, 1986

Photos taken of debris believed to have come from the *Challenger* were flown to NASA's Marshall Space Flight Center to allow engineers a chance to identify them as coming from the Shuttle's right solid rocket.[62] If confirmed, the wreckage would then be recovered and brought to Kennedy Launch Center. Copies of the photos were delivered to the *Challenger* Commission to aid in their investigation.

Tuesday, February 18, 1986

The navy's NR-1 nuclear submarine arrived at Cape Canaveral, equipped with a mechanical arm, floodlights and cameras.[63] The ship has the capability to stay submerged with its crew of seven for up to a week. The sub is also equipped with wheels which allow it to roll along the ocean floor. Meanwhile, NASA reported that the recovery operation could continue for three more months. (In fact, the operation would be completed on August 28, after seven months of searching.)

Wednesday, February 19, 1986

A section of *Challenger*'s right solid rocket booster is identified on the ocean floor and is again photographed by *Sea Link II* and then recovered.[64] The part number on a right SRB hydraulic reservoir was photographed before the stainless steel sphere was recovered by the *Sea Link II*.[65] As yet, no sighting had been made of the SRB section where flame was seen to have come from on the launch video. The head of the Shuttle search-and-recovery operation, Air Force Colonel Edward A. O'Conner reported, "I am confident we will find all of the right-hand SRB."[66] Search teams concentrated on locating the rest of the suspect booster.

Thursday, February 20, 1986

Press corps rumors circulated that the president had pressured NASA to put the *Challenger* crew on the phone to him during the scheduled State of the Union address. This rumor was never confirmed and indeed was flatly denied by White House officials.

As reported by Senator Ernest Hollings (D–SC), astronaut Robert Crippen asked Allan McDonald, who would not sign off to launch *Challenger*, why the flight procedure was reversed. Before, contractors always had to justify why they were ready to fly, but in this case the contractor had to justify why not to fly. McDonald did not know why the change had occurred, but indicated that he was also concerned about the change. Senator Hollings demanded the resignation of NASA's acting chief[67] and stated that the Shuttle could have been launched in the afternoon at its regular time, when it was warmer.[68]

In the search area, rough weather was complicating recovery operations off and on, as high waves and brisk winds swept through the area.

Friday, February 21, 1986

The search was now being conducted by nine surface ships and two submarines, focusing on an area 43 miles east of Cape Kennedy.[69] The searchers wanted to recover as much of the right

SRB as possible, because it was believed to hold the key to the disaster. Weather continued to hinder operations in the general area as more radar targets were searched.

Saturday, February 22, 1986

A fourteen by fifteen foot piece of sheet metal, believed to be from the external tank, was located by *Sea Link II* and brought to Port Canaveral. *Sea Link II* also photographed the left SRB on the ocean floor. At the same time, the navy announced that its unmanned submersible Deep Drone had located and photographed pieces of the main engines. The engines were located 18–20 miles east of Cape Canaveral. The navy's NR-1 nuclear submarine continued operations on the ocean floor in the area around the right SRB, reporting contacts but no positive identification of Shuttle debris.

Navy Captain Joseph P. Kerwin, a former Skylab astronaut and director of the Space and Life Sciences Laboratory at the Johnson Space Center in Houston, replying to press questions, stated that he would not discuss any reports on the pathological examinations which were being conducted on possible crew remains.[70] It would be Kerwin's job to develop the medical report on the deaths of the astronauts, and his team was not about to publish any premature statements.

Sunday, February 23, 1986

Seven surface ships and two manned submarines along with three robotic subs continued to scan the ocean floor for wreckage,[71] as they investigated targets identified by radar and sonar. Working from the USAF *Independence*, the Deep Drone continued to photograph pieces of the main engines on the ocean floor nearly twenty miles northeast of Cape Canaveral.[72] In another search area, a different unmanned sub retrieved a 15-foot-long section that was believed to be from the external tank.[73]

Monday, February 24, 1986

The search on the ocean floor was expanded to cover 350 square miles, as NASA reported it had completed a sonar search of an area of nearly 100 square miles.[74] The search indicated many possible wreckage sites which would have to be identified by remote subs. The USS *Preserver* unloaded parts of the three engines,[75] which had worked perfectly for 73 seconds, at Port Canaveral. The three main engines were found very close to each other and were little damaged by the destruction of *Challenger*. The engines had been seen on the mission video still firing even after the explosion.

Tuesday, February 25, 1986

As the search continued, hampered by high seas, engineers began to step forward and say that they were pressured to okay *Challenger*'s launch. Allan McDonald, Morton Thiokol's booster manager, said, "There's no doubt in my mind.... I felt pressure."[76] He testified that he had stated before the launch that, "If anything happen[s] to this launch I wouldn't want to be the person standing before a board of inquiry."

NASA Administrator James M. Beggs, on leave from his post due to unrelated legal problems, resigned on this date as space agency chief.[77] Beggs had been NASA administrator since July 7, 1981, and was a strong supporter of the Shuttle program.

Wednesday, February 26, 1986

The Associated Press reported that Rocco Petrone, president of the Space Division of Rockwell International, builders of the Shuttle fleet, had expressed fears before the launch, that ice could damage the orbiter if it was launched during the cold temperatures indicated at the launch pad.[78] Rockwell's position was that it was unsafe to launch, but NASA managers overruled their objections. In the search area, several sonar targets were inspected by robot subs and found to be wreckage unrelated to the *Challenger*.

Friday, February 28, 1986

News reports confirmed that a 69-year-old NASA engineer named Elmer Thomas, who was watching the launch from a viewing room at Kennedy, collapsed with a heart attack after viewing the explosion and died two days later. In the recovery area, rain and high winds whipped up rough seas making operations impossible. None of the recovery fleet made it to the search area on this day.

Saturday, March 1, 1986

Countdown magazine quoted John T. Philbin, a coast guard searcher, as saying, "Every time we pick another piece [of *Challenger* wreckage] out of the water, it hurts a little more."[79] The same issue printed several photos showing major pieces of wreckage being unloaded at Cape Canaveral. Ten foot waves, rain and high winds kept salvage ships in port this day further delaying recovery efforts.[80]

Sunday, March 2, 1986

The U.S. Navy's *Stena Workhorse* moved to the recovery area 42 miles northeast of Cape Canaveral in an effort to retrieve a large piece of the Shuttle's left SRB.[81] It was considered a practice run for an upcoming recovery operation in deeper water for the critical right SRB. NASA reports that the Shuttle will be grounded for twelve to eighteen months.[82]

Monday, March 3, 1986

Astronaut Henry Hartsfield, in a meeting with reporters along with fellow astronauts Joe Engle, Gordon Fullerton and Vance Brand, stated, "I was very much upset. It's amazing this [O-ring problem] has been going on and we didn't know about it. It caught me completely by surprise. I don't know where the fault lies, but nobody made a conscious effort to tell us."[83] His statement was in response to the astronaut corps learning that NASA upper management knew about the O-ring problem but did not inform them or correct the situation. From the recovery area no new discoveries were announced as operations continued.

Tuesday, March 4, 1986

The Associated Press reported that the Reagan administration was debating whether or not to resurrect unmanned rockets as a temporary replacement for the now grounded Shuttle fleet in delivering its military as well as civilian satellites into space. Later, it would be decided to end the launch of commercial satellites from the Shuttle fleet entirely. The Shuttle would concentrate in the future on Spacelab missions, space station construction, and deploying scientific payloads.

Wednesday, March 5, 1986

As the search operations continued, the White House press office announced that President Reagan would ask Congress for funds to replace *Challenger*.[84] The new Shuttle would be built using spares already constructed, with an overall replacement cost estimated to run nearly $2.2 billion. On site, a broken winch on the *Stena Workhorse* delayed the retrieval of a large section of the left Solid Rocket Booster.[85]

Friday, March 7, 1986

After several failed attempts, divers located *Challenger*'s crew compartment in 100 feet of water, after side-scanning sonar discovered it nearly eighteen miles northeast of Cape Canaveral.[86] The crew cabin had traveled approximately ten miles downrange after the breakup. The discovery, however, was not announced at the time to the general public. NASA wanted time to confirm the discovery, as well as an opportunity to brief the families of the lost crew members. NASA had reason to believe that the crew members' remains would be located very near the crew compartment. NASA knew that the cabin had probably survived the destruction of the *Challenger* for the most part intact.

Contract divers Mike McAllister and Terry Bailey, working for the air force, located the *Challenger* crew cabin. A blue NASA flight suit 87 feet below the surface in the debris field marked the area.[87] McAllister stated, "The suit was in the pile [of wreckage] and it was pinned down underneath some debris. It was full of air and the legs were floating up towards the surface."[88] This was one of the extra flight suits which had been stored in one of the crew lockers on the mid deck and not one worn by a crew member during the flight. The divers stated that nothing else was identifiable. No crew remains were reported to have been found at the time. It was later reported that some of the divers working around the crew cabin were astronauts, including Jim Bagian.[89]

More pieces from the main engines were recovered by the USS *Preserver*, as two other ships involved in the search returned late in the day to Port Canaveral with sections of the external tank, the rear fuselage, and *Challenger*'s left wing.[90]

Saturday, March 8, 1986

Navy divers from the USS *Preserver* made positive identification and took photos of *Challenger*'s crew cabin with close-up inspections. Crew remains located in the wreckage of the compartment were brought to Port Canaveral by the *Preserver* late in the night under cover of darkness, with the ship's running lights turned off. The remains were in containers with flags draped over them. NASA did not announce the cabin confirmation and made no statement concerning the containers.

Chief astronaut John Young's internal memo was released and printed in papers around the country. In that memo, Young stated his concern about the flawed performance of the solid rockets and NASA's continued use of that system. "There is only one driving reason that such a potentially dangerous system would ever be allowed to fly — launch schedule pressures!"[91] Young's memo did not state the source of the launch schedule pressures. NASA Interim Administrator William Graham stated that the full cost of the *Challenger* disaster would run at least $3.2 billion, including recovery operations and replacement of hardware.

Sunday, March 9, 1986

The NASA public affairs office announced that *Challenger*'s crew compartment had been located and positively identified, and crew remains found 16 miles northeast of Cape Canaveral, 100 feet under water.[92] NASA officials stated that no bodies were found, only remains which were being worked on for identification by personnel from the Armed Forces Institute of Pathology. When the USS *Preserver* came into port, without running lights,[93] astronaut representative Bob Overmyer met the ship. Overmyer would later be killed in the crash of an experimental aircraft.

A navy representative reported that "neither the crew compartment nor the bodies were intact. We're talking debris, and not crew compartment and we're talking remains, not bodies."[94] NASA Associate Director for Space Flight Richard Truly stated, "Local security measures are being taken to assure that the recovery operations can take place in a safe and orderly manner." NASA reported that the families had been informed before the public announcement of crew members' remains, yet the parents of both Jarvis and McNair told reporters that no such notification had been received. Was this just one more NASA error?

NASA sources close to the operation reported that crew members' personal tape recorders had also been recovered. These were the battery-operated recorders that payload specialists Jarvis and McAuliffe were to have used to record their impressions of the flight. They would certainly have been used during the launch phase of the mission, but it is unlikely that their contents will ever be released to the general public by the space agency. Earlier, personal property from crew lockers had been recovered, including material for McAuliffe's Teacher-in-Space Program, which had been found floating on the surface near the crew cabin.[95] It was not reported if a watch worn by McAuliffe which had belonged to her grandmother, or her son's stuffed green frog had been recovered. However, it was later reported that an American flag carried by Onizuka was recovered and presented by his family to the United States Air Force Academy. NASA would continue to say very little about astronaut personal effects, and nothing about the recorders.

Monday, March 10, 1986

Strong winds up to twenty knots and high seas with ten-foot waves interfered with recovery operations.[96] By this time 15 percent of *Challenger* and its component parts, including the external tank and SRBs, had been located and recovered. This was far more than NASA officials had expected would be recovered. The recovery team reported that the section they were seeking showing the possibly burned section on the right solid rocket booster may have been blown away by the destruct order.

Tuesday, March 11, 1986

Calmer seas allowed divers from the USS *Preserver* to continue the recovery of astronaut remains.[97] This had become the divers' prime objective since the crew cabin had been located. The NASA public affairs office refused to comment on how long they would be looking for astronaut remains.[98]

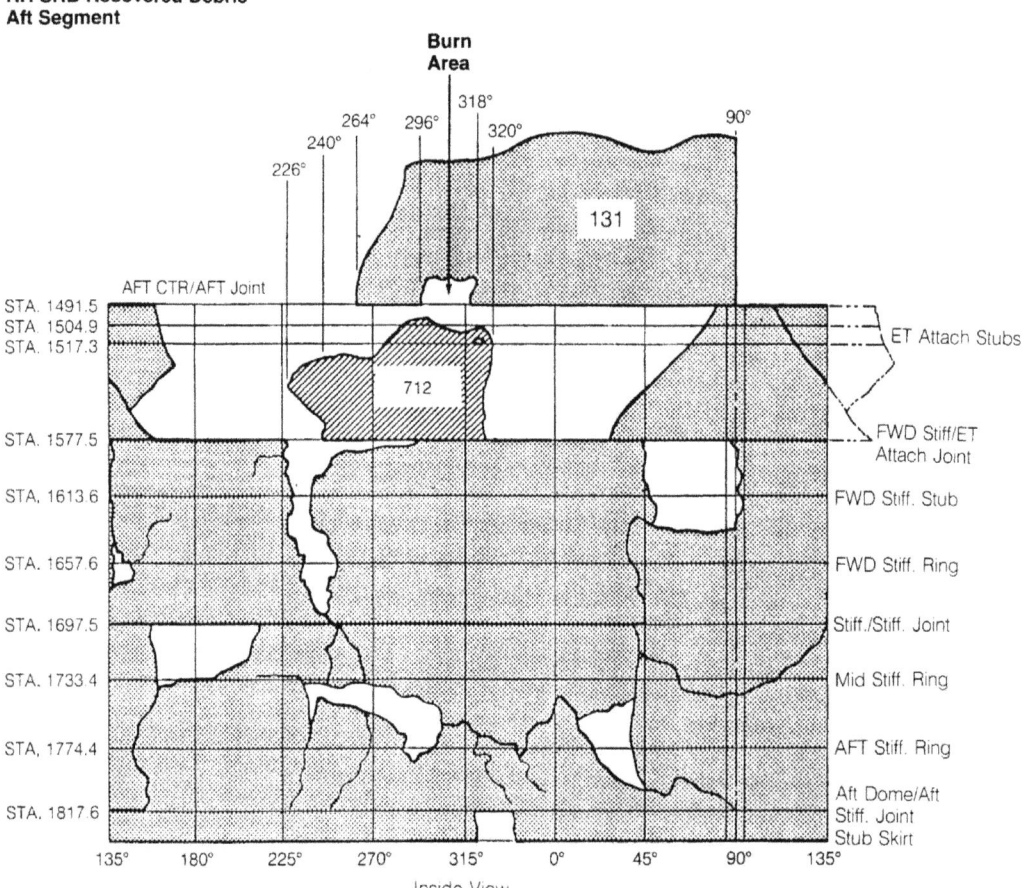

Right solid rocket booster, recovered debris aft segment (graphics from *Challenger* report, page 68).

Wednesday, March 12, 1986

Search crews working late into the night unloaded eight containers of *Challenger* crew remains into four ambulances which were driven to Hangar L, the Life Science Facility at Cape Canaveral Air Force Station. *USA Today* reported that military pathologists were working to identify some remains brought ashore secretly on March 8.[99] NASA refused to give any information about the remains except to say they were found at sea on the 8th.[100]

The computers onboard *Challenger* were also brought ashore late this night along with pieces of the flight deck and two space suits, as well as some personal crew effects.[101] McNair and Onizuka would have used the space suits in the event an EVA was required to help deploy the TDRS. When the USS *Preserver* pulled into port, several crew members were wearing dress blue uniforms and standing by the flag-draped containers as an honor guard.[102] The crews also unloaded several mangled sections of the crew compartment, including four general-purpose computers.[103]

Thursday, March 13, 1986

A NASA official reported that divers recovered all four flight recorders and "some" of the five computers aboard *Challenger*.[104] However, an Atlantic storm kept most of the recovery fleet in port awaiting calmer seas.[105] Astronaut Jon A. McBride, who had been scheduled to command the next Shuttle mission, said in response to reporter's questions, "Personally, I'm sorry that they found it. I think it would have been better for them to have been buried at sea. That would have been my personal desire."[106]

In answer to questions during an interview, President Reagan responded to suggestions that White House pressure was brought to bear on NASA to launch. "We have never from here suggested or pushed them for a launch of the Shuttle."

Saturday, March 15, 1986

Strong winds and large thunderstorms in the Atlantic made recovery efforts extremely difficult. Most of the recovery fleet stayed in port for a second day in a row.[107] NASA and the navy feared the rough weather in the recovery area could bury important wreckage in ocean floor silt. They were also concerned that crewmembers remains could be lost or scattered before recovery divers could retrieve them.

Sunday, March 16, 1986

The *Stena Workhorse* pulled a six-by-eight foot 3,250-pound chunk of debris from a depth of 400 feet, which was reported to be a section from one of the solid rockets.[108] It would later be shown to have come from the right SRB.[109]

Monday, March 17, 1986

The *Stena Workhorse* pulled a four-by-five foot 500 pound section of debris, thought to be another section of the right solid rocket, from the ocean floor,[110] as rescuers checked one sonar target after another. Operations also continued around the crew-cabin area. NASA was unsure if remains had been located for all seven crew members and wanted work completed in the crew-cabin recovery area as soon as possible. Storms continued to be a problem.

Wednesday, March 19, 1986

The USS *Preserver* continued to retrieve crew-cabin debris and crew remains from 100 feet of water, 18 miles northeast of Cape Canaveral.[111] At the same time a 500-pound section of the right solid booster was unloaded at the Cape.[112]

Thursday, March 20, 1986

Press reports indicated that NASA phone logs had been turned over to presidential investigators.[113] White House aides were reported to be anxious to lay to rest reports that any pressure to launch came from anyone at the White House. The logs showed no calls were made to or from the White House staff in connection with the launch of *Challenger*.

Friday, March 21, 1986

Rough seas forced search vessels to suspend operations early and return to port with no additional wreckage recovered. Rumors begin to circulate that military doctors working on the recovered remains had identified all seven astronauts. NASA would not make any comment on these reports. A NASA official did repeat that no information on crew remains would be forthcoming before notification of the families.

Saturday, March 22, 1986

Operations continue to be hampered by bad weather in the recovery area. High winds and choppy seas had even reached down to the ocean floor where visibility was reported to be less than one foot in the recovery areas.[114]

Sunday, March 23, 1986

Operations continue to be hampered by bad weather in the recovery area for a third day in a row, and no ships attempted to enter the recovery area. NASA reported that it would not be necessary to reconstruct the *Challenger*, as investigations already showed that no failure occurred on the orbiter. However, NASA was not in charge of the investigation, and that determination would be made by the *Challenger* Commission and developed by the NTSB.

Monday, March 24, 1986

Using sonar images of the ocean floor, searchers reported that *Challenger*'s crew cabin might have hit the ocean relatively intact.[115] Sonar maps indicated that the debris field was over a relatively small two-square-mile area, seventeen to eighteen miles northeast of Cape Canaveral. Some sources within NASA reported that the cabin "tumbled wildly as it fell some nine miles before its impact on the ocean surface." There was no indication as to what area of the cabin made first contact with the ocean. *Time* magazine reported that, "The lower mid deck ... apparently absorbed the full force of the blast ... and was nearly obliterated. The upper deck ... was still partly intact." The astronaut who was seated closest to the right side of the *Challenger*'s cabin on the mid deck was Greg Jarvis.

Tuesday, March 25, 1986

The families of the *Challenger* crew released a statement, dated March 11, to the world's press: "The spouses and families of the *Challenger* flight 51-L crew gratefully acknowledge your expressions of sympathy and support. We thank the world for sharing in our pride of the *Challenger* crew, and it is with continued pride that we will cherish the memories of their accomplishments. From all of you we draw the strength to bid the crew of 51-L our love and joyful wishes for an extended and exciting exploration of that dimension in space that so intrigued them."[116]

Wednesday, March 26, 1986

The Associated Press reported that remains of six of the seven astronauts had been identified.[117] No word was given as to which crew member's remains had yet to be identified. However, other reports stated that was Greg Jarvis. Bad weather continued in the recovery area.[118] It was possible that Jarvis had not been with the crew cabin when he hit the surface of the ocean, but he was probably still strapped to his seat.

Thursday, March 27, 1986

High waves and strong winds covered the recovery area. Several debris-marker buoys had been moved by the wind and waves in the crew-cabin area. Navy Captain Charles Bartholomew, supervisor of salvage operations, was quoted as saying, "This weather is killing us."[119] He was concerned that bad weather could cause debris to be buried by the shifting sands on the ocean floor, as well as delay an already costly operation.

Friday, March 28, 1986

Bad weather continued to hamper operations[120] as the navy sent the USS *Preserver* out to the recovery area in case of a break in the weather. No debris would be recovered on this day, however, and the USS *Preserver* was forced to return to port.

Saturday, March 29, 1986

High winds and poor visibility down to one foot on the ocean bottom hampered operations in the crew-compartment area as divers searched for crew members' remains.[121] Divers reported that shifting ocean-floor sands were starting to cover wreckage in several areas.

Monday, March 31, 1986

This was the first good full day on site for diving in ten days, after high winds and storm conditions which caused poor visibility on the ocean floor as well as poor surface conditions. The search now concentrated on a 480-square-mile area from fifteen to fifty miles extended downrange east of Cape Canaveral.

Tuesday, April 1, 1986

Sections of the right SRB were recovered as operations concentrate on the now suspect booster.[122] Beyond the recovery of *Challenger*'s crew, this booster's recovery had become the primary task for the fleet.

Wednesday, April 2, 1986

The USS *Preserver* completed operations in the crew-cabin debris field and was assigned to another section of the search area.[123] No further crew remains would be located in the crew-cabin debris field. It was speculated at this point that the remains of Greg Jarvis had not yet been located.

Friday, April 4, 1986

The USS *Preserver* brought its final load of wreckage to Port Canaveral, consisting of a piece of engine nozzle from one of the main engines.[124]

Saturday, April 5, 1986

A large piece of Shuttle fuselage was located and photographed on the ocean floor. The 34-foot-long piece had a United States flag painted on it.[125]

Monday, April 7, 1986

A ten-by-twenty foot section of *Challenger*'s fuselage weighing 5000 pounds was recovered. This section, with the flag on it, was photographed, and the picture published in newspapers around the country.[126]

Wednesday, April 9, 1986

The *Stena Workhorse* brought eight sections of the right solid rocket into Cape Canaveral, as fewer and fewer pieces were being located in the search area.[127]

Friday, April 11, 1986

Several sections from the right SRB were recovered, including a 38-foot-round section, which went completely around the booster. This weighed 11,000 pounds.[128]

Tuesday, April 15, 1986

A navy official reported that a ten-by-twenty-foot piece of the right solid booster (labeled no.131) had been recovered some forty-six miles northeast of pad 39B.[129] The recovered section had a two-foot-square hole burned through the steel casing. The piece was recovered by the *Stena Workhorse*. "It is the piece of evidence that we have been looking for in our total search-and-salvage operation," stated Tom Moser, NASA deputy associate administrator for space flight. This piece would be the final confirmation needed to show that the right SRB had failed, dooming the crew. The remains of Greg Jarvis were recovered, as reported in *Disasters and Accidents in Manned Spaceflight*.[130]

Wednesday, April 16, 1986

As the *G. W. Pierce* returned to Port Canaveral, it was met by ambulances, reportedly to offload crew remains with an astronaut escort.[131] However, the search for crew remains within the crew-cabin area had ended two weeks earlier. Was it possible that at least one of the crew members had indeed been located in a second area? This could point to one of the crew members being thrown out of *Challenger* at the time it was destroyed. The best guess would be Greg Jarvis who was seated closest to the starboard wall of the Shuttle in the lower deck when it was heavily damaged by the explosive burn of spilled fuel.

Thursday, April 17, 1986

Returning to Port Canaveral with the Shuttle's right wing, the USS *Opportune* offloads the wreckage along with a set of Shuttle landing gear. Most of the word "Challenger" can be seen on the scorched wing.[132]

Saturday, April 19, 1986

With a statement from NASA's new associate administrator for space flight, former astronaut Richard H. Truly, the search for the crew was over.[133] The former Shuttle commander reported, "Remains of each [astronaut] have been recovered.[134] Final forensic work and future planning in accordance with family desires are to be completed, and an announcement made in a few days."[135] Greg Jarvis's remains had been found. NASA also reported that several pieces of the right SRB were rigged for recovery. Even though the search for remains had been completed, this work was yet to be finished.

The search for and recovery of the remains of *Challenger*'s crew had been successfully completed. On April 29, the remains were released to the families.[136] By April 30, the final critical section from the right solid booster (labeled no. 712) had been recovered. This piece with a hole burned through it was being brought back to Cape Canaveral for study. The search for Shuttle wreckage would continue until the end of August when the final three ships pulled into port on the 28th. It was seven months to the day after *Challenger* was destroyed. The search had involved 31 ships, 52 aircraft, 3 submarines, 5 unmanned subs, and 6000 individuals, including 115 divers.[137] The search had covered 429 square miles of ocean floor, and nearly 93,000 square miles of ocean surface.[138] Of the 3,367 sonar contacts, an amazing 3,197 turned out to be *Challenger* wreckage. Total debris recovered was 245,000 pounds which represented 90 percent of the crew cabin, 45 percent of the orbiter, 95 percent of Spartan-Halley, 35 percent of TDRS-B, 50 percent of the SRBs, and 50 percent of the external tank. Also, some of the Shuttle crew's personal belongings were returned to family members, but not identified by the space agency. It was now time for NASA, and the nation, to take a hard look at what had gone wrong on the mission, and find answers that would prevent this type of accident from happening again. That would be the job of the Presidential Commission on the Space Shuttle *Challenger* Accident, and it would prove to be a difficult and much debated task.

> *For any contingency they [know] what to do, so I feel very, very comfortable.*
> — Payload Specialist Gregory Bruce Jarvis

Final Resting Places of *Challenger*'s Last Crew

FRANCIS RICHARD SCOBEE	Buried at Arlington National Cemetery with full military honors on May 19, 1986, near the Tomb of the Unknowns. The poem "High Flight" was read at the ceremony.[139]
MICHAEL JOHN SMITH	Buried at Arlington National Cemetery with full military honors on May 3, 1986. Located on a small hill near the Tomb of the Unknowns.[140]
JUDITH ARLENE RESNIK	Remains were cremated and her ashes buried in Akron, Ohio, during a private ceremony.[141] The family will make no further comment.
ELLISON SHOJI ONIZUKA	Buried in Hawaii at the Memorial National Cemetery of the Pacific at Honolulu's Punch Bowl, Puowaina Crater, with full military honors on June 2, 1986.[142]
RONALD ERWIN MCNAIR	Buried in his hometown of Lake City, South Carolina, at Rest Lawn Cemetery. He was buried with full military honors on May 17, 1986.[143]
SHARON CHRISTA MCAULIFFE	Buried in the private Calvary Cemetery on May 1, 1986, on a hillside overlooking Concord, New Hampshire, where she taught high school classes.[144]
GREGORY BRUCE JARVIS	His widow scattered his ashes over the Pacific Ocean from an aircraft on May 3, 1986.[145]*

On May 20, 1986, the unidentified remains of *Challenger*'s crew were buried at Arlington National Cemetery in a common grave after cremation. The crew of *Challenger*'s last mission would remain together in a place of national honor.

---

*Disasters and Accidents in Manned Spaceflight, published in 2000, reported that Gregory Jarvis was buried in Mohawk, New York.

SECTION IV

INVESTIGATIONS AND RECOVERY

CHAPTER NINE

The *Challenger* Investigation

If we die, we want people to accept it.... The conquest of space is worth the risk of life.
—Astronaut Virgil I. ("Gus") Grissom, January 1967

NASA Loses Control of Its Agency

On February 3, 1986, while NASA and the nation focused on the search for *Challenger* and her seven astronauts, President Reagan informed the nation that he had formed a presidential commission to investigate what had gone wrong. President Reagan had been urged by his chief of staff, Donald Regan, to form a panel outside of NASA soon after the accident had been viewed on television by the White House staff. Perhaps the memory of the Apollo One fire on the pad which killed three astronauts 19 years earlier, and the less than sterling investigation conducted by NASA, had something to do with the decision. Before anyone from the space agency could even call the White House, the wheels were in motion to take control of the investigation and remove the space agency from the process. However, the President was not to make his final decision until he was on Air Force One flying back to Washington from the *Challenger* memorial services on January 31, which had been held in Houston, Texas.

One individual who was totally opposed to any outside commission was former (and soon to be reinstalled) NASA administrator James Fletcher. Fletcher had been NASA's administrator when many of the decisions had been made on the orbiter's design. He had also been on hand when it was decided which companies would construct the Shuttle's components, including Morton Thiokol's SRB design, which he knew would come under close scrutiny. If an outside group conducted the investigation, Fletcher would not be able to control which way the investigation would go, and he knew he would be one of those on the hot seat. The president knew that NASA could not be allowed to investigate itself. Not even a glimmer of a cover-up could be allowed to show. NASA had spoon-fed a willing press for years on the safety and reliability of the Shuttle system, and that had to end with this investigation.

William Rogers, former secretary of state under Nixon, a logical choice to chair the commission, was handpicked by Donald Regan because he was a safe Washington insider. Most of the commission's other members were from NASA or the aerospace industry, and could be expected to generally soften any criticism of problems being uncovered. However, Richard Feynman was another matter. This maverick, Nobel Prize–winning physicist would bring to the commission insight into how the system had failed, and how those tasked to guard the system had themselves failed at just the point when they were most needed. As an outsider, he got to the truth quickly and with sharp definition. (Two years later, in February 1988, Dr. Feynman would die of cancer.)

As stated in the commission's final report, the panel was required to: (1) Review the circumstances surrounding the accident to establish the probable cause or causes of the accident; and (2) Develop recommendations for corrective, or other action, based upon the commission's findings and determinations. Even before the commission had formed, NASA administrators and managers were ready to speak about what they thought had caused *Challenger* to be destroyed, but only on condition that they not be identified. These individuals were pointing toward a plume of flame seen on the launch video, which either burned through the external tank or touched off a destruct package, destroying the orbiter. However, only Acting Administrator William Graham would comment for the record, and he would only state that NASA was still looking into the causes of the accident. "We haven't yet finished the analysis and measurements on film to identify the exact point at which the plume appeared."

In the hours and days following the disaster, the nation closed ranks around NASA in a show of grief and solidarity. The disaster was seen by most as a terrible accident which could not have been avoided. Before long, however, that feeling would change to one of horror, as more and more evidence was discovered showing a badly flawed system, a failed design and a supplier who seemed to care more for profit and less for delivering a safe and reliable product. For many the shock would turn to anger and then suspicion at the entire space program and at those who had been tasked to guide it. Many Americans felt they had been betrayed and lied to.

> *It's time now to assemble a group of distinguished Americans to take a hard look at the accident, to make a calm and deliberate assessment of the facts and ways to avoid repetition.*
> — President Ronald Reagan

EXECUTIVE ORDER

PRESIDENTIAL COMMISSION ON THE SPACE SHUTTLE CHALLENGER ACCIDENT

By the authority vested in me as President by the Constitution and statutes of the United States of America, including the Federal Advisory Committee Act, as amended (5 U. S.C. App. I), and in order to establish a commission of distinguished Americans to investigate the accident to the Space Shuttle *Challenger*, it is hereby ordered as follows:

Section 1. Establishment. (a) There is established the Presidential Commission on the Space Shuttle *Challenger* Accident. The Commission shall be composed of not more than 20 members appointed or designated by the President. The members shall be drawn from among distinguished leaders of the government, and the scientific, technical, and management communities. (b) The President shall designate a Chairman and a Vice Chairman from among the members of the Commission.

Sec. 2. Functions. (a) The Commission shall investigate the accident to the Space Shuttle *Challenger*, which occurred on January 28, 1986. (b) The Commission shall:

(1) Review the circumstances surrounding the accident to establish the probable cause or causes of the accident; and

(2) Develop recommendations for corrective or other action based upon the Commission's findings and determinations.

(c) The Commission shall submit its final report to the President and the Administrator of the National Aeronautics and Space Administration within one hundred and twenty days of the date of this Order.

Sec. 3. Administration. (a) The heads of Executive departments and agencies shall, to the extent permitted by law, provide the Commission with such information as it may require for purposes of carrying out its functions.

(b) Members of the Commission shall serve without compensation for their work on the Commission. However, members appointed from among private citizens of the United States may be allowed travel expenses, including per diem in lieu of subsistence, to the extent permitted by law for persons serving intermittently in the government service (5 U.S.C. 5701–5707).

(c) To the extent permitted by law, and subject to the availability of appropriations, the Administrator of the National Aeronautics and Space Administration shall provide the Commission with such administrative services, funds, facilities, staff, and other support services as may be necessary for the performance of its functions.

Sec. 4. General Provisions. (a) Notwithstanding the provisions of any other Executive Order, the functions of the President under the Federal Advisory Committee Act which are applicable to the Commission, except that of reporting annually to the Congress, shall be performed by the Administrator of the National Aeronautics and Space Administration, in accordance with guidelines and procedures established by the Administrator of General Services.

(b) The Commission shall terminate 60 days after submitting its final report.

RONALD REAGAN
THE WHITE HOUSE
February 3, 1986

Commission Members[1]

Chairman William P. Rogers	Former secretary of state in the Nixon administration
Vice Chairman Neil A. Armstrong	Former astronaut and commander of Apollo 11, first man to walk on the moon
David C. Acheson	Washington attorney and formerly senior vice president of the Communications Satellite Corporation
Dr. Eugene E. Covert	Professor of aeronautics at MIT and a consultant to NASA for rocket engines
Dr. Richard P. Feynman	Professor of theoretical physics at Cal Tech and winner of the 1965 Nobel Prize in Physics
Robert B. Hotz	Member of the general advisory committee to the U.S. Arms Control and Disarmament Agency and past editor of *Aviation Week* and *Space Technology* magazine
Donald J. Kutyna	Director of space systems for the air force
Dr. Sally K. Ride	Astronaut and first U.S. woman in space and holder of a Doctorate in physics
Robert W. Rummel	Member of the National Academy of Engineering and former vice president of Trans World Airlines
Joseph F. Sutter	Executive vice president of Boeing and a member of the National Academy of Engineering
Dr. Arthur B. C. Walker, Jr.	Professor of applied physics at Stanford University and a consultant to Rand Corporation
Albert D. Wheelon	Senior vice president with Hughes Aircraft Company and a member of the president's Foreign Intelligence Advisory Board
Charles E. Yeager	Former test pilot who became the first man to fly faster than the speed of sound (attended only one meeting)

With the formation of the President's Commission, NASA's interim investigation panel ceased to exist, and with it NASA's control over its own destiny.[2] Following the members' swearing in on February 6, the commissioners called their first hearings, in which NASA administrators outlined how the space agency operated in general and specifically how the Shuttle program was managed. NASA also reported on the results of its internal investigation of the accident.

The nation was continuing to react to the loss of *Challenger*, and a private fund was established for the eleven children of those killed on board. The U.S. Space Foundation, a private advocacy group, also developed a fund to help pay for the cost of *Challenger*'s replacement,[3] a fund which by February 3 had collected over $30,000 after only four days.

It was not long before the commission and NASA's relationship would become adversarial. After a report in the *New York Times* made reference to possible discrepancies between NASA documents and the testimony of NASA officials, the commission ordered NASA to turn over all internal documents concerning O-ring problems. The next day Jesse Moore, former head of NASA's 51-L Interim Mishap Review Board, delivered a 1.5-inch-thick stack of documents to the commission. The commission then ordered tests to be conducted on the design of the solid rockets as well as on lubricants and O-rings to establish their effectiveness at low temperatures. They were not yet aware that such tests had been conducted by the space agency, and NASA would not come forward until much later with those results. NASA, it would seem, was already trying to cover its tracks. It was clear early on that the decision to appoint an outside investigation group was a good one. The review also looked at the weight of the Shuttle stack as a possible cause of the mission failure. With a total weight of 4,529,122 pounds, including fuel, *Challenger* was the heaviest of any previous Shuttle. The orbiter alone weighed 268,829 pounds including its TDRS cargo.

During its February 10, 1986, closed session held in Washington, D.C., the commission first learned about the many warnings which had been given about the O-rings, and the fact that engineers at Morton Thiokol did not want to launch. At this session, Lawrence Mulloy displayed a cross-sectional model of the solid-rocket-booster seals.[4] Under continued questioning, Mulloy indicated that engineers at Thiokol had warned against a launch.[5] Prior to his testimony, NASA officials had only stated that there had been discussions among engineers and management about the possible effects of cold temperatures on the O-rings. NASA officials stated there had been no disagreement with the final decision to proceed. NASA management had been caught in another lie.

> LAWRENCE MULLOY: There was data presented by Thiokol engineers that there was a suggestion that possibly the seal shouldn't be operated below any temperature that it had been operated under in previous flights.

While commission member Feynman was firing questions at Mulloy about the effects of cold temperatures on the O-rings, Feynman performed an experiment of his own on a small piece of synthetic rubber O-ring he had removed from the model Mulloy had brought to the hearing.[6] Dr. Feynman bent the O-ring in a small vise and then submerged it in a glass of ice water. When he removed the O-ring from the ice water and removed the vise he was able to show the effects of cold on the synthetic rubber. It had moved back to its original shape much more slowly than it had at room temperature. Dr. Feynman said it showed the rings were less resilient in cold temperatures, and could allow hot gases to escape when used on the Shuttle.

> DR. FEYNMAN: Does your data agree with this feature that the immediate resilience in the first few seconds is very, very much reduced when the temperature is reduced?
> LAWRENCE MULLOY: Yes, in a qualitative sense. I just can't quantify at this time.

Mulloy had previously testified that to his knowledge the O-rings had been tested to as low as minus 30 degrees and had operated properly. But these were the phantom tests, which were never performed by Thiokol. (See Chapter Two.) Yet, even Thiokol management, who had deceived NASA, only reported that they had tested to 21 degrees above zero, not an unforgiving 30 degrees below zero. So where did Mulloy get his data? Mulloy even believed, as did many at Marshall Space Flight Center, that the O-rings had redundancy in the secondary seals, despite tests to the contrary. Mulloy later admitted that the rubber O-rings begin to lose their resiliency at 50 degrees Fahrenheit (in a memo dated August 9, 1985, some five months before *Challenger* was launched, Thiokol had informed NASA that the seals did not seal properly at 50 degrees), but that he continued to recommend launching despite this knowledge.

NASA also reported that *Challenger*'s boosters were never physically examined by any NASA engineer for the full 33 days the Shuttle stack sat on the pad. Thomas Utsman, deputy director of the Kennedy Space Center, reported that no Shuttle had ever sat for a longer period of time in colder weather. "Once it is moved to the pad, it is considered a structurally safe vehicle."

To develop a better understanding of testimony just concluded on the O-rings, a special jet was made available to commission members, which carried them to the Kennedy Space Center for two days of closed-door sessions on February 13 and 14.[7] The members interviewed launch-control officials and toured the Vehicle Assembly Building (VAB) to inspect the stacking procedures used with the solids, as well as how the booster segments from the manufacturer were unloaded and stored.

On the 13th, commissioners saw a NASA-produced film, which included video and telemetry of *Challenger*'s launch.[8] This was the first hard evidence the commission was to see of the SRBs' potential failure. On the 14th, dramatic testimony was taken in closed session from Thiokol executives and NASA managers, which would change the direction and scope of the investigation.[9] No longer would commission members be passive collectors of NASA supplied information. They became active participants in the investigation process.

On February 15, 1986, after only a few days of operation, Chairman Rogers announced that the decision to launch *Challenger* "may have been flawed."[10] He spoke for all members when he said, "In recent days the commission has been investigating all aspects of the decision-making process leading up to the launch of the *Challenger* and has found that the process may have been flawed."[11]

Chairman Rogers then asked Acting Administrator William Graham not to permit anyone who was involved with the launch decision to be assigned to any internal (NASA) investigating teams. Charles Redmond from the public affairs office at NASA headquarters in Washington, D.C., stated, "There are an awful lot of people involved in the countdown process who are involved in the investigation into the *Challenger* explosion."[12] NASA was clearly trying to work around the commission request in an attempt to show it would involve too many people at NASA. Once again NASA had missed the point, perhaps intentionally. The operative phrase was "involved with the launch decision."

This was a dramatic shift in the commission's direction, from problems in hardware, to individuals who gave the go ahead for launch. From this point on, management would be considered at least partly to blame for the disaster and suspect in any investigation. News reports were quoting Richard C. Cook, a NASA budget analyst, as saying that engineers had "whispered in his ear"[13] that the O-rings were not a safe design and that engineers had held their breath,[14] when the first few Shuttle missions were launched.

By February 14, the United States Space Foundation had received $63,625 in donations to help build a replacement Shuttle.[15] At the same time, NASA announced the space agency had signed a contract to launch three satellites for Intelsat Corporation, beginning on July 15, 1987.[16] This was the first commercial contract signed since the *Challenger* accident and one NASA would not be able to fulfill. A business-as-usual mentality was already starting to show at NASA only days after the astronaut's deaths.

With developments coming on fast, the commission decided to divide itself into four investigation teams. The areas included development and production, prelaunch activities, mission planning and operations, and accident analysis.

As Chairman Rogers briefed Congress on the new approach developed by the commission, testimony continued. Dr. Feynman reported to the press that he had confirmed, despite NASA's continued denials, that the seven-degree temperatures measured on the pad by infrared readings were accurate when they were taken three to five hours before launch. He further stated that on the outside of the external tank a temperature was recorded at minus 8 degrees, and that a mild breeze moving across the stack could have further cooled the right solid rocket booster. Feynman later stated that the infrared equipment was accurate to within 0.1 degree Fahrenheit. That was something NASA managers did not want to be reminded of because it showed how cold it was, and how these readings were totally ignored by the men responsible for the Shuttle and lives of the astronauts.

In testimony before the commission, Jesse Moore, chief of the Shuttle program and final launch authority, stated that if he had received reports of temperatures less than 10 degrees Fahrenheit on the SRBs he would have asked "more questions about what the readings indicated." However, he was well aware of temperatures below freezing on the pad and he never asked the ice team about the below-zero temperatures. "At this particular juncture, it looks like an avoidable accident rather than an unavoidable one,"[17] said Senator Ernest F. Hollings (D–SC).

In testimony before the commission, Morton Thiokol engineer Allan J. McDonald, a 26-year veteran, said in closed session that he had fought hard to stop *Challenger*'s launch, in discussions which at times became very heated.[18] He also revealed that Morton Thiokol had originally recommended against a launch on the basis of cold temperatures, but that Thiokol management had overridden the engineers. Fifteen of Morton Thiokol's engineers were concerned about the cold effect on the O-rings and none gave any indications that it was a "go" situation.

NASA spokesman Charles Redmond said, "Clearly, the Morton Thiokol engineer has a point of view he is allowed to [express], but beyond that I don't think we have any comment."[19] It would seem from this statement that NASA was unaware that McDonald was not just any engineer; he was director of Thiokol's Solid Fuel Rocket Motor project and an expert in the field. His superior, Joe Kilminster, vice president of Space Booster Programs at Thiokol, eventually overruled McDonald's "point of view."[20] The question is: why would a NASA official attempt to put a spin on what a Thiokol engineer said? What else was going on behind the scenes? And who was really running the show?

Commission members were shocked to hear of a "near revolt" by engineers at Thiokol who had tried to stop the Shuttle launch in talks which had gone on for hours.[21]

> CHAIRMAN ROGERS: Mr. Boisjoly, at the time ... that Thiokol made the recommendation not to launch, was that the unanimous recommendation as far as you knew?
> ROGER BOISJOLY: Yes ... there was never one positive, pro-launch statement ever made by anybody.

When asked about the pressure to launch brought to bear on Thiokol by NASA through Marshall Center, Boisjoly continued.

> DR. WALKER: Do you know the source of the pressure on management that you alluded to?
> ROGER BOISJOLY: Well, the comments made over the [Net] is what I felt.... [W]e were being put in a position to prove that we should not launch rather than being put in the position [to] prove that we had enough data to launch. And I felt that very real.
> DR. WALKER: These were comments from the NASA people at Marshall and at Kennedy Space Center?
> ROGER BOISJOLY: Yes.
> DR. FEYNMAN: I take it you were trying to find proof that the seal would fail?
> ROGER BOISJOLY: Yes.

During testimony at Cape Kennedy, in a closed hearing room, the members of the commission sat stunned after Robert Sieck, launch director; Arnold Aldrich, manager of the Space Transportation System at Houston; and Gene Thomas, director at Kennedy for 51-L, all stated that they were not told about engineers' objections to launching. At that point, Chairman Rogers ordered everyone out of the room and stated, "We must advise the President as soon as possible." The system had fully failed to convey the necessary information to those who had responsibility to launch. Another major flaw had been uncovered. It would not be the last. The commission was doing its work fast and well. It should be noted that this problem would reappear during the investigation into the *Columbia* disaster 17 years later.

On February 19, 1986, a NASA spokesman announced that Jesse Moore, the associate administrator for Space Flight,[22] would "leave his position," being replaced by former astronaut Richard Truly. The next day Senator Hollings called for the resignation of Acting Administrator Graham saying that the administrator was really in over his head.[23] Hollings had accused Graham of "falsely representing the facts"[24] when he told the Senate that there was no evidence that Morton Thiokol had opposed the launch.

The commission began to focus its investigation on Marshall Space Flight Center managers in Huntsville, Alabama. Investigators wanted to determine if they had pressured Morton Thiokol engineers to approve the launch. The commission was also sifting through documents it had requested from NASA on the history of the solid rockets and specifically any problems with the O-rings. At the same time, commission members, who had been at work for little more than 18 days, came under criticism by Senate Democrats who complained that the commission was being too secretive about confirming reports on what the panel had learned.

Meanwhile, *Aviation Week and Space Technology* magazine was reporting that one year earlier NASA tests had indicated that cold weather conditions similar to, but warmer than, those on the night before *Challenger* was launched could freeze the O-rings, causing them to fail. This was critical in the decision-making process which had halted the launch of 51-C, the top-secret defense satellite mission. (See Chapter Three.) The aerospace industry magazine also reported that heaters had been installed on the new SLC-6 Defense Department Shuttle launch site at Vandenberg, California, but due to normally humid and warm weather in Florida they were rejected in the late 1970s for use at Kennedy. Yet, it can often go below freezing in Florida and that rarely happens in California.

As the investigation proceeded, it was found that a small leak in *Challenger*'s huge external tanks holding super-cooled hydrogen might have blown across the right Solid Rocket while it sat on the launch pad. This super cooled cloud of gas could have lowered the aft end of the booster to as low as minus 8 degrees early launch morning. This would give an explanation to the critically low temperatures found by the infrared equipment used by the ice team on pad 39B. The much lower temperatures then admitted by NASA were starting to look all too real.

With more and more information being released to the public by the commission and NASA concerning the solid rockets, Morton Thiokol representative Thomas Russell stated on February 22, 1986, that the company was very comfortable with its decision-making process.[25] He added that it would "take at least five years before a competitor could start taking away business. If the program is going on, we're going to have the contract."[26] The *Los Angeles Times* reported in its February 22 issue, that Thiokol received "between $300 to 400 million a year in revenues"[27] from NASA. (NASA had earlier announced that they would be seeking competitive bids for the next set of solid rocket boosters.)

A NASA official in Washington, D.C., speaking on condition of anonymity, stated that the space agency presently under fire would be ready to launch a Shuttle mission in as little as six months, with a small crew and minimal payload. The official was referring to the possibility of launching a military payload if a national security situation demanded it. Military payloads were beginning to pile up and schedule pressure would be something that would never fully be defeated unless funding was online to fully support seven or eight launches a year.

Indeed, when the program began to launch Shuttles again, two of the first three Shuttle flights would carry top-secret Defense Department payloads. Former secretary of the air force Verne Orr said, "The needs of the Defense Department are not flexible. There are times when we have to get to space." However, with the loss of *Challenger* that left only *Atlantis* at the time capable of carrying the Centaur Upper Stage needed to place the heaviest defense payloads into geostationary orbit.

During this period, the air force reportedly had only one KH-11 photoreconnaissance satellite in polar orbit, but felt that two were required to properly cover targets in the Soviet Union. A launch on August 28, 1985, only four months before *Challenger*'s loss, from Vandenberg Air Force Base, ended when the unmanned rocket was destroyed, taking with it one of the final KH-11 satellites. With this loss, a vast gap in the Defense Department's surveillance capabilities was created. On-orbit resources would need to be stretched as far as possible.

In testimony held in open session on Tuesday, February 25, Allan McDonald stated that he had felt pressure to launch by company officials and that he had refused to sign a Thiokol company memo approving the launch of *Challenger*. McDonald was the first witness to testify since the commission began looking into the flawed decision-making process. "The contractor always had to get up to prove it was safe to fly. In this case we had to prove it wasn't."[28] He further told the commissioners that "conditions for justifying the launch were so different than they had been before."[29] McDonald said he had assumed his and other engineers' objections would be passed all the way up to those who had the final approval for launch. The commission was to find that this was simply not the case and, further, that Thiokol had lied to their own engineers. The decision-making process was indeed flawed inside Thiokol as well as NASA.

It was also during these sessions, held the third week of February, that the commission would discover that Rockwell Space Division President Rocco Petrone had protested the Shuttle launch. Rockwell had built the Shuttle fleet, and Petrone's objections were based on damage which could have occurred if ice chunks were dislodged from the external tank during launch. Ice pieces could have hit the Shuttle's heat tiles and caused serious damage, which was very close to what happened to *Columbia* during its final launch. Dr. Petrone was pressed for details as they related to the *Challenger* flight.

> GENERAL KUTYNA: The question is very simple: are you "go" or are you "no-go" for the launch, and "maybe" isn't an answer. I hear all kinds of qualifications and cautions and considerations here. Did someone ask you are you "go" or "no-go"? Was that not asked?
>
> DR. PETRONE: It had been done at earlier meetings. This was a technical evaluation of a series of problems, and we talked about debris hitting the TPS (Thermal Protection System) and the tiles [which] led us to a conclusion that they were not safe to fly.
>
> CHAIRMAN ROGERS: And your recommendation [is that] it was unsafe to fly?
>
> DR. PETRONE: Correct, sir.

The commission would later determine that the objections raised by Rockwell management were not strong enough to stand as a simple and logical "no-go." Before launch however, Petrone was described as being astonished to learn that a launch was continuing at such cold temperatures, but he did not walk up to a launch controller and yell "No-Go!"[30] He allowed the flight to continue, which was the bottom line and constituted a "Go for launch" from Rockwell.

Roger Boisjoly, O-ring engineer, was next to testify as he described all he had done to try to stop the launch. "I did all I could to stop the launch on the evening before liftoff. I expressed deep concern that the result would be a catastrophe of the highest order — loss of human life."[31] In all, seven Morton Thiokol engineers testified that they had felt a great deal of pressure from NASA management when Thiokol's original decision was to not launch. Thiokol's managers would then send a fax stating they were "go" for launch.

When discussing Thiokol management testimony on the disaster, William Rogers would state, "We've heard the explanation (from NASA management) given. The problem we're having is [that] it's not convincing." Perhaps the true pressure was from an as-yet-unknown outside political source? Or perhaps "go" fever simply ran rampant.

The engineer's testimony was beginning to show a true pattern of pressure to launch. Boisjoly further testified that not a single engineer was in favor of launching in the cold temperatures expected at the Cape that morning. Jerome Lederer, former director of the office of Manned Flight Safety, stated, "These are signs that complacency may have set in, and that is not good for safety."[32]

In response to testimony that pressure was applied to Thiokol, George Hardy, deputy director of Science and Engineering at Marshall Space Flight Center, told the commission that, "I have racked my brain to try to see if there was any conceivable motivation that may have had any contractor representative feel that he was under pressure from anything I said."[33] The motivation may not have been there, but the pressure real or imagined certainly was. It was George Hardy who had said that he was "appalled"[34] that Thiokol did not want to launch due to predicted cold temperatures. Racked brain or not, that was pressure.

Meanwhile, in Alabama, Marshall Director William Lucas responded to a commission report that the launch process may have been flawed by issuing a statement. "I think it was a sound decision to launch. My people did not know of the lingering feeling of some Thiokol engineers."[35]

One fact is certain however: Lucas was fully aware that the design for the solid rockets, approved by Marshall, was flawed. The problem was buried in paper, and its long history of flaws was dismissed as a "feeling of some engineers."

After briefing President Reagan on what was to be announced, Chairman Rogers, in a statement released on February 27, stated, "I believe I am speaking for the whole commission when I say that [the decision-making process] is flawed. The process as it worked in this case was clearly flawed because recommendations made were not fully understood. The commission was also concerned that despite warnings from both the builders of the Shuttle and the SRBs, NASA still decided to launch, but did not have reliable data that would have shown it was safe to proceed."[36]

During his statement, Chairman Rogers was clearly angry and frustrated at the inability of NASA managers to take a solid stand on what he called "life-and-death safety issues." From this point on the commission would, for the most part, concentrate on the technical aspects of the failure. Analysis of launch photos showed that no ice debris had fallen on the Shuttle stack, and thus did not contribute to the Shuttle's destruction. After Rogers's statement, one of the commissioners reported that continuing statements by NASA officials showed that "what they're saying is that they would do it [launch] over again, that it's all right and correct because the paperwork was in order.... The commission was not amused."[37]

On March 6, 1986, President Reagan announced the appointment of James Fletcher to head the National Aeronautics and Space Administration, as members of the Senate were reporting that his confirmation would not be a problem.[38] However, earlier Fletcher had publicly stated that in his opinion the President's Commission was on a "witch hunt." He was perhaps not the best person to lead NASA out of this disaster. It was Fletcher who had convinced then-president Nixon to use solid rockets and not the originally designed liquid-fuel boosters for the Shuttle system, and his statements were, to say the very least, not founded in historical fact. Fletcher would later state that he had "not misled the president."[39]

Two days later, NASA released a 12-page internal memo, dated March 4, 1986, from John Young, a six-time space-flight veteran and chief of the Astronaut Office. In that memo, Young reported that he felt the space agency had exposed the astronauts to numerous "potentially catastrophic hazards"[40] because of NASA's need to push its launch schedule beyond its capabilities. The memo was released as the commission was reporting to Congress that they would examine the launch schedule pressure issue as it pertained to safety. Young's efforts to protect his crews would eventually cost him his last flight — and his job.

As stated in the commission's report, "a general investigative staff began a series of individual interviews to document fully the factual background of various areas of the commission's interest, including the [teleconference] between NASA and Thiokol officials the night before the launch; the history of joint design and O-ring problems; NASA safety, reliability and quality assurance functions; and the assembly of the right Solid Rocket Booster for STS 51-L." In all, more than 160 individuals would be interviewed during more than 35 sessions. These sessions would generate more than 122,000 pages of transcripts and documents in a wide-ranging search for answers to the loss of *Challenger* and her crew.

With a great deal of the press coverage focused on the management of NASA and Morton Thiokol, and how they had failed to properly view launch conditions, Thiokol on March 12, released a press statement. The statement announced that several company employees would be reassigned, but that no one in the management review would lose their job. Joe Kilminster, who faxed and signed Thiokol's go-ahead for launch to NASA, would not be demoted and would retain his position as vice president. The same could not be said for most of the engineers who had spoken up and tried to give warning.

In Congress, Representative Bill Nelson (D–FL), announced that he would conduct separate hearings, not on the *Challenger* disaster, but on the flight which had proceeded it when *Columbia* flew mission 61-C.[41] Nelson had been a passenger/observer on that flight which had landed only days before *Challenger* was launched. He stated that his investigation would not interfere with the commission looking into the *Challenger* disaster. Nelson was interested in uncovering the facts behind the many safety problems, which could have caused the destruction of *Columbia* on its mission. Unfortunately debris hits to the underside of the Shuttles was not a major portion of his investigation in early 1986.

By the second week of March, the commissioners had ruled out the orbiter, the three main engines, as well as the upper-stage rocket that would have been used to place TDRS-B into orbit as having anything to do with the accident. This would leave only the external tank and solid rockets as possible causes, with most of the examination focused on the solids. The commission had also hired an outside expert to examine O-ring tests because they were skeptical of the accuracy and the conclusions reached in tests conducted by Thiokol and NASA. The investigation teams could no longer trust their data due to conflicting testimony. A source close to the commission reported that, "the new tests are to determine at what temperatures the seals begin to deteriorate and what the effect of prolonged cold would be." This news was widely reported in the nation's newspapers.

That week, in a closed-door hearing, commissioners viewed an enhanced film and analysis along with a timeline, which for the most part confirmed that the O-ring seals had failed, most likely causing the *Challenger* accident. At the same time, NASA management began to respond to the commission's reports to the press that the decision to launch had been flawed. Richard G. Smith, director of the Kennedy Space Center, was quoted as saying, "We ought to get some of the facts understood before we start drawing conclusions, that's the normal process and I think we're going about this backwards. The commission got off on a bad foot by searching for bureaucratic indictments rather than technical explanations for the explosion." However, it was because of the failure of management to properly respond to the requirements of their positions that led to the destruction of *Challenger*, and those were the easiest problems to uncover.

By the end of March 1986, NASA was already developing a schedule for a return to space flight within one year.[42] It is difficult to understand what logic was used to put together this new timeline. By no means were investigations by the commission or NASA complete, and the recovery operations were still ongoing and would not be completed for another four months. With little or no real data, NASA seemed to be continuing to make decisions which made no sense. NASA management even went so far, during the investigation by the commission, to sign a new contract with Morton Thiokol for more solid rockets without any changes to those boosters. The commission ended that contract as soon as the members were made aware of the situation, and NASA was told, in no uncertain terms, not to sign any more Shuttle contracts until the investigation was completed. NASA management did not seem to realize that they were no longer in the driver's seat, and they would continue to fight for control.

On March 29, James E. Kingsbury, newly appointed by the space agency to direct the solid-booster fix, responded to NASA's one-year target date for the next launch by saying, "It's going to be very tight, but I think we can make it."[43] With the astronauts' remains not yet identified, NASA was already back to pressing a very tight launch schedule. NASA would continue to push its can-do, post–*Challenger* agenda, announcing that civilians, including a Teacher-in-Space replacement, would still be given an opportunity to fly. The agency did not seem to be learning much from the disaster, even as *Time* magazine in its June 9, 1986, issue quoted astronauts as saying that their work load was "too heavy for such extravagance." Privately, they described their civilian passengers as operational liabilities. One astronaut confided that Senator Jake Garn was a space invalid on his flight; Florida Congressman Bill Nelson, who later became a Senator, was described as a pain; and Saudi Prince Sultan Saud was "a nice guy, but he was a problem for everybody. He was lost, totally lost."

Shuttle pieces were still being recovered off the coast of Florida by a fleet battling high winds and pounding waves. Those were the same conditions in which *Challenger* had flown. It would later be shown that the Shuttle had flown through 75-mile-per-hour crosswinds, hurricane strength, producing an extremely rough ascent. At sea, ships prepared to recover the solid rockets were heading back to port due to high seas and rough conditions. These facts alone should have stopped the launch. Air Force Colonel Edward A. O'Connor briefed commission members on what had been recovered so far and what the prospects were for recovering critical parts of the Shuttle stack, which would conclusively show how events had unfolded.

On April 3, 1986, astronauts John Young, Robert Crippen, Henry Hartsfield and Paul Weitz, all Shuttle veterans, testified before the Presidential Commission. Safety and reliability were on most of the minds of these men who had flown Shuttles. Crippen stated that he felt no escape system would have saved the crew's lives in the situation which had just occurred, but that some kind of system must be developed. Hartsfield said that he would like some kind of "low-altitude bail out capability" developed.[44] He further stated that present procedures called for ditching the Shuttle in the ocean, something he doubted the Shuttle or the crew could survive. A later study by NASA would confirm his feelings when it was discovered that any ocean impact would destroy the orbiter. Any cargo in the cargo bay would be thrown into the crew compartment, killing the crew. As for problems with the O-rings, or the debris hits, they were simply not problems discussed at any length with the crews.

John Young said it for all the astronauts when he reflected on not being informed of O-ring problems. "I don't ever recall anything coming out of the flight-readiness reviews on solid rocket motor seals.... If anyone in the gang had known about this business and said something, we could have done something about it."[45] About the long hours worked by the crews who prepare the Shuttles for launch, Young said, "People are just working long hours, long periods of time. If we had more missions, from an operational standpoint it would have been tough. I think we would have been pushing it."[46]

That day, William Rogers told these astronauts that the commission would, as part of its final report, recommend that a safety group be formed, fully independent of NASA management, to oversee safety on upcoming flights.[47] That promise was kept and it would be the astronauts themselves who would eventually become part of that safety group. The problem with that solution was that, as time wore on, even the safety group would be cut back, all the way up to the time of the *Columbia* disaster. The further away NASA got away from a disaster, the lower the safety standards became.

The *Los Angeles Times* in its April 9, 1986, issue, reported that NASA investigators had filed preliminary findings describing "a two-step scenario in which a leak was created in a joint of *Challenger*'s right Solid Booster a split second after ignition, but did not become a catastrophic problem until the spacecraft hit a heavy, high altitude wind shear." Four other areas were being looked at which could have contributed to the accident, which included, according to the report:

- Poor performance of the putty used as an airtight seal
- Joint rotation at ignition, which may have damaged the O-rings
- Bad alignment of solid rocket segments which had to be forced into place, bending the segments
- Sub-freezing temperatures on the O-rings

The ongoing investigation of Shuttle wreckage was beginning to show surprising results. Despite the fact that wreckage recovered from the right side of the *Challenger* showed many burned areas, the breakup of the Shuttle was found to be due to in-flight overloads, not an explosive event. The investigation also found undamaged forward fuel tanks. This would indicate that the flash seen to destroy the nose of the Shuttle was in fact preceded by aerodynamic overloads on the vehicle which first broke up the front of the Shuttle and this allowed spilled fuels to rapidly burn; this was not an explosive event.

As commission members were interviewing individuals at several space agency centers, a news conference was being held at Marshall Space Center in Huntsville, on April 16. Neil Armstrong and three other commissioners discussed with the press an analysis which had been conducted on a two-ton piece of wreckage from the right solid rocket booster, which had been recovered off of Cape Canaveral.[48] This section, according to Armstrong, confirmed a hypothesis long held, that a failed seal had leaked during the boost stage of mission 51-L.[49]

Investigators were also on hand at Kennedy Space Center when workers disassembled a solid rocket booster to see if anything unusual could be discovered in its assembly which could point to problems in the general assembly process. These were the solid rockets which would have been used to launch *Challenger* on its scheduled May 1985 mission, designated 61-F, to send the Ulysses probe to Jupiter, and eventually into polar orbit above and below our sun. Ulysses would eventually be launched on board *Discovery* on October 6, 1990, on a new mission designated STS-41 from the same launch pad that *Challenger* had used for her final flight, pad 39B.

The final week of April 1986 would not be a good one for NASA, as more reports of problems within the space agency would see the light of day. First, the New York Times News Service in an April 23 report detailed problems with newly appointed NASA Administrator James Fletcher's NASA history. Fletcher was reported to "have misled Congress and the public in the 1970s about essential costs of the Shuttle program.[50] In 1972, in news releases and Congressional testimony, Fletcher predicted a cost of $10.45 million for each Shuttle launching. Today the costs are $279 million for each launch ... and $151 million each time just for operations."[51] According to Fletcher, the cost of building a Shuttle would be $675 million but the cost was actually $1.47 billion — more than double.[52] At the time NASA (read "Administrator Fletcher") announced a flight schedule that would top out at an astronomical 60 flights a year.[53] History would show that the most flights ever launched in a single year was nine in 1985.

The next day the *New York Times* again reported major problems within NASA. Federal audits conducted by the General Accounting Office showed that the space agency had[54]:

- Falsified, through X-rays, faulty welds on *Challenger* to avoid the cost of repairs.
- Failed to detect equipment flaws "so critical that they could cause loss of life or destruction of the spacecraft," due to major reductions in the number of inspection personnel.
- Misled Congress on the overall costs of the Shuttle from development at start to flight costs.
- Withheld critical documents from inspection agencies.
- Violated Federal codes thousands of times.
- Spent billions on Shuttle equipment not thoroughly tested.

During a news conference, the space agency released new enhanced photos taken from the ascent video, which showed the crew cabin breaking free and falling seemingly intact after the explosive burn. This was the first published visual confirmation that *Challenger*'s cabin had survived the breakup as a single unit. Recovery crews—who found a relatively small area on the ocean floor that held the wreckage and most of the crew remains—would confirm this view.

As reported by news agencies across the country, rocket engineers Allan J. McDonald and Roger M. Boisjoly were being pressured by Morton Thiokol management to not come forward with information on the problem-plagued solid rockets. In testimony, McDonald stated that he was approached by Lawrence Mulloy who told McDonald, "You're giving information to the commission without going through your own management; without going through NASA."[55] The question must be asked, was Mulloy ever charged with obstruction of justice in a federal investigation? He was also reportedly questioning McDonald's motivations. Another Thiokol employee, unnamed, stated, "The only thing you can conclude is that someone, somewhere, decided to make an example of Boisjoly and McDonald,"[56] when asked about these men's transfers and demotions.[57]

Both of these Thiokol engineers were later re-interviewed by commission members who simply could not believe what they were hearing.[58]

CHAIRMAN ROGERS: Do you have any reason to think you were given another assignment because of the testimony you gave?
ALLAN MCDONALD: Yes, I do. I feel I was set aside so that I would not have contact with people from

NASA again because they [Thiokol] felt that I either couldn't work with them [NASA] or it would be a situation that wouldn't be good for either party.
ROGERS: So, you were in effect punished for being right?
MCDONALD: I feel I was.

And from Thiokol O-ring team leader Roger Boisjoly:

ROGER BOISJOLY: I, too, have been put on the sidelines.
ROGERS: Do you feel that may be in retaliation for your testimony?
BOISJOLY: I think that is a possibility, a distinct possibility.
ROGERS: In this kind of an accident, where people come before a commission and tell the truth and then are treated as he [McDonald] believes he has been treated, which obviously is in some way punishment or retaliation for his testimony, it is extremely serious. To have something happen to him that seems to be in the nature of punishment [for testifying] is shocking.

Twenty-eight United States senators would later send a letter to newly appointed NASA Chief James Fletcher in response to charges of retribution against Thiokol engineers demoted for testimony at the commission. In part the letter stated, "Should these charges be true, we believe you should re-evaluate your agency's relationship with Morton Thiokol."[59]

Testimony, which had been taken in closed session earlier, was released on May 10, 1986. That testimony indicated that SRB managers at Marshall Space Center had consistently downgraded information about the problems found in development of the SRBs to higher headquarters. It was also the first time that the commission members had been informed that Marshall had placed a "launch constraint" on the solid rockets. At the start of that day's testimony, a very angry Chairman Rogers demanded to know "why we didn't know about it and what gave rise to the launch constraint?" It was later discovered that Mulloy and other Marshall managers did not go out of their way to inform NASA upper management of the launch constraint and perhaps much more. Angry questions were directed at Lawrence Mulloy, who had already testified and had said nothing about the waiver, regarding the fact that he had been the individual who placed the constraint, and then consistently waived it.

CHAIRMAN ROGERS: All you did on these waivers was to waive [them]!

Even though documents clearly showed that the problem was considered closed with no work done (meaning that no waiver was considered necessary), Mulloy testified that this was in error. Chairman Rogers then demanded to know how this closure, which was mysteriously left unsigned and therefore untraceable, could have occurred with no fix performed?

LAWRENCE MULLOY: A failure of the human being within the system.
CHAIRMAN ROGERS: It was a little more than that. It's a failure of the whole system if one letter and one [unknown] human being can close out a constraint that has been concerning you for many years!

Rogers further stated that NASA had "almost covered up, ever so slightly," when speaking of the way the space agency had handled information on the flawed SRB design. Ever so slightly?

What else were NASA managers covering up in 1986? Certainly they did not put forth a major effort to solve the debris problem. Can the American people ever again trust this government agency while any of these men are still part of it in any capacity? And what are they covering up today about the 2003 *Columbia* disaster? Will it lead to another Shuttle disaster?

On May 12, 1986, a source close to the commission reported that the investigation had located the "closest thing we've got to a smoking gun."[60] He was describing a document dating from August 1985, which showed that O-rings "lost contact at 75 degrees for 2.4 seconds and at 50 degrees the O-ring did not re-establish contact in 10 minutes." The commission was then able to show, by that letter from Morton Thiokol to Marshall Space Flight Center, that both groups were well aware of design problems and did nothing. Earlier, NASA officials had testified that there was no clear correlation between lower temperatures and O-ring failure.

Presidential Commission
on the
Space Shuttle Challenger Accident

June 6, 1986

Dear Mr. President:

On behalf of the Commission, it is my privilege to present the report of the Presidential Commission on the Space Shuttle Challenger Accident.

Since being sworn in on February 6, 1986, the Commission has been able to conduct a comprehensive investigation of the Challenger accident. This report documents our findings and makes recommendations for your consideration.

Our objective has been not only to prevent any recurrence of the failure related to this accident, but to the extent possible to reduce other risks in future flights. However, the Commission did not construe its mandate to require a detailed evaluation of the entire Shuttle system. It fully recognizes that the risk associated with space flight cannot be totally eliminated.

Each member of the Commission shared the pain and anguish the nation felt at the loss of seven brave Americans in the Challenger accident on January 28, 1986.

The nation's task now is to move ahead to return to safe space flight and to its recognized position of leadership in space. There could be no more fitting tribute to the Challenger crew than to do so.

Sincerely,

William P. Rogers
Chairman

The President of the United States
The White House
Washington, D. C. 20500

Cover letter to President Reagan by William P. Rogers, Chairman of the *Challenger* Commission (graphics from *Challenger* report).

On the same day, Chairman Rogers sent an FBI agent to interview Marshall management after an anonymous letter was received reporting that NASA management at that space center had ordered documents destroyed. No evidence had by then been located to confirm this report. On May 20, 1986, the Associated Press reported, "No documents necessary for the investigation of the *Challenger* accident are unavailable, even though some were destroyed at Marshall Space Flight Center in Alabama." Copies of those destroyed documents were found elsewhere, at other NASA locations.[61] Later, the commission would report that documents were indeed destroyed after *Challenger* was lost, but that there was no attempt to cover anything up.

As the commission was finishing its work, Morton Thiokol announced that it was reversing itself by appointing Allan McDonald to head the booster redesign team and would return him to his former position.[62] He would later become engineering vice president at Thiokol.

During the first week of June, reports in the nation's papers stated that NASA had ignored potential failure-rate reports and instead had accepted a 1983 Marshall Space Center report, which used made-up, arbitrary results, which had falsely shown a failure rate of 1 in 100,000 launches.

On June 4, 1986, William R. Lucas, director of Marshall Space Flight Center, announced his early retirement from NASA.[63] Lucas had been aware of the potentially deadly effect of cold weather on the O-rings and of the launch debate the night before *Challenger* was launched, but did nothing to stop the launch, in or outside of the chain of command. On this day the president and families of the *Challenger* crew were briefed on the commission's results before the actual release of the report to a waiting nation on June 6, 1986.

> The consensus of the Commission and participating investigation agencies is that the loss of the Space Shuttle *Challenger* was caused by a failure in the joint between the lower segments of the right Solid Rocket Motor. The specific failure was the destruction of the seals that are intended to prevent hot gases from leaking through the joint during the propellant burn of the rocket motor. The evidence assembled by the Commission indicates that no other element of the Space Shuttle system contributed to this failure [Chapter IV, Report of the Presidential Commission on the Space Shuttle *Challenger* Accident].

Later, two hundred members of the House of Representatives would sign a letter urging NASA to hold an open competition so that the space agency could develop a second source for its supply of solid rockets.[64] More than twenty years after the loss of *Challenger* and her seven-astronaut crew, Morton Thiokol (later changed to Thiokol Corporation) is still supplying NASA with solid rocket boosters, and astronauts still take great risks each and every time they fly. These rockets are now being looked at as a possible method to return astronauts to the moon.

In July 1986, NASA created a new Headquarters Office of Safety, Reliability, and Quality Assurance. It did not prevent the *Columbia* disaster.

> *Our objective has been not only to prevent any recurrence of the failure related to this accident, but to the extent possible to reduce other risks in future flights.*
> —William P. Rogers, Chairman, Presidential Commission, June 6, 1986

Recommendations by the *Challenger* Commission

In its report, the commission urged NASA to submit a report to the president one year after its submission on the progress that the space agency had made effecting the commission's recommendations.

- I -

Design. The faulty Solid Rocket Motor joint and seal must be changed. This could be a new design eliminating the joint or a redesign of the current joint and seal. No design options should be prematurely precluded because of schedule, cost or reliance on existing hardware. All Solid Rocket Motor joints should satisfy the following requirements:

- The joints should be fully understood, tested and verified.
- The integrity of the structure and of the seals of all joints should be not less than that of the case walls throughout the design envelope.
- The integrity of the joints should be insensitive to:
    - Dimensional tolerances
    - Transportation and handling
    - Assembly procedures
    - Inspection and test procedures
    - Flight and water impact loads
    - Environmental effects
    - Internal case operating pressure
    - Recovery and reuse effects
- The certification of the new design should include:
    - Tests which duplicate the actual launch configuration as closely as possible.
    - Tests over the full range of operating conditions, including temperature.
- Full consideration should be given to conducting static firings of the exact flight configuration in a vertical attitude.

Independent Oversight. The Administrator of NASA should request the National Research Council to form an independent Solid Rocket Motor design oversight committee to implement the Commission's design recommendations and oversee the design effort. This committee should:
- Review and evaluate certification requirements.
- Provide technical oversight of the design, test program and certification.
- Report to the Administrator of NASA on the adequacy of the design and make appropriate recommendations.

- II -

Shuttle Management Structure. The Shuttle Program Structure should be reviewed. The project managers for the various elements of the Shuttle program felt more accountable to their center management than to the Shuttle program organization. Shuttle element funding, work package definition, and vital program information frequently bypass the National STS (Shuttle) Program Manager.

A redefinition of the Program Manager's responsibility is essential. This redefinition should give the Program Manager the requisite authority for all ongoing STS operations. Program funding and all Shuttle Program work at the centers should be placed clearly under the Program Manager's authority.

Astronauts in Management. The Commission observes that there appears to be a departure from the philosophy of the 1960s and 1970s relating to the use of astronauts in management positions. These individuals brought to their positions flight experience and a keen appreciation of operations and flight safety.
- NASA should encourage the transition of qualified astronauts into agency management positions.
- The function of the Flight Crew Operations director should be elevated in the NASA organization structure.

Shuttle Safety Panel. NASA should establish an STS Safety Advisory Panel reporting to the STS Program Manager. The charter of this panel should include Shuttle operational issues, launch commit criteria, flight rules, flight readiness and risk management. The panel should include representation from the safety organization, mission operations, and the astronaut office.

- III -

Criticality Review and Hazard Analysis. NASA and primary Shuttle contractors should review all Criticality 1, 1R, 2, and 2R items and hazard analysis. This review should identify those items that must be improved prior to flight to ensure mission success and flight safety. An Audit Panel, appointed by the National Research Council, should verify the adequacy of the effort and report directly to the Administrator of NASA.

- IV -

Safety Organization. NASA should establish an Office of Safety, Reliability and Quality Assurance to be headed by an Associate Administrator, reporting directly to the NASA Administrator. It would have direct authority for safety, reliability, and quality assurance throughout the agency. The office should be assigned the work force to ensure adequate oversight of its functions and should be independent of other NASA functional and program responsibilities. The responsibilities of this office should include:

- The safety, reliability and quality assurance functions as they relate to all NASA activities and programs.
- Direction of reporting and documentation of problems, problem resolution and trends associated with flight safety.

- V -

Improved Communications. The Commission found that Marshall Space Flight Center project managers, because of a tendency at Marshall to management isolation, failed to provide full and timely information bearing on the safety of flight 51-L to other vital elements of Shuttle program management.
- NASA should take energetic steps to eliminate this tendency at Marshall Space Flight Center, whether by changes of personnel, organization, indoctrination or all three.
- A policy should be developed which governs the imposition and removal of Shuttle launch constraints.
- Flight Readiness Reviews and Mission Management Team meetings should be recorded.
- The flight crew commander, or a designated representative, should attend the Flight Readiness Review, participate in acceptance of the vehicle for flight, and certify that the crew is properly prepared for flight.

- VI -

Landing Safety. NASA must take actions to improve landing safety.
- The tire, brake and nose wheel steering systems must be improved. These systems do not have sufficient safety margin, particularly at abort landing sites.
- The specific conditions under which planned landings at Kennedy would be acceptable should be determined. Criteria must be established for tires, brakes and nose wheel steering. Until the systems meet those criteria in high fidelity testing that is verified at Edwards, landing at Kennedy should not be planned.
- Committing to a specific landing site requires that landing area weather be forecast more than an hour in advance. During unpredictable weather periods at Kennedy, program officials should plan on Edwards landings. Increased landings at Edwards may necessitate a dual ferry capability.

- VII -

Launch Abort and Crew Escape. The Shuttle program management considered first-stage abort options and crew escape options several times during the history of the program, but because of limited utility, technical infeasibility, or program cost and schedule, no systems were implemented. The Commission recommends that NASA:
- Make all efforts to provide a crew escape system for use during controlled gliding flight.
- Make every effort to increase the range of flight conditions under which an emergency runway landing can be successfully conducted in the event that two or three main engines fail early in ascent.

- VIII -

Flight Rate. The nation's reliance on the Shuttle as its principal space launch capability created a relentless pressure on NASA to increase the flight rate. Such reliance on a single launch capability should be avoided in the future. NASA must establish a flight rate that is consistent with its resources. A firm payload assignment policy should be established. The policy should include rigorous controls on cargo manifest changes to limit the pressures such changes exert on schedules and crew training.

- IX -

Maintenance Safeguards. Installation, test, and maintenance procedures must be especially rigorous for Space Shuttle items designated Criticality 1. NASA should establish a system of analyzing and reporting performance trends of such items.

Maintenance procedures for such items should be specified in the Critical Items List, especially for those such as the liquid-fueled main engines, which require unstinting maintenance and overhaul.

With regard to Orbiters, NASA should:
- Develop and execute a comprehensive maintenance inspection plan.
- Perform periodic structural inspections when scheduled and not permit them to be waived.
- Restore and support the maintenance and spare parts programs, and stop the practice of removing parts from one Orbiter to supply another.

Concluding Thought

The Commission urges that NASA continue to receive the support of the Administration and the nation. The agency constitutes a national resource that plays a critical role in space exploration and development. It also provides a symbol of national pride and technological leadership.

The Commission applauds NASA's spectacular achievements of the past and anticipates impressive achievements to come. The findings and recommendations presented in this report are intended to contribute to future NASA successes that the nation both expects and requires as the 21st century approaches.

Chapter Ten

Unanswered Questions: Beyond the Commission

> *It is my honest and very real fear, that if we do not take immediate action to dedicate a team to solve the problem, then we stand in jeopardy of losing a flight along with all of the launch pad facilities.*
> — Roger Boisjoly, O-ring lead engineer, Morton Thiokol

Dark Passage

Two of the most critical unanswered questions for the future of the U.S. space program are how the astronauts on *Challenger* were killed and how this can be prevented in the future. The question can only be partly answered. Detailed examination of enhanced launch footage and analysis of the wreckage was inconclusive, due to overwhelming forces inflicted upon the crew cabin when it hit the ocean. However, much is known. The Shuttle *Challenger* was not destroyed during the breakup and fast burning of the fuels housed within the external tank upon which the craft was riding. This fast burn did, however, cause a vacuum between the tank and the Shuttle, causing the spacecraft to change its angle of flight. This abrupt change in angle caused the orbiter to break up, but in itself this would not have been sufficient to cause the deaths of the crew members. The aero-breaking effect may have been sufficient enough to incapacitate several of the crew members, but not all. Indeed, the Kerwin letter (see below) states, "The probability of major injury to crew members is low."

It is also possible that some injury to the crew could have occurred as the expanding cloud of fuel was burning, causing the lower section of the *Challenger* to absorb the pressure pulse of the burn. But again, it appears to be unlikely, though not impossible, that any of the astronauts would have sustained life-threatening injuries from this pulse, as the explosive sound moved across what had been the crew cabin. Further, there was no detectable explosive pulse of energy released, because the fuel did not explode. However, some damage to the base and right side of the Shuttle cabin may have been incurred which could have buckled the lower flight deck in which three of the astronauts were seated. This damage would have caused a loss of cabin pressure, which would have caused unconsciousness within a few seconds. The sounds of a powerful, rocketed spacecraft, felt more on the body than actually heard, would have been replaced by a stunning silence, accompanied only by a whisper of very cold wind in the thin air.

Within one and a half seconds of breakup, the forward-reaction-control area in the nose of the Shuttle was ripped apart and the spilled fuels burned immediately. Again, no explosion was seen. Soon after this event, at least three astronauts activated their Personal Egress Air Packs (PEAP), which would give them an emergency air supply. The investigation showed that three out of the four recovered PEAPs were activated.[1] Commander Scobee's was not, indicating that he was unconscious, or otherwise incapacitated, during or soon after the breakup, and probably did not recover consciousness.

Pilot Smith's PEAP was activated, which must have been turned on by Mission Specialist Onizuka after his own, as Smith would have been unable to perform the task himself, due to the way the unit was laid out behind his seat.[2] That would account for two of the activated units. It is logical to expect that if Mission Specialist Resnik were able, she would have activated the commander's PEAP after her own if she could reach it, but since Commander Scobee's was not used, it is possible that Resnik lost consciousness just after breakup and was therefore unable to perform the task.

The final activated PEAP may also have belonged to one of the crew members in the lower crew compartment, but there is no conclusive evidence to show this. However, it is also possible that a piece of flying debris penetrated one of the windows in front of Commander Scobee and either incapacitated or killed this astronaut. A piece of debris was found imbedded between the front windows. The most likely origin for such debris would be from the breakup of the forward-control area on the nose of the Shuttle or the external tank as it shattered.

One other possible situation must be considered. This is the evidence concerning Payload Specialist Greg Jarvis. After searchers finished work on the crew-cabin wreckage area, reports circulated that remains of only six of the seven Shuttle astronauts had been recovered. Two weeks later, it was announced that remains of all seven had been recovered. It is possible that Jarvis, seated on the mid deck and closest to the right side of the Shuttle, may have been ejected from the crew cabin during the Shuttle's breakup. If correct, his remains would have been found apart from those of the rest of the *Challenger* crew remains found with the wrecked crew cabin. A review of recovery charts and ships' logs could confirm this scenario. The space agency has said nothing concerning this theory.

Although not conclusive, evidence seems to indicate that several of the crew were conscious for at least long enough to be fully aware of the situation. It also shows it is possible, but not conclusive, that none of the crew were killed due to the explosive burn seen on the launch video. It is most likely, based on evidence gathered by NASA, that most of the crew lost their lives at approximately 3 minutes and 58 seconds after launch (2 minutes and 45 seconds after breakup) when the separated and wildly tumbling crew cabin hit the Atlantic Ocean.

Can we trust NASA to tell the public the truth about the astronauts' deaths? As reported by the Associated Press on November 13, 1988, "Dr. Ronald Reeves, Chief Assistant to the Medical Examiner in Brevard County, says the public may never learn exactly how the astronauts died because NASA kept the remains from Florida pathologists, whose records would be public." NASA denied the cover-up, but the medical examiners were never allowed to examine the remains. In fact, NASA officials, and pressure from Representative Bill Nelson, who had taken his free ride on the Space Shuttle courtesy of NASA just weeks earlier, physically blocked them from their legal responsibilities.[3]

The Kerwin Letter

On July 28, 1986, Rear Admiral Richard H. Truly, newly appointed associate administrator for space flight, released a letter from Dr. Joseph P. Kerwin, the physician-astronaut who had served on Skylab 2. The two-page letter/report, commissioned by the president's commission on the *Challenger* disaster, related the known medical facts about the deaths of the *Challenger* crew. That report is reproduced here to allow the readers the opportunity to come to their own conclusions as to the fate of the *Challenger*'s final crew.

RADM Richard H. Truly, Associate Administrator for Space Flight
NASA Headquarters
Washington, DC 20546

Dear Admiral Truly:

The search for wreckage of the *Challenger* crew cabin has been completed. A team of engineers and scientists has analyzed the wreckage and all other available evidence in an attempt to determine the cause of death of the *Challenger* crew. This letter is to report to you on the results of this effort. The findings are inconclusive. The impact of the crew compartment with the ocean surface was so violent that evidence of damage occurring in the seconds which followed the explosion was masked. Our final conclusions are:

- the cause of death of the *Challenger* astronauts cannot be positively determined;
- the forces to which the crew were exposed during Orbiter breakup were probably not sufficient to cause death or serious injury; and
- the crew possibly, but not certainly, lost consciousness in the seconds following Orbiter breakup due to in-flight loss of crew module pressure.

Our inspection and analysis revealed certain facts which support the above conclusions, and these are related below: The forces on the Orbiter at breakup were probably too low to cause death or serious injury to the crew but were sufficient to separate the crew compartment from the forward fuselage, cargo bay, nose cone, and forward reaction control compartment. The forces applied to the Orbiter to cause such destruction clearly exceed its design limits. The data available to estimate the magnitude and direction of these forces included ground photographs and measurements from on board accelerometers, which were lost two-tenths of a second after vehicle breakup.

Two independent assessments of these data produced very similar estimates. The largest acceleration pulse occurred as the Orbiter forward fuselage separated and was rapidly pushed away from the external tank. It then pitched nose-down and was decelerated rapidly by aerodynamic forces. There are uncertainties in our analysis; the actual breakup is not visible on photographs because the Orbiter was hidden by the gaseous cloud surrounding the external tank. The range of most probable maximum accelerations is from 12 to 20 G's in the vertical axis. These accelerations were quite brief. In two seconds, they were below four G's; in less than ten seconds, the crew compartment was essentially in free fall. Medical analysis indicates that these accelerations are survivable, and that the probability of major injury to crewmembers is low.

After vehicle breakup, the crew compartment continued its upward trajectory, peaking at an altitude of 65,000 feet approximately 25 seconds after breakup. It then descended striking the ocean surface about two minutes and forty-five seconds after at a velocity of about 207 miles per hour. The forces imposed by this impact approximated 200 G's, far in excess of the structural limits of the crew compartment or crew survivability levels.

The separation of the crew compartment deprived the crew of Orbiter-supplied oxygen, except for a few seconds supply in the lines. Each crew member's helmet was also connected to a personal egress air pack (PEAP) containing an emergency supply of breathing air (not oxygen) for ground egress emergencies, which must be manually activated to be available. Four PEAPs were recovered, and there is evidence that three had been activated. The non-activated PEAP was identified as the Commander's, one of the others as the Pilot's, and the remaining ones could not be associated with any crewmember. The evidence indicates that the PEAPs were not activated due to water impact.

It is possible, but not certain, that the crew lost consciousness due to in-flight loss of crew module pressure. Data to support this is:

- The accident happened at 48,000 feet, and the crew cabin was at that altitude or higher for almost a minute. At that altitude, without an oxygen supply, loss of cabin pressure would have caused rapid loss of consciousness and it would not have been regained before water impact.
- PEAP activation could have been an instinctive response to unexpected loss of cabin pressure.
- If a leak developed in the crew compartment as a result of structural damage during or after breakup (even if the PEAPs had been activated), the breathing air available would not have prevented rapid loss of consciousness.

- The crew seats and restraint harnesses showed patterns of failure which demonstrates that all the seats were in place and occupied at water impact with all harnesses locked. This would likely be the case had rapid loss of consciousness occurred, but it does not constitute proof.

Much of our effort was expended attempting to determine whether a loss of cabin pressure occurred. We examined the wreckage carefully, including the crew module attach points to the fuselage, the crew seats, the pressure shell, the flight deck and mid deck floors, and feed throughs for electrical and plumbing connections. The windows were examined and fragments of glass analyzed chemically and microscopically. Some items of equipment stowed in lockers showed damage that might have occurred due to decompression; we experimentally decompressed similar items without conclusive results.

Impact damage to the windows was so extreme that the presence or absence of in-flight breakage could not be determined. The estimated breakup forces would not in themselves have broken the windows. A broken window due to flying debris remains a possibility; there was a piece of debris imbedded in the frame between two of the forward windows. We could not positively identify the origin of the debris or establish whether the event occurred in flight or at water impact. The same statement is true of the other crew compartment structure. Impact damage was so severe that no positive evidence for or against in-flight pressure loss could be found.

Finally, the skilled and dedicated efforts of the team from the Armed Forces Institute of Pathology, and their expert consultants, could not determine whether in-flight lack of oxygen occurred, nor could they determine the cause of death.

[Signed] Joseph P. Kerwin

As to how such a disaster can be prevented in the future, this is a question which can also only be partially answered. Certainly, much has been done to examine and correct the hardware responsible for the disaster. Procedures have been rewritten and new levels of safety and inspection have been initiated. Many of those directly involved in the flawed decision to launch *Challenger* have been removed from the process. Yet, as long as Shuttles fly and as long as Thiokol solid rocket boosters supply 80 percent of the initial liftoff power needed to push the Shuttle into space, no one can state with any certainty that another *Challenger* accident will not occur again.

This is not news to the men and women who fly spacecraft for a living. Risk is, and always will be, a part of the job. Astronauts willingly risk their lives each and every time they step into a Shuttle. The only measure that may be taken to reduce the possibility of an accident recurrence to zero is to never fly again, and that is something this nation and the astronaut crews who fly Shuttles are unwilling to do.

Criminal Responsibility for *Challenger*?

In September 1988, the U.S. attorney's office in Birmingham, Alabama, brought forth indictments against the A. O. Sammons Company. These indictments were for allegedly falsifying documents sent to NASA which stated that bolts produced by the company for the Spacelab module had passed specified tests for safety. A grand jury brought 17 counts of mail fraud against the company, and 26 counts of making false statements to NASA.[4]

It has been well documented by the *Challenger* report that both NASA and Morton Thiokol lied, not only to Congress and the American people, but to each other. NASA had informed Congress that Shuttles would be able to fly up to 100 missions and at a far lower cost. The space agency informed the public that an accident could occur in 1 in 100,000 flights and that information was pure fantasy. With this in mind, Congress weighed the possibility of criminal charges in the death of *Challenger*'s crew.[5]

Morton Thiokol management for their part were not honest in their dealings with NASA when it came to SRB design and testing problems.[6] When NASA requested data on low-temperature testing of the solid rockets, Thiokol managers were less than truthful when they informed Marshall managers that the cold weather tests had been conducted, a clear federal violation. When three of seven mandatory government O-ring inspections were not made, Edward Dorsey, general manager of Thiokol, stated that it was just a mistake in paperwork. Perhaps, but mistake or not, the tests were not made, and the results are clear.

Internal NASA studies on safety issues showed that at times some paperwork on safety inspections were completed before the actual inspections were done, if the inspections were done at all. During hearings before Congress held after the accident, Representative Robert Walker (R–PA) in May of 1986, stated, "If the big ones [safety issues and procedures] can be routinely ignored we have every reason to think that the process is probably being ignored in some other instances. That transcends the whole organization."

Just before the *Challenger* commission released its report, which clearly showed fault in Thiokol's solid rockets, the chairman of Morton Thiokol, Charles S. Locke, clearly expressed his concerns about the results of the report: "This Shuttle thing will cost us this year ten cents a share."[7, 8]

It was left to Senator Donald Riegle (D–MI) to state, "Every single person that didn't behave and function properly has got to be identified and some kind of disciplinary action has to be taken."[9] What the senator failed to state was that the major responsibility for the Shuttle disaster must fall on NASA managers and administrators for allowing a flawed design to go forward without being corrected. Under the direction of Marshall Space Flight Center, Thiokol was allowed to change the approved O-ring design, allowed to continue onward without being properly managed or even challenged in areas which would prove to be a matter of life and death. Thiokol did the work, but NASA controlled events and the funding. Clearly there is enough blame to go around.

On April 4, 1987, fifteen months after the accident, it was revealed that Morton Thiokol was the subject of an FBI criminal investigation.[10] Susan Schnityer, an FBI official, reported that a court document with information on the investigation had been "inadvertently released" and was supposed to have been secret. That document and other files were resealed. Days later, a federal judge, Harold Greene, ordered the FBI to release all court documents describing its investigation of Thiokol.[11] Within those documents were charges that the astronaut's deaths were being investigated as criminal homicides. Yet Chairman Rogers, in testimony before Congress, stated that he felt it would not be in the national interest to prosecute anyone for the loss of *Challenger*. This statement came after some U.S. senators had complained that the commission had not gone far enough in its investigation of the accident. These senators wanted the report to place blame, by name, on the individuals responsible for the loss.

At the same time, Roger Boisjoly filed a suit which claimed that NASA used its wildly optimistic safety figures to "pressure the government to pressure Thiokol's unlawful monopoly as the sole source provider"[12] of solid rocket motors to NASA. Under the False Claims Act that Boisjoly used, U.S. citizens can file claims against any government contractor on behalf of the taxpayers. It was an attempt to hold NASA and Thiokol responsible, but not much was to become of the suit, nor, for that matter, the FBI investigation.

On September 7, 1987, the newly formed four-member Space Flight Safety Panel, headed by astronaut Colonel Bryan D. O'Conner, reported to NASA with its first recommendations. Among other things, the panel asserted that, "information from witnesses should not be released to the public"[13] after an accident. It further recommended that NASA should find a way to prevent information from being disclosed to the general public, using any legal method it could, including rejecting requests made under the Freedom of Information Act. Is this NASA policy today?

A Question About Solids

From the beginning, solid rockets used on manned spacecraft raised questions. Many engineers and managers experienced with manned flight and unmanned uses of solid rockets knew that using SRBs to launch a manned vehicle was far too dangerous. They could not be turned off once they were ignited. Despite many arguments to the contrary, the original Shuttle design which used liquid fuel rockets was somehow allowed to be changed to an untested solid-rocket system. Even managers at NASA's Marshall Space Flight Center, tasked with the booster's overview, knew it was a bad decision.

Only a full investigation of the contract and design history could begin to answer questions about the SRBs' dangers. (The long history of problems with the solids was well documented years before the first launch.) It should be remembered that all U.S. manned spacecraft, before the Shuttle had used liquid fuel rockets, and with great success. The United States had launched one-man Mercury, two-man Gemini and three-man Apollo crews on liquid-fuel rockets for 14 years. These flights culminated with six successful moon-landing missions using the 7.65-million-pound thrust, three-stage Saturn V rocket. With this much background and successful engineering development, it made no sense to risk development of a completely new system for human space flight. Research should have continued to develop for use of solids on unmanned systems, but not placed on a new manned system until NASA had a successful track record.

During a period in NASA's history when the agency was desperately trying to keep its manned flight capabilities intact, why did it pick a solid-rocket design which was rated last among four contenders? Was this a sound decision based on engineering excellence and a history of success? As one astronaut stated after learning about the *Challenger* disaster, "A lot of people [in the astronaut group] are saying; I don't trust those bastards anymore. What they are really saying is that we had a system that was supposed to function and it didn't."[14]

Shuttle commander Jon A. McBride said, "A lot of people feel punch-in-the-nose angry. We feel ... betrayed or taken advantage of by what we sense happened to us."[15] McBride was scheduled to command the next Shuttle mission after *Challenger*.

After all the developmental problems with the solids, and after two decades since the destruction of *Challenger*, why has NASA yet to develop reliable liquid-fuel rockets to replace the patched together, flawed system now being used? Within a relatively short period of time, NASA could design and build a liquid-fuel rocket booster to replace the SRBs. There is even a model to work from. The now-grounded Russian Space Shuttle *Buran* was test flown in November 1988 using liquid-fuel boosters. It was grounded, not because the boosters did not work, but because the old Soviet system collapsed, and along with it the funding for any kind of operational manned space program. Perhaps it is time to take a lesson and return to a more reliable system of placing manned spacecraft into orbit.

In July 1986, Aerojet General Corporation, whose solid rocket design had beaten Thiokol's during the original SRB proposals, announced that they could be ready with a single-cast solid rocket booster in time for the next Shuttle flight.[16] The company informed NASA management that its design would eliminate the segments and O-rings which had failed on Thiokol's design. Further, they would be prepared to test the new boosters on old existing Apollo/Saturn launch pads 34 or 37 at Kennedy, thus taking advantage of NASA's existing facilities. This type of vertical testing was recommended by the *Challenger* Commission, but would not be conducted by Thiokol and would not be demanded by NASA.

NASA's only response came in September 1986 when they awarded five contracts for what the agency called "block 2" concepts for new SRB designs. The 120-day, $500,000 contracts were awarded to Aerojet Strategic Propulsion Company, United Technologies, Hercules Aerospace, Atlantic Research Corporation and Morton Thiokol.[17]

Later it was revealed, in a Congressional study released in November 1986, that Thiokol was continuing to receive incentive pay. "The seal problem was serious enough to lead both [Thiokol and NASA] to briefings at Headquarters and to establishment of a redesign task force. Yet, in spite of all these problems, Thiokol was eligible to receive a near-maximum incentive fee of approximately $75 million."[18]

The General Accounting Office reported that of all government agencies it investigated, NASA was the most difficult agency to work with in terms of obtaining information and getting cooperation. In letters to the GAO, Dr. Fletcher and other upper NASA management "consistently refused to correct many of the problems cited by the inspectors and auditors, including management errors, spending abuses, delays in hardware testing, lax monitoring of contractors, and optimistic estimates of cost and technical problems." As John Glenn said, "A can do attitude was replaced by a can't fail attitude."[19]

Spare Parts?

Soon after the release of the *Challenger* report, two-time Shuttle commander Robert Crippen admitted during congressional questioning that due to a critical shortage of spare parts, NASA would never have been able to complete the flights scheduled for 1986. Fifteen flights had been scheduled, but continued cannibalization of Shuttles to meet those demands would not have been sufficient — the flight schedule was unattainable.

Where did the money go to fund the spare-parts program which Congress had funded year after year? Representative Don Fuqua said, "I know every year this committee has added ... money in for spare parts; we've had to twist nose[s], kick, [and] scream to get NASA to use it, and most times they have not used it for spare parts as it was intended. I don't think that's good policy when we do that. That caused a lot of pirating of one vehicle to another vehicle." During congressional hearings, this question was posed to Richard Truly and James Fletcher, the new NASA chief. It was not expected that Truly would have an answer, as he was new on the job, but he did state that he would come back with an answer. However, Fletcher was the past NASA administrator and would have been well aware of this discrepancy, yet all he too could say was that he would try to find an answer.

What Truly stated was that NASA wanted to take a look at all hardware spares across the board and find where best to place new funding to increase the safety of future Shuttle operations. But he further stated that NASA was not interested in completely eliminating the cannibalization of one Shuttle to fly another. The rationalization given was that it would cost too much and that even aircraft squadrons do this. But these are spacecraft, not aircraft, and NASA is a research program not an airline. Either way, the question of spare-parts funding was not answered, and NASA continues parts cannibalization to this day.

Rockwell International, builder of the Shuttle fleet, reported that due to a shortage of flight spares and hardware, coupled with the time required to process Shuttles for flight, no more than nine flights per year could ever be supported. Indeed, the most flights ever launched by NASA during the Shuttle program has been nine (in 1985).

The bottom line is that NASA was not allowed enough time to develop a safer system and not enough money to solve the problems which come to the forefront in its Shuttle research program. Along with meeting safety concerns, there must be a properly funded spares program, one which is closely monitored by Congress or other outside agencies to insure that funds are correctly allocated for programmed needs.

Why Did NASA Launch *Challenger*?

Cold, unforgiving temperatures; ice on the pad greater than ever before; O-ring engineers fearing a catastrophic failure; hurricane strength crosswinds within the launch profile; high winds at the Shuttle emergency landing area; 25-foot waves in the Solid Rocket Booster recovery area; and marginal landing conditions at the transatlantic emergency runways in Africa. Why did NASA launch *Challenger*?

Challenger had been scrubbed the day before due to high crosswinds. Winds were also high on the ground, causing so many high-wave problems that the SRB recovery ships had to return to the Cape. When the 51-L mission was launched, the orbiter punched through the highest recorded crosswinds ever experienced by a Shuttle during ascent. The winds were even discussed by the crew during their short flight.

The easy answer, much discussed, is pressure from a political source such as the White House, but this explanation does not ring true. Even if the president wanted to speak with members of the crew during the State of the Union address, that would not have required a morning or early afternoon launch. The Shuttle would have been in orbit for at least a few hours no matter when it was launched within the required launch window. And a presidential speech is easy to postpone for another day.

On the day of the launch, flight rules were badly bent if not completely violated; launch constraints were ignored or glossed over. Major engineering concerns were ignored or pushed aside, as managers risked launching an experimental vehicle well beyond its known design capabilities. The push-to-launch answer lies in the overall history of the Shuttle program, and dates back to NASA's original projected costs and projected launch rates.

During his confirmation hearing before Congress, James Fletcher was questioned on projected costs for the Shuttle and achievable flights. "We may have overestimated the number of flights for the complete five–Orbiter fleet," he said. At the time, Administrator Fletcher testified that the cost per mission would be $10.45 million, but in reality it ended up being $279 million, nearly 26.7 times more expensive. When the General Accounting Office reported that his figures were misleading, all Fletcher would say was, "I don't recognize the second figure at all. We never stated the $10.45 million per launch as an absolute figure; it was an estimate." With a thin smile he added, "Something happened on the way to the bank." It was during 1972 that these figures were originally quoted, and they reflected an amazing projected flight rate of sixty Shuttle missions a year. Better than a launch a week was the NASA headquarters' estimate, yet there were not facilities enough to handle even one-third of that flight rate.

As projected costs per launch began to increase, NASA managers and administrators knew that the only way to reduce this per-launch figure was to launch as often as possible. In that environment of internal launch pressure, *Challenger* was flown when all logic indicated a no-go. No outside pressure was ever required. As Chief Astronaut John Young stated in his memo written before the 51-L *Challenger* mission: "Launch schedule pressure."

Within weeks of discovering what had destroyed *Challenger*, Richard Truly had already put together a flight schedule that had as many as 15 flights scheduled in one year, and this was projected when no one at NASA or anywhere else had any idea when the Shuttle system would again become operational, if ever. NASA was continuing to take on more than it could reasonably be expected to handle.

It should also be noted that the Shuttles had continued to receive debris hits to their Thermal Protection Systems, though this was unacceptable according to flight rules. The acceptance of this risk, well outside design requirements, and the fact that Shuttle management at the highest levels refused to order a simple photographic survey of the Thermal Protection System would doom *Columbia*'s final crew.

If I want to do something, I will do it. I don't usually see any two ways about it.
— Mission Specialist Ronald Erwin McNair

Chapter Eleven

Recovery for Flight? STS-26

It's very important for this nation to get back into space as soon as practical. You've got to balance that with wanting to make sure you don't risk these very valuable assets that the country has.
— Frederick Hauck, Commander, *Discovery* STS-26

1986 — Pressure from All Sides

Even before the *Challenger* commissioners had finished their work, Congress began a series of hearings on Shuttle safety, as well as overall NASA operations. The House Science and Technology Subcommittee looking into NASA quality assurance conducted one such hearing in May 1986. The chairman of that panel was Representative Bill Nelson (D–FL), who had flown on *Columbia*'s mission 61-C as an observer, and he was not about to let anything NASA managers said pass unchallenged. NASA managers were about to come face to face with an individual who had come very close to losing his own life because of NASA's recklessness.

Representative Nelson, who would later become a senator, came out shooting when he asked NASA Chief Engineer Milton Silveira about a bad weld which allowed a sensor on ground equipment to flow through the fuel lines and become lodged in a main engine pre-valve. If the launch countdown had not been stopped because of bad weather in Africa, the Shuttle would have been launched. After the main engines had shut down eight minutes later, the stuck valve would have caused the engine to explode, destroying the Shuttle in space, before it had reached orbit. The 61-C temperature probe sensor was found to be old and badly welded, which was a manufacturer's defect, along with the fact that NASA design specifications had not been met. Representative Nelson was attached to that rocket during the countdown, and now he wanted answers. But even with advanced notice, Silveira was not prepared for the questions. Nelson went right to the attack:

REP. NELSON: We told you guys in advance that we were going to be talking about this and we expect some answers. Now I want to know when it was inspected and by whom it was inspected? Now, do you have records?
MILTON SILVEIRA: Yes, sir.
REP. NELSON: You have those records here?
MILTON SILVEIRA: No, I do not.
REP. NELSON: We would like you to send us copies of those inspection records.
MILTON SILVEIRA: Yes, sir.
REP. NELSON: I don't know ... why we should be having this hearing then, unless we can have some of

these answers. This is why we asked you to have specific answers in this area. So we may as well adjourn the hearing.

NASA was not about to be let off the look, as the hearings continued with Representative Robert Walker (R–PA) continuing to press for answers to questions about safety and reliance. Tough as nails, Walker chose to use an incident which had occurred at Kennedy Launch Center only one month before *Challenger* was launched. The incident involved a continuing violation of rules which had resulted in a solid rocket booster section being damaged in handling as the SRB was being assembled for an upcoming flight (November 8, 1985).[1] Safety personnel had been looking the other way. Milton Silveira responded for NASA.

> MILTON SILVEIRA: They had been doing this a number of times. In their defense, they found out that it was very difficult to get some of the bolts out.
> REP. WALKER: You said "in their defense." There is no defense for the fact that they were violating operation procedures. Why would there be a defense for them in that [incident]? Shouldn't there have been a safety officer at that point that was saying, "You can't do it; I don't care how easy it makes it, you can't do it"? And that did not happen, did it?
> MILTON SILVEIRA: That did not happen.
> REP. WALKER: It didn't happen here and it didn't happen several times before this.
> MILTON SILVEIRA: Yes, sir. That's right.
> REP. WALKER: The fact is that in this particular incident the safety officer wasn't even around to observe it. He was around getting sign-offs wasn't he ... for things that hadn't even been done yet, so that the paperwork would look good, but the procedures were being violated?
> MILTON SILVEIRA: Yes, sir.
> REP. WALKER: Has he been fired?

Colonel Marvin Jones, director of the Kennedy Space Center Safety Office, was then questioned, and responded in a halting voice. These members of Congress wanted blood and the NASA officials were on the firing line.

> COLONEL JONES: I can't comment whether he was specifically fired.
> REP. WALKER: Should he have been fired?
> COLONEL JONES: [If] you consciously and deliberately violated procedure without some safety concern or constraints or waiver to it then certainly you should not be in a position to be making judgments on those procedures.
> REP. WALKER: I'd like to know how many people have been fired out of this. [It] is your quality assurance program. You know, bottom line is [that] you're responsible. Now what did we do about getting people fired, getting people disciplined as a result of this? What did NASA tell the contractor had to be done?
> MILTON SILVEIRA: As we indicated earlier, the matter of following the procedure, the matter of safety, the matter of quality has to start at the top of the organization and work its way down. We became very concerned....

The House subcommittee members were not buying the excuses. It was later found that the individuals responsible had requested a change in procedure which would have allowed them to handle the solid booster in exactly the way in which it was damaged. Higher headquarters denied their request, but the people at Kennedy went ahead and violated the procedure anyway. No one was ever fired for this incident, and Lockheed, the company who had the contract to handle the SRBs, continued to process the solid rockets for the space program.

Soon after the president's commission issued their report, James Fletcher and his new space flight chief Richard Truly were called before Congress to review the report and advise the Congress on specifically how NASA would implement its recommendations. They did so, as President Reagan told the nation, "I don't believe there was any deliberate, criminal intent in any way on the part of anyone."

Both Fletcher and Truly fell short of fully endorsing the commission's report, saying that some of its areas were "merely recommendations" and that NASA would not necessarily take all of them.

The next day, President Reagan held a meeting with NASA Administrator Fletcher, informing him point blank that Fletcher was to implement all of the commission's recommendations as soon as possible and report back to the president in 30 days on "how and by when the recommendations will be implemented." The president informed Fletcher that this was "essential to resuming effective and efficient operations." This was a presidential directive, and a not-very-pleased NASA administrator had gotten his marching orders. But would this be sufficient to get the job done?

Not all commission members believed that the Shuttle program should be allowed to continue at all. David C. Acheson said in an interview just after the report was issued, "I believe we should have abandoned the Shuttle and gone on to something else." But it was the system the nation was stuck with for many years, the only one going for those who wanted Americans in space. Twenty-two years later, given an ever-aging program and after a second Shuttle disaster, the time may well have come to move on to a new program.

NASA may not have been prepared for the *Challenger* report, but the families of 51-L crew members knew exactly what the report said and what it meant. Mike Smith's widow Jane, in an interview with the *Washington Post*, stated, "The report reflects incredibly terrible judgments, shockingly sparse concern for human life, instances of officials lacking the courage to exercise the responsibilities of their office and some very bewildering thought processes." She had expressed the feelings of many family members, as it came to be known that lives were lost not so much due to a mechanical failure as to "human gross neglect and greed." Later, all of the families would receive large compensations from NASA and Morton Thiokol.

At this point, a fix was needed fast, and cost was not allowed to become an obstacle. Both Congress and the executive branch understood that the nation had backed itself into a Shuttle-only hole and would have to fight hard to get out. With the loss of Shuttle orbital capabilities, the government and private industry would have to rely on ELVs (Expendable Launch Vehicles) to place payloads into orbit. It would not be long before these would also fall to the wayside. In quick succession, the United States experienced launch failures of Titan 34D and Delta rockets. These were the workhorses of America's unmanned fleet. During the same period, Europeans experienced a failure of their new Ariane booster. For all intents and purposes, the United States and Europe were out of the space business for a while, leaving only the Soviet Union and China with systems online and launching. At the time, news reports came out that claimed that agencies such as the CIA and FBI were investigating possible sabotage of the U.S. and European launch systems. Nothing would come out of those investigations, however.

On July 3, 1986, Marshall Space Flight Center's John Thomas was asked his opinion of the new solid rocket booster design the agency was unveiling. "This design will preclude ... disaster from happening again," he said. "It will be a system that will be extremely resistant to any failure.... We are designing a joint that will be safe."[2] For more than eight years, NASA and Thiokol had known that the solid rocket booster design was flawed and did little if anything to correct the problem. Now, only five months after *Challenger*, Marshall officials endorsed a new design which had yet to be tested or flown.

That same month, Richard G. Smith, director of the Kennedy Launch Center, announced his retirement, making clear his displeasure with the work of the commission, even though no mention of personnel blame ever came out of the investigations as many felt it should have. In his final interview as director he stated that the *Challenger* commission "needlessly damaged the reputations of NASA officials." "Every time there was a delay the press would say, 'look, there's another delay ... here's a bunch of idiots who can't even handle a launch schedule.' If you think that doesn't have an impact ... you're stupid."[3] He made no reference to the lost astronauts or their families.

By the end of the month, the Kerwin report would be released, which said that at least three of the crew had survived long enough to use their personal emergency air packs. Astronaut Robert Crippen speculated during an interview, "They were probably breathing all the way to water impact, which they would be doing if they were unconscious." It had not been a good year for NASA, and with lawsuits piling up, there were more problems to come before the space agency would get back on the track to human space flight.

On August 12, 1986, NASA released data on the engineering fix for the redesigned solid rocket booster. It included something called a "J feature," named for how it looked, as well as a third O-ring.[4] The "J feature" was designed to stop the rocket sections from rotating in different directions at ignition, and the third O-ring was added more as a publicity feature than a true engineering requirement. No one in or out of NASA was willing to say that flying with only two O-rings would ever be sufficient for a manned vehicle again. The projected cost of the fix was reported to be $300 million and the taxpayers would bear the full cost.[5]

At the same time, NASA was again refusing to release to the press and public copies of audio portions of the crew comments during the ascent. This was reportedly out of respect for the families, but that excuse was questionable. News agencies wanted an opportunity to review the tapes in case NASA missed anything.[6] Six days later President Reagan signed a directive approving the construction of a replacement Shuttle for the lost orbiter, which would bring the fleet back up to four. He also ended private-corporation satellite launches from the Space Shuttle.[7] One month later, Congress would approve NASA's budget by a vote of 407 to 8, which included funds for the new Shuttle. It was the largest margin of approval for a NASA budget in the space agency's history, well beyond even the historic days of Apollo.

By October, the National Research Council had completed its review of the booster design and endorsed NASA's new limited redesign of Thiokol's original SRB. The council cited what it called "a need to get Shuttles back into space"[8] and speculated that sooner, rather than later, would be the best way to go. However, the space agency, already pushing as fast as it could, had announced that they would allow Thiokol to test its new design in a horizontal stand instead of the commission-recommended vertical testing configuration. This would speed up the process, but would make it difficult to simulate the launch profile of a flying solid rocket.

As 1986 came to an end, the space agency was again feeling pressure to put together a new launch schedule and get back into the manned-space-flight business as soon as possible. NASA announced a tentative solid rocket booster test program, which would be a road map back to space. The agency also began to develop methods to help astronauts escape from a crippled Shuttle. Although no one in the program believed such a system would have saved the crew of 51-L, any chance would be better than no chance at all. The end of the year also saw reports being published of a new Soviet Shuttle named *Buran* being prepared for its upcoming first unmanned orbital flight. U.S. reconnaissance photos, taken from space, had shown the Soviet Shuttle being transported piggyback on its carrier aircraft as well as on the launch pad for a series of prelaunch tests. Intelligence sources believed that in early 1987 the large new booster would be tested with the first Soviet manned Shuttle flight coming in the first part of 1988. It was anyone's guess which country would place the next Space Shuttle into orbit. The race for space was on again.

1987 — A Crew, but No Booster Yet

On January 5, 1987, NASA tentatively approved one of the proposals being developed for a Shuttle escape system. The system would consist of a hatch which would explosively separate from the spacecraft, combined with a rocket pod system which would pull individual astronauts away from the Shuttle in the event of an emergency.[9] The astronauts would then parachute to a probable water landing. It was reported that the system would only be used if the Shuttle were in some type of level stable flight. However, it would be hard to imagine that astronauts would not use it in just about any situation in which they felt the Shuttle would not make it to a safe runway landing.

Three days later, NASA workers began the job of storing *Challenger*'s wreckage in two old Minuteman silos located at Cape Canaveral Air Force Station. Later, in May 1990, NASA announced that the Apollo One spacecraft (the remains of capsule 204, destroyed in the January 1967 launch pad fire) would be placed in one of these same silos. The silos would then be sealed with tons of concrete. The next day, NASA announced the names of the astronauts who would fly the Shuttle *Discovery* on the next mission then being planned for sometime in 1988: Frederick H. Hauck, commander; Richard O. Covey, pilot; and John M. Lounge, George D. Nelson, and David C. Hilmers, mission specialists.

On January 15, the new crew gave their first press conference.[10] It was businesslike and subdued. This all-veteran Shuttle crew would fly mission STS-26, and in doing so would once again place Americans in space. It would become the most closely observed Shuttle mission ever flown by NASA. If there was ever going to be a relatively safe Shuttle flight, this would be it. At least, that is how the press and many NASA public relations people reported it. However, mission commander Hauck had a more realistic and experienced point of view. "Let's be very clear. There is risk in flying in space — there should be no doubt about that. There will never be a space flight that does not have risk associated with it."[11] Indeed, this flight would be the first test of a completely redesigned Shuttle system, not just the solid rockets, and as such was as much a risk as any other test flight. The Shuttle crews knew that they were flying a craft with over 800 "criticality 1" items. This meant that if any one of those flight items failed it would be catastrophic because none of them had any backup.

At the same time, the space agency had established what it called a "scrub team."[12] This was a standby team of controllers and technicians who would take over in the event that a launch was scrubbed (cancelled or postponed for any number of reasons with the expectation of launching in the very near future) allowing the launch team enough rest between launch attempts. The commission report had been very critical about the long hours required of launch team members, something that was thought to have caused unsafe conditions during several Shuttle launches, including the almost disastrous 61-C *Columbia* flight. Astronaut chief John Young had earlier pinpointed this as a major problem, doing so in a memo which at the time was ignored.

On January 23, 1987, as reported by the Associated Press, NASA was prepared to pay Morton Thiokol $350 million to correct their flawed design on the solid rockets, although new director of Marshall Space Flight Center, J. R. Thompson, stated he was, "not in love with this design."[13] If he did not think this was such a good design why did he not require Thiokol to go back and do the work again? Flight schedule pressure? It was also announced that Thiokol could be fined up to $10 million because of the O-ring failure[14] which had destroyed the Shuttle. It was made clear, however, that no such fine had yet been implemented.

One week later, Allan McDonald, one of Thiokol's engineers who had fought so hard to stop the *Challenger* flight, now heading the redesign, announced that the new effort would be "several orders of magnitude more reliable than the old design." That would indicate that this design was from 100 to 1000 times more reliable.[15] In other words with 400 flights, using two solid rockets each, this system would not be expected to ever fail. "I think it is as good a joint as you can make," said McDonald. But why use a joint in the first place? Why not design a new single-case solid rocket with no joints and thereby eliminate an entire area of possible failure? Thiokol had no such design, nor production capability online to make a single-case solid rocket, but there were other companies who did, and those companies wanted to do the work.

By the end of January, O-ring engineer Roger Boisjoly had applied for, and had accepted, long-term disability leave from Thiokol. A week later he filed a $1 billion defamation and antitrust suit against the booster maker.[16] It was going to be a long year, as the next target launch date of February 18, 1988,[17] seemed to move farther away.

As redesign efforts continued, according to a NASA report to Congress, Thiokol had agreed to accept a $10 million reduction in its fees and the company would also agree to "take no

The crew of STS-26, the first post–*Challenger* Shuttle flight, were Mission Specialist David C. Hilmers, Pilot Richard O. Covey, Mission Specialist George D. "Pinky" Nelson, Commander Frederick H. "Rick" Hauck, and Mission Specialist John M. Lounge (NASA).

profits from the now $409 million worth of work that is required to fix future rockets."[18] The only comment from Thiokol about this agreement, which was also reported to have shielded the company from any future lawsuit, came from spokesman Thomas Russell who affirmed "neither admission by Morton Thiokol nor determination by NASA as to Morton Thiokol responsibility or liability for the accident."[19] Apparently Thiokol's spokesman never received his copy of the *Challenger* report. On page 72 it reads, "In view of the findings, the commission concluded that the cause of the *Challenger* accident was the failure of the pressure seal in the aft field joint of the right Solid Rocket Motor." This was a part designed, built, tested, assembled and delivered by Morton Thiokol. While the commission's report never explicitly put the blame on Thiokol, no one else's name was on the rocket. It is one of the major failings of the commission that they did not name the company and individuals who were truly responsible. That has yet to come. Will proper responsibility ever be assigned or admitted regarding the *Columbia* disaster?

NASA's Advisory Council, which reviews all of the space agency's operations and advises on future programs, announced on March 12, 1987, that it had advised NASA to stop using the Shuttle as its principal launch vehicle, due to high costs and dangers inherent in any Shuttle flight.[20] It further advised the agency to acquire a fleet of ELVs (Expendable Launch Vehicles) as a way of reducing reliance on the Shuttle and to clear out some of the now growing backlog of satellites waiting for launch. A week later, astronaut Robert "Hoot" Gibson, speaking as a member of the

NASA/Thiokol redesign team, and as a representative of the astronaut office, stated that he was still very concerned about the new design and wanted NASA to add two more tests to the five already planned by Thiokol.[21] "The current number of tests doesn't give us a good data base." The test program would actually end up completing six full-scale rocket tests, but only five before *Discovery* flew on the next Shuttle mission.

Due to pressure from within NASA, much from the astronauts themselves, and also from Congress, James Fletcher very reluctantly decided to put out an open-competition procurement contract for the new advanced solid rocket motor (ASRM).[22] This new contract would be valued at more than $2.4 billion over a five to seven year period, and would be expected to replace the old SRBs then being used. At the same time, NASA announced that they would add countdown tests to the launch schedule, including a so-called wet test.[23] The wet test would allow technicians at Kennedy the opportunity to fully load the Shuttle's external tank with fuel and oxidizer. NASA managers knew that a long break in the launch schedule could easily harm readiness and a practice would be required. Also added would be a 20-second flight readiness firing of the Shuttle's three main engines, similar to tests conducted before the first Shuttle was launched in April 1981.[24]

Always looking for more funds, the politically savvy Fletcher informed Congress at the end of April that due to a lack of funding, the start of relaunching Shuttles might be slowed. "We just don't want to start the first flight and then have a big gap in the program."[25] Fletcher, who died on December 23, 1991 of lung cancer,[26] knew how to play Congress, and more funds were soon on the way. Congress was mindful of criticism that the program had been underfunded all the way through, and that lack of needed funds may have contributed to the accident. For the most part, NASA would find the Congressional checkbook open, at least until Shuttle operations were back online.

On May 15, 1987, the Soviet Union successfully tested its new Energia Shuttle booster. It failed, however, to place its dummy test payload into orbit.

By the end of May, Thiokol was ready to fire the first of six scheduled full-scale solid rocket tests in a series designed to requalify the solids for space flight. This would not be a test of the new O-ring design. It would only test the newly developed O-ring material, as well as the new heaters which had been designed to wrap around the field joints to keep them a toasty 80 degrees or above. On hand to view the May 27 firing of Engineering Test Motor number 1A (ETM-1A) was the full crew of STS-26. There was not much to actually see, but it did make for a good photo opportunity. It was also an opportunity for the crew to come face to face with the SRB engineers. The test, which proved to be successful, was also a check of the new, external, graphite reinforcing bands, which wrapped completely around the field joints to cut down on the joint rotation, a problem that had compounded the O-ring problems on older designs. NASA and Morton Thiokol had cleared the first hurdle.[27] For NASA it was a move toward launching spacecraft; for Thiokol it was a move toward profits and higher stock figures.

Only days after this victory, however, NASA was to lose in court. For over a year the space agency had been attempting to stop the release of recorded flight tapes from the *Challenger* crew cabin. On June 4, U.S. District Judge Norma H. Johnson ordered the space agency to release the voice tapes of the crew.[28] The ruling rejected NASA's view that somehow this would violate the privacy of the dead astronauts. As the judge eloquently argued, NASA is owned by all U.S. citizens who have a right to know as much as possible what is going on in their names. The next day NASA selected 15 more candidates for training as new Shuttle astronauts.

On June 8, NASA tested its new safety net for Shuttle landings. The Shuttle *Enterprise* was rolled slowly into the new net which had been designed to stop any Shuttle from going off the end of the runway at Kennedy. The astronauts viewed this as a last-ditch effort to stop an orbiter as it came in for a landing, but they had no real confidence in the system, and it was not tested at any great speed.

At the end of June, as part of its continuing oversight function, the White House received a

191-page report on what the space agency had so far accomplished to implement the commission's recommendations. The report, which was released to the public, stated that the first post–*Challenger* flight would have the new "jettisonable" hatch in place. Later, all orbiters would be reworked with the same quick-release hatch. The report also listed twenty other major changes required by the commission, which were implemented to insure greater safety when the main engines were being fired.[29] It had been one full year since the *Challenger* report had been issued and much had been accomplished, but the published February 18, 1988, launch date for the next flight was beginning to no longer appear realistic, if it ever had. When NASA added three more solid rocket tests to an already hard-pressed schedule in July,[30] the new launch schedule was going to be dropped. Following requests from Congress, the astronaut office, and the National Research Council to add the tests, the published launch date was all but forgotten. All concerned knew that it had become unobtainable. Yet, once again outside pressure was needed to get the space agency to do the right thing and fully test the system. On August 1, 1987, NASA awarded Rockwell International the $1.3 billion contract to build the new Shuttle. "Essentially, we've gone back to square one with *Discovery*. Every system ... has been meticulously examined [and] retested," Maria Metcalf, Lockheed's manager of flow planning said in the August 6, 1987, issue of *Lockheed Stargazer*.

On August 10, the STS-26 crew made a visit to Cape Kennedy to speak with the workers who would be assembling their flight hardware. Commander Hauck made a motivational speech in front of the Orbiter Processing Facility, but this was more than an astronaut pep talk. It was part of a newly valued face-to-face communication between the Shuttle crews and the individuals who were responsible for their safety. It was starting to look like the old Apollo days, with astronauts all over the place.

It was during August that a problem was discovered in the solid rocket motor then being prepared. Thiokol had discovered that an O-ring leak check had failed during assembly, causing reassembly of the booster to be required. Later in the month, during a scheduled 58-second test of the newly designed solid rocket motor, a major failure occurred. Just five seconds into the test a steel forward dome failed, sending the rocket flying along the ground for 100 yards![31] No one was injured, but it was a major setback in the development program. However, Thiokol was able to test fire another Solid Rocket Booster only days later. John W. Thomas, manager of the Solid Rocket Motor Design Team, was quoted by *Aviation Week and Space Technology* on September 14, saying, "No hot gas penetrated the new insulation J-seal feature, and the O-ring seals looked good." Demonstration Motor No. 8 (DM-8) had passed its trial by fire. It was two down, four to go.

While NASA was testing for the near future, it was also looking at possible paths the United States could take in the years and decades to come. Astronaut Sally Ride, who had been the astronauts' representative on the *Challenger* commission, and later on the *Columbia* investigation team, delivered a 63-page study to NASA on where the agency should be going.[32] The Ride Report had developed recommendations based on a wide and diverse set of programs rather than a space agency doing one major project at a time, as was the case when NASA went to the moon. The report stressed developing a diverse program of goals which would complement each other. By reaching for many goals at the same time, the costs would be dramatically reduced for each program while the benefits overall would greatly increase. Goals foreseen by Ride included the establishment of a manned orbital space station, a lunar research outpost on the surface of the moon, and exploration of Mars. The report pointed out that research done and problems solved in one area would directly affect the others, which would, as such, reduce the overall costs of such a wide-reaching program. The report for the most part gathered dust on the shelf. A decade later the United States had yet to orbit the first module for its space station, although the agency has stepped up its exploration of the red planet with sets of unmanned probes expected to fly to Mars every twenty-six months for the next decade. It is a program of exploration which will in the future form the basis for manned exploration.

Space Shuttle Demonstration Motor-8 (DM-8), August 30, 1987 (Morton Thiokol, Inc.).

Back on earth, in September 1987, NASA announced that *Discovery* would deploy a replacement TDRS for the relay satellite that had been lost with *Challenger*.[33] The TDRS is such a powerful relay station in orbit that it can process as much information from multiple sources as can be found in 100 encyclopedias in a single second. The flight to launch this was tentatively scheduled for June 2, 1988, and it would also carry two student experiments which had previously flown on *Challenger*.[34] Earlier in September, NASA released a small news booklet entitled *Crew Egress/*

Escape System.[35] The booklet outlined two methods being worked for crew escapes in emergency situations. The crews would either be rocketed out of the hatch or would bail out using an extended pole system, which would ensure that they would not hit the wings or tail of the orbiter.

On October 9, 1987, Thiokol test-fired sections of a solid rocket booster. The test, which was quickly reported to have been a success, was conducted to discover if deliberately placed flaws in the troublesome nozzle joint would fail, causing the hot gases to leak and interact with the O-rings which were using the new J redesign feature. According to project chief Allan McDonald, "There was no evidence of any such leak."[36] NASA managers were crossing their fingers since this was the first "hot" test of the new features, and the agency wasted no time in publicizing this apparent success.

Two weeks later, NASA headquarters, through Richard Truly's office, announced an ambitious 19-Shuttle flight schedule for the next three years beginning in 1988.[37] The manifest also included 49 flights using so-called expendable rockets. It was solid confirmation of the agency's desire to put the program back into space using as many options as they could. "The new manifest is a demonstration that we are serious about a mixed fleet," Truly said.[38] The dates and primary payloads of the first ten scheduled Shuttle flights included:[39]

June 2, 1988	TDRS-C
September 8, 1988	Department of Defense/Lacrosse
December 1, 1988	Department of Defense/KH-11
February 2, 1989	TDRS-D
April 27, 1989	*Magellan* spacecraft to planet Venus
June 1, 1989	Hubble Space Telescope (earth orbit)
June 29, 1989	Astro Ultraviolet Telescope (earth orbit)
August 24, 1989	Department of Defense
October 9, 1989	*Galileo* spacecraft to planet Jupiter
November 9, 1989	NAVSTAR

There were no flights scheduled for the now-mothballed Shuttle launch facility at Vandenberg Air Force Base. Soon after the announcement, NASA admitted that the first launch could easily be pushed back again due to delays in manufacturing the new solid rocket boosters and the newly reworked orbiter main engines. NASA was also making a great many modifications to the launch-pad facilities, as well as improving the hardware which would be used in any emergency escape situation causing the crews to leave the pad. NASA was concerned that many sections of the launch structure did not allow sufficient clearance between it and the Shuttle stack during liftoff, so modifications were being completed on both pads 39A and 39B, but these were not expected to affect the new launch schedule.

On October 16, at Hurricane Mesa in Utah, NASA tested the new tractor-rocket-powered escape system. Astronaut George "Pinky" Nelson was on hand to evaluate the tests and came away unimpressed with the complicated system. Some of the astronauts would state in private that the bail-out systems were only being developed to alleviate public fears brought on by the *Challenger* disaster. During the first part of December, the first flight tests were conducted on the new tractor-rocket escape system over China Lake Naval Weapons Center in California. Using a modified Convair 240 fitted with a simulated Shuttle hatch, test dummies were rocketed fifty feet out of the side of the aircraft flying at 200 miles per hour, then parachuted to a dry lakebed. It was a successful test of the new system under development since June 1986.[40]

That same week, managers at Cape Kennedy who had responsibility for processing the Shuttles for flight, reported to NASA directors that they did not have enough qualified individuals working. They felt that even with a reduced fleet of only three orbiters, more workers would be required to keep the orbiters online for launch schedules as they were presently published. The feeling was that if and when the fleet got back up to four orbiters, it would not make much difference in the flight rate. They could not support the flow requirements to meet a four-orbiter fleet.[41] NASA and Lockheed, the process contractor, would need to train at least two more complete crews to keep the orbiters on schedule. Funding once again was a primary concern, as launch pressure was again allowed to creep into the system. As part of the flow-process checks on December 15, workers at the Cape conducted a practice mating of the SRB segments. Astronauts Fred Hauck and John "Mike" Lounge were on hand to view the test.

Earlier, on November 17, Hauck, Richard Covey and David Hilmers ran the first test of the new flight software in the Shuttle flight simulator at Johnson Space Center. It was a successful test run, but the astronauts were using 1970s technology which held far less than half the memory of the average home computer at the time. Because of this limitation, the memory was always full, which meant that if anything needed to be added it was done at the loss of other programs. It was a risky guessing game of trying to put as much as possible into the system and hoping nothing critical had been left out. Lack of proper funding was the major cause of this problem.

On December 23, Morton Thiokol fired its second full-scale test of the solid rocket redesign at its Utah facility. It would be a good test of the new heaters designed to keep the field joints warm, because the local temperature had fallen to 22 degrees above zero, with a wind-chill factor which brought that temperature down to 25 degrees below. Instruments checking the test firing showed that the joints on Demonstration Motor 9 (DM-9) had been kept at a safe 82–90 degrees throughout the entire test. Allan McDonald reported that the O-rings were "Nice, warm and toasty, as we intended them to be."[42]

The O-rings may have been toasty, but the test was not the home run that Thiokol had hoped for. When the booster was examined after the two-minute static test, it was found that part of the booster nozzle had somehow torn apart.[43] A section of the outer boot ring, something new to the design, had been somehow sucked into the motor. This was a part of the system which does the majority of the Shuttle steering for the first two minutes of powered flight. The Associated Press reported that "nearly four feet of the eight foot diameter ring was gone, and more than a foot of that material was discovered inside the motor." A spokesman for Thiokol could only state, "Clearly we have missed something."[44] Earlier, the National Research Council, a group formed to review NASA's recovery, had looked at the new joint design and reported that it might "fail to meet the agency's new safety requirements." This malfunction caused a two-month delay, which set the launch date back to August 1988 for the next Shuttle mission.

One week later, Representative Manuel Lujan, Jr. (R-NM), who had been pressing for a new SRB contract to be given to another company, said, "We've been insisting that we have a second source, but NASA has been fighting that.... [M]aybe now we can convince them. We discovered the problem early enough in the test series that we can do something about it without what we think would be a major schedule problem."[45] It was a bad end to a problem-filled year, and NASA did not know if this booster would ever fly again. Some within the agency wondered if a Shuttle would ever get off the ground and the possibility of a cancelled program was a real fear.

1988 — Fire in the Trenches

As the new year began, NASA received even more bad news. While the space agency was reporting the failed nozzle and stating that the nozzle would have to be redesigned before the Shuttle would be cleared to fly, Thiokol was calling the test a general success. Thiokol official Rocky Raob refused to call a full-blown news conference and stated, "It's not anywhere near as bad as has been suggested."[46] He reported that the company considered the problem so small that they would not even answer a formal request from NASA.

The next day, Thiokol reported that a second problem had been uncovered in its second full-scale test. Despite the use of joint bands to hold together the field joints and the new heater for the O-rings, hot exhaust gases had indeed reached the first O-ring. With NASA's original and post–*Challenger* design criteria that hot gases should never reach an O-ring, this test was a failure. David Winterhalter, NASA's solid rocket booster manager, would only say, "We do not believe it is harmful."[47] A week later an industry-review team looking into safety and quality assurance at Kennedy Space Center found that workers handling the Shuttle had a business-as-usual attitude again. Within two months these very same workers would receive and begin to process the first segments of the newly modified solid rockets from Thiokol's plant in Utah. This shipment would occur before most of the qualification tests had been completed on the new design, including three full-scale tests. NASA was starting to pick up speed toward a launch, but all was not right with the system.

In late January, NASA reported that it would use Northrup landing strip at White Sands, New Mexico as its new alternate in case the Shuttle could not land on Edwards lakebed in California.[48] Concerns about landing on the concrete runway at Edwards and Kennedy, coupled with continuing problems with the orbiter braking systems, were given as reasons for the selection. The Shuttle runway at Kennedy would be the last choice for a landing, at least for the first few flights. At the same time the space agency was in the middle of a $700,000 rework of the Shuttle runway itself. The surface was found to be too hard on tires, which were consistently damaged the five times the Shuttle had launched on the strip. The 3,500-foot end sections would be generally smoothed while the 8,000-foot center section would be reworked, with shallower grooves being ground lengthwise to lessen the amount of tire damage on future landings.

On January 28, 1988, the Associated Press reported that NASA had set a launch date of August 4, 1988, for Next Shuttle Flight. At a news conference given at the end of February, the *Discovery* crew was asked about the progress NASA had made for their upcoming flight and the new safety bailout procedures. The press were concerned with the patch together makeup of the two new systems. Answering for the crew, Pinky Nelson stated, "The next spacecraft anyone designs will have an escape system designed into it, but for now, if we want to fly, we've got to live with what we've got."

Three days later, over Edwards Air Force Base, six navy parachutists successfully tested the second astronaut-escape system. On hand to observe the test were Pinky Nelson and Carl Meade from the astronaut office. The tests consisted of six passes at 7,500 feet over the drop area, as one by one parachutists slid down a 12-foot-long pole protruding out the left side of a modified C-141 aircraft.[49] The tests were considered a complete success and the system was endorsed by the astronaut office as the safest method developed so far. One month later, on April 7, NASA announced that the pole system would be adopted for the Shuttle. The agency would install the 9.8-foot telescoping pole on *Discovery* before it would be recertified to fly. The 241-pound pole would be used at 22,000 feet or lower if the Shuttle was in some type of stable, level flight, or so NASA

Redesigned field joint used by solid rocket boosters (NASA).

said. As far as the crews were concerned, they would use any method any time they deemed it necessary for their own survival.

On February 1, 1988, *Time* magazine reported that once again the space agency was "putting schedule over safety," that despite the *Challenger* accident, the Shuttle program was ignoring whistle-blowers. The article went on to report that NASA had formed a blue-ribbon committee to examine safety issues and procedures after *Challenger*. The committee was blunt in its evaluation, stating, "When NASA rated its program managers, safety was conspicuous by its absence. There was also disturbing evidence that schedules were given priority over safety." Whistle-blowers had been harassed and fired, and some had received threatening phone calls and letters. In one particularly worrisome case, a former Rockwell engineer tasked with quality assurance reported that he had conducted an audit of the Shuttle hardware and found that only 12 percent had met NASA's contract specifications. The day after submitting his report, his supervisor told him that "he had to change the figure to 96 percent or better." He refused. Five weeks later he was suspended and then later fired. NASA's only response came from Administrator James Fletcher who said, "We will fly only when we are ready. And readiness means that the Shuttle will fly only when it's as safe as we can make it." It had been a little more than two years since *Challenger*'s destruction and NASA was pushing hard for a launch.

Five months later, during an investigation by the House Science and Technology Subcommittee, these charges and others regarding safety and quality control would be confirmed. It was shown that even as early as a few months after the disaster, NASA and its contractors had cut corners in many areas of development throughout 1986 and 1987.

A major milestone was achieved when, on March 9, Thiokol shipped by rail the SRB segments which were to be used on Shuttle *Discovery*'s next flight. The train left Utah for Cape Kennedy, but in Biloxi, Mississippi, it struck a car at a road crossing, killing the two passengers. Later, after X-rays were taken, it was found that no damage had occurred to the solid rockets during the crash.

On April 20, as Kennedy Space Center workers were assembling and checking the first set of solid rocket boosters for the Shuttle's return to space, Thiokol was testing its new design with its Qualification Motor No. 6 (QM-6).[50] The test would prove successful, and NASA could breathe a sigh of relief, as well as continue to aim for its August launch date.[51] It would be a short breath. On May 5, 1988, the *Los Angeles Times* reported that in Nevada "three searing explosions of orange flame [had] destroyed a plant that makes rocket fuel oxidizer for the Space Shuttle and many of the nation's nuclear missiles," adding that "one person was killed and 210 were hurt, one critically."

The explosions, which had leveled the plant owned by State Senator James Gibson, had occurred at Pacific Engineering and Production Company located 15 miles south of Las Vegas. Two of the three explosions had been powerful enough to be recorded on Cal Tech seismographs in Pasadena, California, some two hundred miles away. These instruments, normally used to record earthquakes, recorded the force as nearly equivalent to a 1-kiloton nuclear explosion. Flying nearly overhead was American West Airlines Flight 46 out of Burbank, California. The jet was on final approach to McCarran International Airport in Las Vegas when it was hit by the shock wave from the most powerful of the blasts. One passenger reported, "The second explosion looked like a mini–atomic bomb.... There was a tall column of smoke going up 20,000 feet." Workers and others who were nearby ran in all directions, as far away as three miles, scattering all over the desert. With this one disaster, the space program had lost half of its capability to produce oxidizer for solid rockets. It was going to be another long year, as NASA would struggle to find the resources to keep flying.[52] They would also be competing directly with the Defense Department for the limited oxidizer resources that both agencies needed to keep space accessible.[53]

On June 7, one week after mock-emergency-landing tests were conducted at both Kennedy and Johnson Space Centers, Thiokol ran its next Qualification Motor Test (QM-7). This full-scale test was the first conducted with all of the modifications required by NASA in place. It proved to be an unqualified success. However, one week later, as bids were being accepted for the updated advanced solid rocket motor (ASRM), Thiokol announced that the company would not submit a bid. This would have effectively taken Thiokol out of the solid rocket business for manned flight within a few years. The ASRMs were then scheduled to phase out the redesigned Thiokol Motors by 1994–97. Perhaps Morton Thiokol management knew that the advanced rockets would never be purchased by the government, and would never launch a Space Shuttle.

On July 4, 1988, the nation's 212th birthday, NASA rolled Shuttle *Discovery* out to launch pad 39B with great fanfare, surrounded by astronauts and families of Kennedy Launch Center workers. The trip to the pad took 7 hours and 55 minutes, and it was begun under giant spotlights in the early morning. John Pike, from the Federation of American Scientists, viewed the rollout and stated, "If it's anything short of picture perfect, the Shuttle program is going to be at an end. NASA will be chopped up into little pieces."[54] Two days later, a House of Representatives Task Force found that NASA had "deficiencies in people, skills, management systems and independent safety oversight functions." No solutions to these problems were given other than to request another report from NASA. Roger Boisjoly said of the upcoming return to space, "My own personal feeling is that it is a flip of the coin."

Later that month, a Shuttle flight-safety study contracted for McDonnell Douglas Corporation by Pickard, Lowe and Garrick, Inc., was released on Shuttle flight safety. The report concluded that based on critical items and failure rates the Shuttle could be "expected to crash in one out of 70 flights."[55] When the study was released to the space agency, NASA responded that orbiter fixes then being worked on would double the expected life span of the Shuttle, but the safety report's conclusions were not specifically addressed. NASA was then hoping to bring the Shuttle up to forty missions each. Originally the orbiters were sold to Congress and the American people as being flyable up to 100 times each. Among other things, NASA was suffering from bad memory.

On August 10, NASA test fired Shuttle *Discovery*'s three main engines as part of their Flight Readiness Firing (FRF). The 21.8-second test had been postponed five times in the previous weeks because of mechanical problems. Six days earlier a stuck valve had been detected by a sensor, and just 6.6 seconds before the test was to begin the firing had to be halted. This time, however, the test was a go and a complete success.[56] NASA now had three tested, flight-ready main engines on the pad ready to help lift *Discovery*, and the Shuttle program, back into space. But they did not yet have a certified solid rocket.

Eight days later, Thiokol conducted its fifth and last preflight test of its new solid rocket booster design. This test of the over $470 million redesign effort had 14 intentional flaws imbedded to certify all of the new changes and safety features. With this qualification motor test successfully completed, manager Allan McDonald said, "We're ready to go launch. We built in a joint that is clearly, in my mind, the safest thing on the space vehicle."[57] Only time and test flights would judge the validity of that statement.

One week later, the new Tracking and Data Relay Satellite (TDRS-C) was loaded on *Discovery* and the payload bay doors were cleared and closed out. On hand for the loading were several of *Discovery*'s new crew. When asked about the new SRBs he was about to ride into space, astronaut John Lounge said, "We should get off the solid rocket boosters. We shouldn't use them, they're just inherently dangerous."[58]

On September 26, only three days before *Discovery* was scheduled to lift off from Kennedy, three engineers from a 25-member redesign team at Thiokol risked their jobs and their security and came forward in Washington, D.C., to report that the new design might be dangerously full of flaws. They stated that Morton Thiokol had consistently withheld critical reports from NASA that had identified potential problems in the design. Steven Agee, one of the engineers, said, "I would never ride one [spacecraft] that had a Morton Thiokol booster strapped on it.... I think the entire system is very unsafe."[59] A second engineer, Aroon Murthi, was quoted as saying, "Cost, schedule and performance was driving the design, not safety and reliability." Finally, a third engineer on the project, Ron Clary, said that "reports documenting potential dangers, called hazard analysis, were lost or were not allowed to appear ... hundreds of them must have been lost." Clary had further stated that when he discovered a problem with the SRB's nozzle, which had failed, he was transferred to "another part of the rocket program considered to be trouble-free." Asked for a response by the press, Thiokol vice president Richard Davis stated, "There was no attempt to suppress any hazards that have been identified by anyone."[60] The FBI would continue its investigation into this and many other areas Thiokol was involved in. Strangely, although the engineers' concerns were covered in the European press and other parts of the world, the story was not covered in the U.S. press.

During an interview, astronaut George Nelson, speaking about the risks involved with space flight, reflected, "Certainly all of us as crew are aware that this is a risky business and the crew of the *Challenger* were no different that way.... [W]e have to live with what happened and keep going."

The day before *Discovery*'s launch, Richard Truly was asked about the possibility of another accident occurring. This time the response was straightforward and honest, as he informed the press that NASA had done everything it possibly could to prevent such a recurrence, but he added, "Somewhere in the future of our country I think we are going to have another accident. It is inevitable, it's in the cards." He also reminded the press that the space agency had overhauled the entire Shuttle system, and in a very real sense this was a first test flight of that new system. Much on *Discovery* had never flown before and the only way to test all of the fixes was to go fly.

Go!

On September 29, 1988, eighteen members of the 1730th Para-rescue Squadron stood ready for rescue operations onboard two C-130 Hercules cargo planes in Morocco and Gambia in Africa. Nine paramedics from the New York Air National Guard were on station at Patrick Air Force Base operating off of another C-130, with eighteen others onboard rescue helicopters. And 225 miles down the Atlantic missile range, two rescue teams sat ready on two ships in case they were needed.[61]

At 11:37 A.M. EDT, Space Shuttle *Discovery* (STS-26) with its crew of five, lifted off from launch pad 39B. It was the same pad used by *Challenger* 32 months earlier. Calling the flight for NASA was Steve Nesbitt who had been the individual narrating *Challenger*'s final ascent. He had requested this assignment because he wanted to end his commentary career with a success. The launch had been held for 98 minutes while very light winds were allowed to build to predicted levels. Eight minutes and 37 seconds later, Americans were once again orbiting the earth. Below, 156 miles down, two SRB recovery ships, which had been used for recovery operations for *Challenger*, steamed toward the splashdown locations of the two used solid rocket boosters which had, at least this time, performed exactly as planned. That was the way it looked before the boosters came back for a closer inspection.

Forty-seven days later, the Soviet Union launched their new unmanned Space Shuttle *Buran* on its first and only two-orbit mission. Insiders would later explain that the vehicle had been severely stressed during launch and would no longer be flightworthy. Yet, even if the vehicle could fly there was simply not enough money to keep the program operable.

We are all explorers and the race to space was back on. The year 1988 would be a much better one for NASA than 1986.

The ascent probably excites me more than anything because it's so much happening in so short a period of time.
—*Challenger* Commander Francis Dick Scobee

Launch of *Discovery* on STS-26 on September 29, 1988 (NASA).

The Courts

July 2, 1986 — The family of Michael Smith filed a $15.1-million suit claiming negligence by NASA.[62] The suit stated that NASA should have known that "a catastrophic accident would likely occur." Named in the suit was Lawrence Mulloy. Two weeks later Mulloy announced his retirement from the space agency.

September 5, 1986—The family of Ron McNair filed suit against Morton Thiokol claiming gross negligence.[63] The suit named Joseph Kilminster, Robert K. Lund, Jerald Mason, and Calvin Wiggans as defendants.

December 29, 1986—The U.S. government settled all claims against NASA with the families of Scobee, Onizuka, Jarvis and McAuliffe with "confidential financial settlements." Under a separate claim worked out by the government, Morton Thiokol also settled all claims with these families.[64]

January 28, 1987—Former Thiokol engineer Roger Boisjoly filed a $1 billion anti-trust and defamation suit against Morton Thiokol.[65] In the suit he claimed that Thiokol had tried to present him as a "disgruntled or malcontented employee whose views should be discounted and whose professional expertise should be doubted." The suit also reported that NASA managers who had testified before the commission gave "false and misleading" testimony. The suit would later be dismissed.

January 30, 1987—The father of Greg Jarvis filed a $5-million suit against NASA claiming wrongful death and negligence.[66]

May 6, 1987—The wife of Michael Smith filed a suit in Orlando Federal District Court against Morton Thiokol for $1.5 billion.[67] The suit filed in federal court stated that the crew was "in imminent peril of death. The motive was money ... a conspiracy of silence and deceit with reckless disregard for human life." NASA was also named in the suit, along with Lawrence Mulloy.

May 8, 1987—The family of Ron McNair settled their suit against Morton Thiokol. The amount was not disclosed. "My family has amicably and equitably concluded legal proceedings against Morton Thiokol," said Cheryl McNair.[68]

February 17, 1988—Judith Resnik's parents reached a settlement with Morton Thiokol after a year of negotiations. The amount was estimated to be from $2 to 3.5 million.[69]

February 27, 1988—U.S. District Judge Patricia Fawsett set down a ruling that Mike Smith's widow Jane could not sue NASA (the government) because he was on military duty at the time he flew on *Challenger*. The ruling stated that she was only entitled to the standard death benefit given to service members' families.[70] The suit against Morton Thiokol stood.

March 7, 1988—The federal government and Morton Thiokol purchased $7.7 million in annuities in order to settle claims brought by the families of Scobee, Onizuka, Jarvis and McAuliffe. Thiokol purchased 60 percent of the annuities, with the U.S. government picking up the other 40 percent.[71]

August 23, 1988—Mike Smith's widow Jane settled her wrongful death suit against Morton Thiokol for an undisclosed amount.[72]

Section V

From *Challenger* to *Columbia*

Chapter Twelve

Beyond the Recovery: Was It Really Fixed?

There was a high level of schedule pressure; they wanted us to go from eight, nine launches a year to 14 to 18, while reducing costs at the same time.
— Robert B. Sieck, Director of Shuttle Ops, Kennedy

Before the Shuttle program could get back into space with its newly redesigned solid rocket booster, NASA had to do a little housekeeping. It was not done, however, in a way that might have been expected after the fatal accident. No one who had anything to do with the deaths of the astronauts was ever fired, fined or put in jail. Most of the managers and directors who had direct responsibility were either reassigned to generally equal or better positions, or were allowed to retire with little or no fanfare. Employees of both NASA and Morton Thiokol escaped criminal indictment. No one was named as being responsible in any of the thousands of pages released by the *Challenger* commission, and that was one of the great failings of the investigation.

Years later, reporting on the NASA cultural problems, the agency's *Columbia* accident report stated, "By the eve of the *Columbia* accident, institutional practices that were in effect at the time of the *Challenger* accident — such as inadequate concern over deviations from expected performance, a silent safety program, and schedule pressure — had returned to NASA."[1] The report further stated, "The Apollo era created at NASA an exceptional 'can-do' culture marked by tenacity in the face of seemingly impossible challenges. This culture valued the interactions among research and testing, hands-on engineering experience, and a dependence on the exceptional quality of the workforce and leadership that provided in-house technical capability to oversee the work of contractors. The culture also accepted risk and failure as inevitable aspects of operating in space, even as it held as its highest value attention to detail in order to lower the chances of failure."[2]

Who then was to pay the price for the *Challenger* accident? In March 1988, Commander John Young was removed as commander of the Hubble Space Telescope deployment mission. The most experienced member of the astronaut team, Young had six flights to his credit beginning with the first manned Gemini mission in 1965, and including two missions to the moon, plus the first Shuttle mission. Young had been one of four Shuttle astronauts who testified before the *Challenger* commission. He had also written memo after memo to NASA's upper management, voicing his concerns over many safety issues, which would only see the light of day after *Challenger* was destroyed. Young said, "If they don't want me, all they [have] to do is say that; I can't get straight answers from them." Young was also removed as chief astronaut and assigned as "special advisor" to Aaron Cohen, director of Johnson Space Center. It was punishment for an outspoken veteran space commander who was not afraid to tell the truth, someone who only wanted to do his job and fly in space.

Veteran Shuttle astronauts Henry Hartsfield, Robert Crippen and Paul Weitz had testified along with Young. None of these Shuttle commanders would ever be allowed to fly in space again. It is easy to suspect that these individuals, who were in a position to best know what was needed to fix the management system, were not going to be allowed to fly because they had told the truth and blown the whistle on NASA. The truth had a price, and these and other astronauts who were willing to tell that truth would be the ones to pay it instead of the managers who made the decisions. Where was the outcry from the press and commission members on this? And did the astronaut office suffer once again after the *Columbia* investigation finished their work?

The *Columbia* accident report would criticize NASA culture, saying, "Apollo successes created the powerful image of the space agency as a 'perfect place,' as 'the best organization that human beings could create to accomplish selected goals.' In the aftermath of the *Challenger* accident, these contradictory forces [i.e., 'perfect place' and the demise of the Cold War space race with the Soviet Union] prompted a resistance to externally imposed changes and an attempt to maintain the internal belief that NASA was still a 'perfect place,' alone in its ability to execute a program of human space flight.... [NASA managers] lost their ability to accept criticism, leading them to reject the recommendations of many boards and blue-ribbon panels, the Rogers Commission among them."

Recognizing NASA's reluctance to change, the Office of Technology Assessment gave NASA a dire warning in 1989. "Shuttle reliability is uncertain, but has been estimated to range from between 97 and 99 percent. If the Shuttle reliability is 98 percent, there would be a 50–50 chance of losing an Orbiter within 34 flights.... The probability of maintaining at least three Orbiters in the Shuttle fleet declines to less than 50 percent after flight 133."[3]

The Augustine Committee report from 1990 states clearly, "NASA has not been sufficiently responsive to valid criticism and the need for change." However, the committee understood that NASA was overcommitted "in terms of program obligations relative to resources available — in short ... trying to do too much, and allowing too little margin for the unexpected."[4]

It's the Booster Once Again

After the picture-perfect launch of *Discovery* on mission STS-26, its solid rocket boosters were returned to Morton Thiokol to be unstacked and inspected. During this inspection, engineers found that soot had penetrated beyond a filler material which was designated to help protect the booster's nozzles against heat. Once again a flight motor had shown a problem which had not occurred during horizontal ground tests. The O-ring had not been damaged during this flight, but material next to it had. NASA would once again push aside any concerns when NASA booster manager Royce Mitchell at Marshall stated, "It's a non-problem." Non-problem or not, the design was such that no soot would be expected in that area of the joint. That made it a problem.

On December 2, 1988, NASA flew *Atlantis* on STS-27, which came very close to becoming a disaster. During launch, debris had fallen off the stack and knocked off a thermal tile. This tile loss resulted in structural damage to the Shuttle and a near burn-through. NASA managers did not set up a team to solve the debris problem. They did not consider it to be a "safety of flight issue." After the *Columbia* disaster it would be.

On January 20, 1989, Morton Thiokol ran its final booster test in NASA's ever-growing $600 million redesign effort. It was a full-scale rocket test designated Qualification Motor 8 (QM-8), fired at 40 degrees Fahrenheit with all redesigned systems onboard. In NASAese, the test had "no anomalies seen." Again a solid motor had functioned as designed — on the ground. As far as

Morton Thiokol and NASA were concerned, the program had been successfully completed and the redesigned solid rocket motor (RSRM) was now part of a safe system. Morton Thiokol changed its name to Thiokol Corporation and moved some of its offices to Chicago.

As NASA began to increase its scheduled Shuttle flights, the space agency once again started to look like the NASA of old, at least on the surface. On May 4, 1989, Shuttle mission STS-30 deployed the *Magellan* Venus Orbiter for spectacular close-up views of Venus with a clarity never before seen. On October 17 of that year, the Shuttle *Atlantis* was launched on mission STS-34 with the Jupiter-orbiting spacecraft *Galileo* in its cargo bay. Performing a perfect deployment in earth orbit, the craft began its six-year journey to the Jovian planet. However, it would not be long before problems with the booster would cause NASA to take another look at the Thiokol Corporation.

At the end of November, it was announced that a federal grand jury had met in Birmingham, Alabama, to hear charges that Thiokol had placed inaccurate sensors on the solid rockets and had improperly inspected them. The charges were made by a design engineer who had worked for 17 years at Thiokol, Dan S. Joos. Joos, who had worked on solid rockets at the Thiokol Utah facility, told the grand jury that the new sensors which had flown on the first three flights after *Challenger* generated a great deal of data on how the new design was operating. However, he further stated that up to 50 percent of the information was inaccurate and that "we could lose another Shuttle today." He testified that booster managers could be lulled into a false sense of security about booster performance and therefore once again push the system beyond its capabilities. NASA had been relying heavily on these sensors. As Allan McDonald, new vice president for engineering at Thiokol, stated after the flight, the sensors would "verify the integrity of the booster and its performance." No indictment would come out of this grand jury, but that was mostly because there was no one identified to charge. Meanwhile, the solid rocket data would not be considered valid, as engineers from NASA and Thiokol continued their oversight of a system no one fully trusted.

Despite continuing problems with booster hardware, space agency managers at Kennedy continued to push for an accelerated launch schedule. Former astronauts then in management positions knew problems could arise if NASA pushed too hard again. Robert Crippen, then director of NASA's Space Transportation System said, "We knew [the increased flight rate] might be a problem when we established the manifest. We think we can keep three Shuttle Orbiters in the processing flow simultaneously. We are massaging the workforce skills right now to be sure we have the right ones in the right places."

NASA was trying to bring the number of days it took to completely process an orbiter for flight down to 81. This was about half the time it had been taking. Things would become even more complicated when the fourth orbiter came online. Skills or not, NASA was running out of workers. At the same time, payloads were again backed up, waiting for launch, including several for the Department of Defense. NASA was still attempting to launch 12–14 times per year, a rate it would never achieve. In 1989, NASA was only able to launch five Shuttle missions, which was a reflection of just how difficult the process could be, yet the agency was pushing to double that rate for 1990. Before that dream could become a reality, the space agency would be forced to delay the launch of its space telescope due to problems discovered in its already stacked solid rockets at Kennedy.

In late January 1990, NASA announced a delay of up to four weeks for the launch of *Discovery*, which would carry the billion-dollar Hubble Space Telescope. Kennedy Launch Center found that they would have to replace one of the shuttle solid boosters because it had incomplete joint-leak-test information. Several "unexplained anomalies" were found in the test data after the segments were mated, and some of the stacking documentation had not been completed. Once again, booster processing would push back a launch schedule. This time, however, NASA would err on the side of safety and remove the questionable boosters and delay the launch. The questionable data involved leak tests which had been conducted on the O-ring which was part of the joint between the right booster's center and aft booster segments. It was an area known for its problems. After the booster switch, the space telescope was successfully deployed into earth orbit, but it would soon be shown to have a problematic mirror, which would require a major space-walking repair a few years later. After more than sixteen years, the telescope is still sending magnificent images of the universe back to earth.

On January 9, NASA launched *Columbia* on STS-32. This flight would be the second time that insulating foam from the bipod ramp would break off, allowing debris to fall dangerously close to *Columbia* as it reached for space.[5] NASA and *Columbia* had dodged another bullet.

By mid–May, a new report was issued by the Office of Technology Assessment, an oversight arm of the Congress, which showed that NASA's stated goal of 14 launches in any single year would pose increased risks to astronauts. The report recommended that Shuttles be flown only on flights which required human crews, and that the space agency should schedule a sustainable rate of eight to ten launches a year. It also recommended that the agency seek funding for a fifth orbiter, which would place less stress on the vehicles and add launch capability. Despite the report, a fifth orbiter would never be built nor would a third Shuttle launch facility. With the loss of *Columbia*, the agency would be down to only three Shuttles.

The rest of 1990 was mostly uneventful, unless one counts the cocaine found under a desk in one of the orbiter processing facilities at Kennedy and the launch of STS-35. NASA was not clear about whether the drug was found in a general search or whether someone had reported it. NASA, as would be expected, denied that there was a major drug problem at the Cape. The launch of STS-35 was another matter entirely. *Columbia* had launched on December 2 and for the first time NASA management would label foam debris as a "safety of flight issue."[6] The thermal protection system was taking hits but none had been deadly so far.

The year ended with the 12-member Advisory Committee on the Future of the Space Program recommending the fast development of a replacement for the lost Shuttle, as well as major cutbacks in the flight rates for the orbiters. The White House panel appointed by Vice President Dan Quayle reported that NASA was still depending too much on the manned Shuttle system to place payloads into orbit, something which could be done more safely and cheaply by unmanned rockets. More than a decade and a half later, the space agency still does not have a Shuttle replacement nor anything approved on the drawing boards. A later White House advisory committee would report in 1992 that the cost of the *Challenger* disaster had reached $12 billion[7] which included the cost to replace *Challenger* with *Endeavour*. Once again the program was living up to its reputation as being the most expensive one NASA had ever embarked upon.

During 1990, the space agency was able to launch six Shuttle missions. In 1991, six more flights got off the ground, and in 1992 a total of eight missions were flown. It was a sustainable launch rate, one which NASA could use as a planning base. It was more realistic as budgets began to drop and personnel were taken off the payroll. It was also a period in which there was no "major" problem found in the solid boosters. A leak had been found in a booster in December 1992 (STS-55), due to a one-inch-long human hair being pressed into an O-ring seal.[8] However, the seal leak had been spotted and corrected.

The year 1992 was also when problems would be discovered on the Reinforced Carbon-Carbon (RCC) on the leading edges of the wings. Pinholes were discovered on *Columbia* after STS-50 was flown. Subsequently these penetrations were found on all orbiters. It was soon learned that these small holes were caused by zinc oxide contamination from a primer. It was also discovered that *Atlantis* had suffered what was called a "low-velocity impact" during its STS-45 March-April 1992 mission.[9] The impact on the RCC panel number 10 on the right wing had caused a 1.9 by 1.6-inch gouge on the surface. The impact had been enough to damage the outer and inner surface of the RCC. The event had occurred during launch, but NASA would report "most likely orbital debris." After STS-50, the third known "bipod ramp foam event" would be reported as an "accepted risk" on Hazard Report 37.[10] The forth ramp foam loss was recorded on STS-52,[11] which launched *Columbia* on October 22, 1992. Again the risk was "accepted." NASA's luck was still holding.

Shuttle *Discovery* (STS-56) was launched on April 8, 1993, for an Atmospheric Laboratory mission named ATLAS-2. Eighteen days later, *Columbia* was launched on its Spacelab mission for West Germany. When the solid boosters were inspected after *Discovery*'s launch, a pair of nine-inch long pliers, which had been lost by one of the booster technicians, was found wedged in an outer skirt near the nozzle of the Booster Rocket. One of the Thiokol technicians had noticed his pliers were missing after working on the rocket a week before the launch.[12] When the pliers were found, they still had the safety tether attached. NASA's report does not mention whether or not the technician was fired, but it is extremely unlikely. In a review of the incident, a 45-page study was written which found that workers felt that if they reported errors they would be possibly fired. However, a review of mistakes reported showed that no one had yet been disciplined after making a report. Another disturbing reality of the *Discovery* flight of STS-56 was the very large tile area damaged by debris during launch.[13] Despite the large area, the report filed reported this extensive damage as being "within experience base."

In December, a much more disturbing report would surface showing that NASA had been quietly investigating what were called "alarming variations" in the thrust being produced by the solid rocket boosters. All year long the solids were showing variations which critics had claimed could become even larger and, as such, could cause an orbiter to be torn apart during the first two minutes of its powerful ascent. When the report surfaced, not before, NASA issued a statement which said that the thrust variations were within the limits set by the space agency for this type of situation. However, it was later found that NASA had lowered the safety standards to accommodate the problem so it could continue to fly Shuttles rather than demand a fix by Thiokol. Once again, NASA managers had become comfortable with a problem and rather than fix it had papered over it with a new type of waiver. When questioned about the new lower performance standards, Larry Williams, manager of Shuttle integration, admitted that the space agency had lowered the safety standard from what he called "plus-plus" to "plus." Frank Jordan, an aeronautical engineer working for Aerojet Solid Propulsion Company, was quoted as saying, "It's death-threatening." With all of its new paperwork complete, NASA managers continued to fly with the "plus" boosters. Will a new presidential commission in the future look into this? Only time and Shuttle launches will tell. In August 1993, Rocketdyne received a "poor" rating on the Space Shuttle main engines from NASA.

In the meantime, NASA continued to cut operating costs for the Shuttle. From 1991 to 1994, NASA cut costs by 21 percent, which amounted to a great loss in personnel working directly on the Shuttle program both inside and outside the space agency. "Contractor personnel working on the Shuttle declined from 28,394 to 22,387 in these three years and NASA Shuttle staff decreased from 4,031 to 2,959." From the *Columbia* accident report we further learn that "a 1994-1995 NASA 'Functional Workforce Review' concluded that removing an additional 5,900 people from the NASA and contractor Shuttle workforce — just under 13 percent of the total — could be done without compromising safety. These personnel cuts were made in Fiscal Years 1996 and 1997. By the end of 1997, the NASA Shuttle civilian workforce numbered 2,195, and the contractor workforce 17,281."[12] Too few people were minding the shop and Shuttle safety was bound to suffer.

On June 27, 1995, NASA launched *Atlantis* (STS-71) on its mission to dock with the Russian Space Station Mir. Sixteen days later, *Discovery* (STS-70) would fly on a mission to deploy America's TDRS-G satellite. When the solid rockets were recovered, first from *Atlantis* and then from *Discovery*, it was found that hot gases had singed primary O-rings on both missions' boosters.[15] The problems had occurred in an area around the end nozzle, which directs most of the Shuttle's thrust for the first two critical minutes. A failure in that area could easily cause the Shuttle to violently veer off course and possibly to break up and crash. A week later, the space agency announced that the orbiter fleet would be grounded until NASA could understand how the problem had occurred. The agency stressed that this situation was unrelated to the O-ring problem on the *Challenger*, but once again the solids were causing problems. This time gas bubbles had formed in an insulator which had been put into place to protect the O-rings; the bubbles had formed a path which had allowed hot gases to directly affect the O-rings. Former astronaut Brewster Shaw, then director of Shuttle operations, reported that these bubble paths had been seen on more than ten previous missions, but this was the first time they had caused a problem. The fix would be to develop a method by which the bubbles could be sucked out so that the O-rings would maintain a tight fit. Within two months the fleet was declared operational and orbiters were again flying into space.

It will only be a matter of time before the next solid rocket booster fails. The only fail-safe method to avoid that is to stop using solid rockets, and no one in NASA management is going to call for that. Funding has not been allocated for the change. Funding, or the lack of it, was a prime concern addressed by the *Columbia* accident report. With a flat budget that was not going to fully fund NASA programs, the agency was being forced to take funds from one program to feed another. "The flat budget at NASA particularly affected the human space flight enterprise. During the decade before the *Columbia* accident, NASA rebalanced the share of its budget allocated to human space flight from 48 percent of agency funding in Fiscal Year 1991 to 38 percent in Fiscal Year 1999, with the remainder going mainly to other science and technology efforts. On NASA's fixed budget, that meant the Space Shuttle and the International Space Station were competing for decreasing resources. In addition, at least $650 million of NASA's human space flight budget was used to purchase Russian hardware and services related to U.S.-Russian space cooperation.[16] This initiative was largely driven by the Clinton Administration's foreign policy and national security objectives of supporting the administration of Boris Yeltsin and halting the proliferation of nuclear weapons and the means to deliver them."

The year 1995 also saw a new problem crop up. NASA launched *Columbia* on October 20 for mission STS-73 commanded by Kenneth D. Bowersox. The Spacelab mission went well, but upon return NASA engineers noticed that a gap filler in the belly of the orbiter had pushed out around 1.4 inches. These ceramic coated gap fillers are as thin as a credit card and are placed between thermal tiles. With one of the gap fillers pushed out it caused at least some "noticeable additional heating" downstream of the protrusion. The overheating could be as much as 600 degrees Fahrenheit which could cause problems on reentry. So far NASA had been lucky and no real problems had developed.

Review After Review

On April 3, 1989, NASA's nine-member Aerospace Safety Advisory Panel released a report to NASA Administrator Fletcher, advising him to cancel or vastly cut back funding on the new advanced solid rocket motors (ASRM). The group based its recommendation on the reliability and safety of such a system, which it thought would not be much greater than the presently used solid booster, if at all. The panel, chaired by *Challenger* commission member Joseph Sutter, endorsed a plan to replace the solids entirely with a new liquid booster similar to ones used on the Soviet Shuttle *Buran*. The study indicated that a liquid booster with four engines would perform better, be safer, and could be ready to use within eight years or less. Fletcher accepted the study, but refused to cancel the projected $1.5 billion design program for the ASRM. That money would have gone a long way towards a new liquid-fuel booster design. As long as he was administrator, the Shuttle was going to continue to fly on Thiokol solid rockets. Fletcher had announced his resignation days before, leaving the way open for former astronaut Richard Truly to lead NASA for the next three years, but Truly would not go for the liquids either.

In August 1989, the Congressional Office of Technology Assessment issued its report to Congress on Shuttle reliability. That report found the Shuttle was 98 percent reliable. With those numbers it could be expected that there would be a 50 percent chance of losing a second Shuttle in the next 34 missions and a 72 percent chance before the Space Station was assembled. By the year 2000, there was an almost 95 percent chance of losing a Shuttle, even if it was flown completely within all design limits.[17]

One year after its 1989 report, the Aerospace Safety Advisory Panel again issued a review of Shuttle operations and this time it warned NASA that another accident was "likely to occur" unless systems were redesigned to cut back on unnecessary risks.[18] Further, it asserted that "the Space Shuttle is still very much a research and development activity with significant chances for accidents and failures." Chairman Joseph Sutter reported to new NASA Administrator, Richard Truly, "To conduct the hundred or so flights required to achieve the planned NASA programs including the construction of the space station — without further reducing Shuttle risk — will probably entail the loss of another Space Shuttle." He was right

The panel had reviewed engineering and management areas and had been impressed by many of the changes, but it stressed that the Shuttle should never be considered operational. Areas of greatest concern were long-term maintenance of the now-aging fleet, the Shuttle main engines and of course the solid rocket boosters, specifically the new igniter, new case-to-nozzle joints and "improved" nozzles. The panel also recommended placing stronger windows on the Shuttle, noting that 25 windows had already been hit by orbital debris. The Shuttle's large external tank was also studied, but was not considered to be a major safety problem.

Before the *Columbia* disaster, NASA was well aware that the Reinforced Carbon-Carbon (RCC) panels on the leading wing edges were taking hits. When *Columbia* returned from STS-87 in November 1997, three RCC panels were found to have been damaged by debris impacts large enough to expose carbon substrate.[19] In January 2000, *Discovery* had returned from STS-103 with a damaged RCC panel that caused NASA to replace the unit, and in April 2001, *Columbia* returned from STS-102 with a .2 inch by .3 inch by .018-inch-deep dent in an RCC panel.[20] This panel was repaired and *Columbia* flew one more mission with the fix but the panel would later be scrapped.

In May 1991, NASA reported that it had narrowly averted a possible disaster by removing one of Shuttle *Columbia*'s fuel-temperature sensors that past September. The space agency had flown *Columbia* seven times with the sensor badly cracked all the way around, and it could have easily broken off and flown into the main engine, causing it to explode. This was not the first time this had happened. Dan Germany, Shuttle flow manager, admitted that they had "dodged a bullet on that one; it was just a matter of time."[21] NASA Chief Truly named an investigative panel to discover why it took eight months to locate the problem. He also ordered tests to be conducted on all other sensors. These tests found one more problem sensor on *Columbia*, two on *Atlantis* and two more on *Discovery*.[221] Every Shuttle in the fleet was affected.

Two years later, in January 1993, an internal NASA safety-review panel issued a report on the Shuttle's main engines, referring to them as being extremely temperamental. The report calculated the chance of a catastrophic failure during powered flight at 1 in 120. An earlier report had placed the odds at 1 in 171. That meant that with three engines per flight NASA could expect a major failure to occur once in forty to forty-five flights. These numbers would indicate that Space Shuttles were the most dangerous craft on earth to fly. The panel called for an immediate enhancement of engine reliability and safety. Indeed, we have yet to lose a Shuttle due to engine failure, but there are no guarantees. If an engine fails on the launch pad in an explosive shutdown, it will destroy the Shuttle along with the entire launch pad and the crew will not survive. At that point the Shuttle program will be over and so will the International Space Station which relies almost exclusively on Shuttles for servicing and construction.

In February 2003, it was found that NASA had been flying Shuttles for about a year without proper inspections being completed before launch. This led to several missions being flown with old retainers on the main engines which could have come off during powered flight. If they had broken off during a flight, they could have damaged the turbine blades, causing a catastrophic, explosive shutdown.

In June 1993, a study conducted by the National Research Council reported that the space agency had such a lax program of safety that it was now "infecting" computer programs being developed for the Space Shuttle. The report was prepared by Nancy Leveson, a computer science professor at the University of Washington. It warned that NASA had gone back to a can't-fail attitude and was heading toward a major disaster, unless the agency spent a great deal more time and money on its safety programs. It must be remembered that all of these reports and reviews were conducted years after the *Challenger* was destroyed.

In recent years the space agency budget for Shuttle operations has gone from $4 billion a year to $3 billion. The Shuttle work force has been reduced from 34,000 to 24,000 without a substantial reduction in the workload, with plans in work for 10 percent reductions in the years to come. The Shuttle system is simply not being funded at a safe level. With NASA's flat budget for human flight programs, monies originally devoted to Space Shuttle have time and time again been diverted to the International Space Station and elsewhere:[23]

- Fiscal Year 1997 — $190 million transferred to International Space Station (ISS)
- Fiscal Year 1998 — $65 million transferred to International Space Station (ISS)
- Fiscal Year 1999 — $34.3 million transferred to International Space Station (ISS) and $31 million transferred to other programs
- Fiscal Year 2000 — $15.3 million transferred to International Space Station (ISS) and reduced an additional $11.5 million
- Fiscal Year 2001 — $40 million transferred to Mars initiative
- Fiscal Year 2002 — $7.6 million transferred to fund NASA Headquarters

In 1994, before an audience at the Jet Propulsion Laboratory, NASA Administrator Dan Goldin would state, "When I ask for the budget to be cut, I'm told it's going to impact safety on the Space Shuttle.... I think that's a bunch of crap."[24] Goldin's priorities were not with the Space Shuttle as he had his eyes on the exploration of Mars and the completion of the Space Station. What was lacking was a full understanding that both goals were decidedly tied to the safe operation of the Shuttle.

The bottom line was easy to read. One: that the Space Shuttle will never become the safe, reliable earth-orbital system once envisioned by planners decades ago. Two: that there was no doubt that this nation would suffer the loss of another Space Shuttle. The only questions were which orbiter it would be and which crew would take that long ride down. Those questions were answered on February 1, 2003.

Replacing the Shuttle

After the *Challenger* disaster there was a great deal of discussion about how the space agency could extend the life of the Space Shuttle system, as well as a small voice crying out that extension was not the answer and that replacement was the way to proceed. After the *Columbia* disaster, a sea change occurred. The Shuttle system is now viewed, and rightly so, as an experimental program never able to accomplish the goal of regular, on-call, orbital flight on a cost-effective basis. From the start of the program, NASA propagandized the Shuttle program as a space "truck." That was a deeply flawed vision which has now been clarified by twin disasters. It has become well beyond the time to replace the Shuttle program with a new low-earth orbital vehicle which can support the Space Station and push the U.S. human space-flight program back to the moon and beyond, perhaps to Mars.

The Shuttle program was never fully funded for what NASA was tasked to accomplish. It can only be hoped that the Shuttle replacement will be funded enough to accomplish its goals safely and as certainly as possible. When should we begin the process? The underfunding of the Shuttle has shown that you cannot vacillate between building a new system and then canceling before the new program has a chance to leave the launch pad. A program must be agreed upon, funded and supported.

Before the *Challenger* accident, the National Commission on Space released a report in January 1986 stating that "the Shuttle fleet will become obsolescent by the turn of the century." Due to this report, and not the January 28, 1986, destruction of *Challenger*, the White House announced the approval of a new, second-generation, reusable launch vehicle soon known as the National Aerospace Plane (X-30). It was hoped that the human-piloted craft would be online and operating by early 1990.[25] This proved to be far too ambitious, as technology for this type of vehicle would take many years to develop. After spending $1.7 billion on its development, far below what would have been required to make the X-30 a reality, the program was cancelled in 1992.[26] It is not known whether the X-30 project could have overcome the difficulties involved in developing a whole new technology, but without proper funding the project never really had a chance.

With the Aerospace Plane gone, NASA spent the next two years vacillating between upgrading the existing Shuttle fleet and funding a new effort to replace the aging Shuttle system.[27] In 1994, NASA developed a list of alternatives. First on the list was to upgrade the Shuttle fleet to operate until 2030. Second: develop a new expendable launch vehicle along the lines of the Saturn V which had taken Apollo astronauts to the moon. Third: replace the Space Shuttle with a new, advanced, reusable vehicle that would be cost-effective. By 1996, NASA would seek and find White House approval for the third option which would reflect NASA's history of taking on bold new initiatives. The X-33 and X-34 projects were born.[28]

NASA selected Lockheed Martin to develop the X-33 and the Orbital Sciences Corporation to develop the X-34 as joint ventures with the space agency. NASA hoped that these two proposed vehicles would pave the way for private-sector involvement in the manned space program. It was hoped, as reported in the *Columbia* accident report, that "NASA could replace the Shuttle through private investment, without significant government spending."[29] Once again the space agency was attempting to keep America in the manned space business on the cheap.

These two programs were to incorporate advanced technologies, including a new aerospace engine, composite lightweight materials to replace metal propellant tanks, and new thermal protection systems. After these two test programs had demonstrated the new technologies, what was learned would be used to design and build a fleet of advanced Shuttles (the prototype was called *Venture Star*) to completely replace the Space Shuttles by 2006.[30] However, before the X-33 and X-34 could even get off the ground, they were both cancelled because of technical problems.[31] By 2001, NASA ended funding for these programs after spending $1.3 billion,[32] effectively ending any hope of replacing the aging Shuttle program with the *Venture Star* or any other new craft for at least a decade. By this time NASA had spent several billion dollars which could have been used to upgrade the Shuttle system and complete the Space Station, and had wasted 15 years in the process. They would also cancel plans for the X-38 Crew Return Vehicle[33], which had been designed around the Air Force X-24A Lifting Body, which effectively removed any American rescue vehicle from the Space Station and lowered the station crew to the three who could fit on board the Russian Soyuz.

In 2000, NASA once again changed plans and developed a $4.5 billion, multi-year research program referred to as the Space Launch Initiative, expected to build on the past to develop new space flight technologies and programs. After spending at least $800 million on the initiative, NASA concluded that no new technologies existed that could "revolutionize space launches." So once again the agency shifted direction to develop an orbital space plane which would use existing technology. The new spacecraft was to carry only crew, not cargo, into earth orbit. Any cargo flights would still need to be flown by a Shuttle, but not necessarily one with human pilots. NASA at

National Aero-Space Plane (X-30) (NASA).

least had the ability to remotely fly Shuttles to the Space Station where a crew could already be stationed, having arrived on the new orbital space plane. But this meant that the Shuttle would need to stay in operation until 2020 or beyond. Not many at NASA felt that flying a fleet of 40-year-old Shuttles made much sense, but the options were becoming fewer every year.

In the meantime, NASA flew STS-112 with *Atlantis* on October 7, 2002. It would be the sixth known "bipod foam loss" from the left bipod. It would be the first time a major "debris event" had not been assigned an "In Flight Anomaly" report number.[34] For the most part, NASA management had become very comfortable with debris loss from the external tank and were ready to "close out" the problem just before STS-107. The *Columbia* accident investigation would discover 14 Shuttle flights which had "major Thermal Protection System damage or major foam loss." STS-107 was number 14.[35]

Mission	Date	Data
STS-1	April 12, 1981	Lots of debris damage; 300 tiles replaced
STS-7	June 18, 1983	First known left-bipod-ramp foam-shedding event
STS-27	December 2, 1988	Debris knocks off tile, structural damage, near burn-through
STS-32	January 9, 1990	Second known left-bipod-ramp foam event
STS-35	December 2, 1990	First time NASA calls foam debris "safety of flight issue," and "re-use or turn-around issue"
STS-42	January 22, 1992	First mission after which the next mission (STS-45) launched without debris "In-Flight Anomaly" closure/resolution
STS-45	March 24, 1992	Damage to wing RCC Panel 10-right (Unexplained Anomaly, "most likely orbital debris")
STS-50	June 25, 1992	Third known bipod-ramp foam event (Hazard Report 37: an "accepted risk"

STS-52	October 22, 1992	Undetected bipod-ramp foam loss (fourth bipod event)
STS-56	April 8, 1993	Acreage tile damage (large area); called "within experience base" and considered "in family"
STS-62	March 4, 1994	Undetected bipod-ramp foam loss (fifth bipod event)
STS-87	November 19, 1997	Damage to orbiter thermal protection system spurring NASA to begin nine flight tests to resolve foam-shedding (foam fix ineffective; "In-Flight Anomaly" eventually closed after STS-101 as "accepted risk")
STS-112	October 7, 2002	Sixth known left-bipod-ramp foam loss; first time major debris event not assigned an "In-Flight Anomaly" (external tank project was assigned an "Action"; not closed out until after STS-113 and STS-107)
STS-107	January 16, 2003	*Columbia* launch; seventh known left-bipod-ramp foam-loss event (loss of crew during reentry)

Only after NASA had flown *Discovery* on STS-114, the first flight after the *Columbia* disaster, was it made clear that the once-again grounded Shuttle fleet would need to be retired as soon as possible. For all intents and purposes, the Shuttle would have outlived its usefulness as soon as the International Space Station was completed.

This Shuttle thing will cost us this year ten cents a share.
— Charles Locke, Chairman, Morton Thiokol

Chapter Thirteen

The Fall of Space Shuttle *Columbia*

Columbia, Houston. We see your tire pressure messages, and did not copy your last.

Columbia's Dark Passage

On January 28, 2003, the seven-member crew flying *Columbia*'s final mission came together on the Shuttle's lower deck to honor and remember the final crew of *Challenger*, who had been killed during their launch seventeen years earlier, with a moment of silence.[1] They could not know that a similar fate awaited them less than four days later as they attempted the always-risky reentry into the earth's atmosphere. And even though manned U.S. spacecraft have performed that demanding task with success for 42 years, professional astronauts and others who work for our space agency know that the reentry must be performed with great skill every single time and that success is never guaranteed no matter how skilled or well-trained the crews are.

When Mercury astronaut Gordon Cooper was asked how he felt when he learned about the latest Shuttle loss he reflected on the closeness of the astronauts. "When we were up there we tried to think of everything we needed to do. It's like losing members of your family."

Fifteen minutes before a landing would have successfully ended the *Columbia* mission at the Kennedy Space Center in Florida, people everywhere were once again dramatically reminded of the human cost of exploring the wonders and mysteries of space. On February 1, 2003, we all heard the name of the Shuttle and learned the names of the seven crew members who died during *Columbia*'s dark passage. Before this flight very few knew their names.

As before, a president addressed the nation and the world to officially recognize the loss, and to again state that once the problem was found (and it would be found), Americans would once again ride powerful rockets into the unknown.[2] What President Bush failed to say was that once again a dangerously under-funded and under-inspected Space Shuttle program was being operated, while the people in charge knew that unacceptable risks were being taken. In a very real sense, the problems that helped bring down *Challenger* had continued and the result was the loss of *Columbia*. Once again "can do" became "can't fail." These internal and external problems have yet to be completely solved as of this writing.

Forty percent of the Shuttle fleet is now gone. The fourteen astronauts lost means one crew member death, on average, every eight flights. History shows that NASA launched six Mercury, ten Gemini, eleven Apollo, three Skylab and one ASAP manned mission without losing a single astronaut in flight. Is the Shuttle program as safe as we are told or can it be made safer? NASA

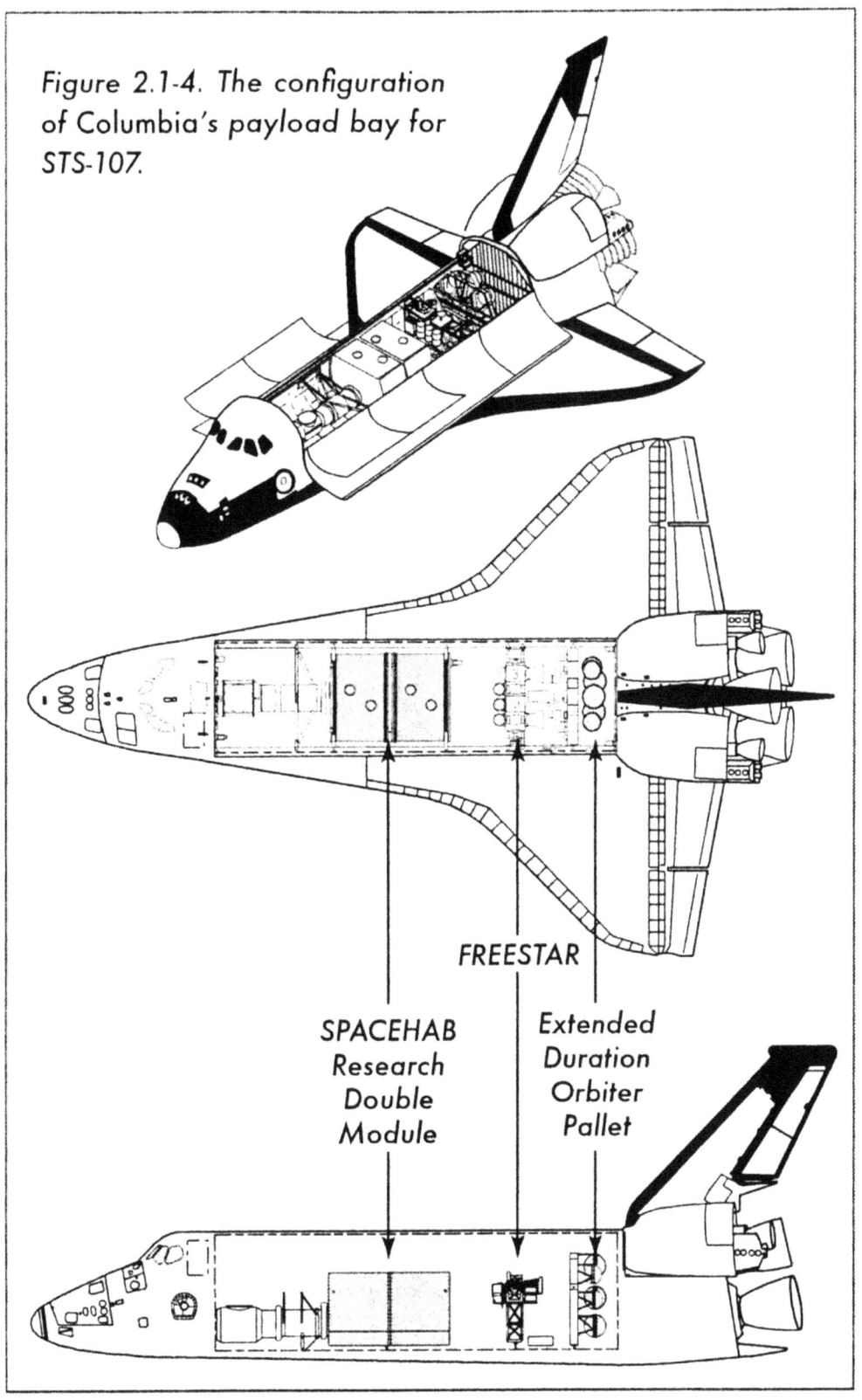

Columbia's payload bay configuration for mission STS-107 (graphics from *Columbia* Accident report, page 31).

had launched Space Shuttles 113 times over a period of twenty-two years losing two spacecraft during these flights. This shows a per-launch safety/reliability record of 98.2 percent,[3] which we are told is well above any other system used to launch crews or satellites into space. Is that good enough for America's manned space program, or could we be doing much better? If not, then the American people need to be prepared to accept the probability that we can expect to lose at least one more Shuttle before NASA closes out our earth orbital Shuttle program. We can expect to lose one more Shuttle crew in order to complete the International Space Station.

The Final Mission of *Columbia*

Under the tightest security ever developed for a human-crewed space flight due to concerns about terrorism, the *Columbia* with her seven-member crew left launch pad 39A at the Kennedy Launch Center on January 16, 2003, for a planned 16-day science mission designated STS-107, featuring more than eighty experiments from many different nations. On board were Shuttle Commander Rick Husband, a 45-year-old air force colonel on his second Shuttle flight; Pilot Commander William C. McCool, a 41-year-old navy pilot on his first flight; 43-year-old air force lieutenant colonel Michael P. Anderson, a mission specialist on his second flight; Dr. Laurel Clark, a 41-year-old navy commander on her first flight, also serving as a mission specialist; as well as David Brown, a 46-year-old navy captain. Filling out the crew was 41-year-old Dr. Kalpana Chawla, a mission specialist born in India, flying her second mission, and 48-year-old Israeli Air Force colonel Ilan Ramon, on his first flight to oversee some of the experiments.

As the mission got underway, fighter jets and combat helicopters patrolled the skies as warships scanned the ocean off of the Cape, enforcing a 35-mile perimeter around the launch pad.[4] With the full recognition that any Shuttle carrying an Israeli astronaut would be a tempting target for terrorists, the government was ready to bring down even a small single-engine aircraft with F-15 fighters if it became necessary to protect the *Columbia* and her crew. Ilan Ramon had earlier said, "I feel very safe here. I mean there's no real risk here. Being a pilot in the air force is a risk position."[5]

As it turned out, the security measures did the job very well, as there were no attempts on the Shuttle; however, the launch itself would prove to be more of a problem than anyone outside of NASA realized at the time. Eighty-one seconds into the flight, a 20-by-16-inch section of insulation broke away from the huge external tank, striking the Shuttle's left wing along the heat-resistant carbon panels and then disintegrating along with at least two other insulation pieces. The impact punched a hole in the leading edge of the wing. The impact may have been severe enough to have damaged the under-metal of the Shuttle's aluminum skin. It was a fatal hit.

Early reports would interpret the problem as "a shower of debris"[6] and point to the possibility of three separate pieces coming into contact with sections of the Shuttle's underbelly or wing at nearly the same time. The heat-resistant Carbon-Carbon panel absorbed the impact and that information would not be discovered until the launch videos were reviewed the next day. No one on the Shuttle or on the ground suspected any problems with the launch at the time. It was later calculated that an area as large as 32 inches long and 7 inches wide could have been damaged by the impact. The impact theory was later boosted by the recovery of a Shuttle data recorder, which had been left on *Columbia* since her first test flights in the early 1980s. That recorder, which had pressure and temperature measurements from at least 420 sensors, would show that *Columbia* was "mortally wounded" during launch. The spokesman for the Accident Board would later say, "This is the one we really wanted to get our hands on."[7] No other Shuttle had this data recorder on it.

On January 22, meetings were held to determine what, if any, damage could have been caused by the launch incident. Two scenarios were viewed as possible results of an impact, first that a

Shuttle *Columbia* on the launch pad before STS-107 mission (NASA).

Launch of *Columbia* on STS-107, January 16, 2003 (NASA).

wide section of the left wing's leading edge had been damaged or tiles knocked off or secondly that the impact had knocked off at least one six-inch-square tile from an area on or around the left landing-gear door. Either result could be life threatening to the crew if the damage was substantial, and no one on the ground knew whether that was the case. Later, investigators would also focus on the possibility that pinholes which had formed on the leading-edge panels, caused by zinc oxide which had dripped off of the launch towers during rainstorms, could have contributed to the Shuttle's loss. The corrosive material could have caused from thirty to forty such holes, weakening the leading edges and making them more susceptible to impact damage as they aged. (Most were more than twenty years old.)

By January 27, mission management teams and NASA engineers met and came to the conclusion that the effect of the impact "pose[d] no threat to the safety of the Shuttle or crew."[8] This conclusion was made without ever taking a close look at the area on the Shuttle which had possibly been hit. When NASA Administrator Sean O'Keefe was later asked about the possibility of taking a photo of the area to help in the decision-making process, he would state, "I'm not ever engaging in shoulda, coulda, woulda. We'll never end up settling this debate."[9] But he soon added that NASA would be taking such photos on future missions. Debate over. It was later reported that an unnamed "federal intelligence agency" had agreed to do the work with one of its spy satellites, upon request by NASA.

Upper NASA managers never made that simple request; in fact, they blocked it. The U.S. Air Force Strategic Command and the National Reconnaissance Office who operate the advanced KH-11 spy spacecraft began to prepare to use the system to image the underbelly of *Columbia*. NASA's N. Wayne Hale, Jr., launch integration manager at Kennedy, had contacted the air force, but the request was halted by upper management who thought that no "safety-of-flight problems" could be identified.[10]

On February 1, without being informed of any concerns about the reentry, but armed with information about possible impact during launch, the *Columbia* and her seven-member crew reentered the atmosphere. At 200,000+ feet, the Shuttle disintegrated, killing all crew members. There is a possibility that the crew knew they were doomed for as long as one terrifying minute. Investigators would later learn that "an attempt may have been made to override *Columbia*'s autopilot in the final seconds of its doomed flight."[11] Although a NASA official would refer to the information as "really suspect," one or both of the two pilots may have tried to take control of the Shuttle before all was lost.

Investigators would later piece together a nearly complete video of the breakup, thanks to the many amateur space photographers who got up early to record the reentry using hand-held video cameras. NASA engineer Douglas White later reported, "To our great surprise, people are still very interested in the space program. These folks got up before sunrise and went out on their own and stuck their camera up in the sky. Most of them knew where to look in the sky because they are amateur astronomers. Without these folks, we wouldn't know any of this [the sequence of events]. These people are definitely our heroes."[12] Portions of some of these videos were broadcast all day long to a shocked nation.

Columbia's Final Reentry*

8:10:00 A.M.	Crew is given a "go" for de-orbit burn.
8:15:30 A.M.	Pilot McCool executes the de-orbit burn on the 225th orbit for 2 minutes, 38 seconds. The vehicle was then maneuvered by Commander Husband into a right-side-up, forward-facing position — nose pitched up.
8:44:09 A.M. (EI+000)	Entry interface at 400,000 feet, arbitrarily defined as the altitude when the orbiter interacts with the discernable atmosphere.
8:48:39 A.M. (EI+270)	A sensor on the left-wing leading-edge spar shows strains higher than seen on previous *Columbia* reentries. Data stored, later recovered but not transmitted.
8:49:32 A.M. (EI+323)	At Mach 24.5, *Columbia* executes a right roll to begin a banking turn to control lift, rate of descent and heating.

Columbia's final re-entry timeline is taken from live NASA Select video, supplemented by the *Columbia* Accident report and reports by *Aviation Week & Space Technology*.

> -----Original Message-----
> **From:** STICH, J. S. (STEVE) (JSC-DA8) (NASA)
> **Sent:** Thursday, January 23, 2003 11:13 PM
> **To:** CDR; PLT
> **Cc:** BECK, KELLY B. (JSC-DA8) (NASA); ENGELAUF, PHILIP L. (JSC-DA8) (NASA); CAIN, LEROY E. (JSC-DA8) (NASA); HANLEY, JEFFREY M. (JEFF) (JSC-DA8) (NASA); AUSTIN, BRYAN P. (JSC-DA8) (NASA)
> **Subject:** INFO: Possible PAO Event Question
>
> Rick and Willie,
>
> You guys are doing a fantastic job staying on the timeline and accomplishing great science. Keep up the good work and let us know if there is anything that we can do better from an MCC/POCC standpoint.
>
> There is one item that I would like to make you aware of for the upcoming PAO event on Blue FD 10 and for future PAO events later in the mission. This item is not even worth mentioning other than wanting to make sure that you are not surprised by it in a question from a reporter.
>
> During ascent at approximately 80 seconds, photo analysis shows that some debris from the area of the -Y ET Bipod Attach Point came loose and subsequently impacted the orbiter left wing, in the area of transition from Chine to Main Wing, creating a shower of smaller particles. The impact appears to be totally on the lower surface and no particles are seen to traverse over the upper surface of the wing. Experts have reviewed the high speed photography and there is no concern for RCC or tile damage. We have seen this same phenomenon on several other flights and there is absolutely no concern for entry.
>
> That is all for now. It's a pleasure working with you every day.
>
> [MCC/POCC=Mission Control Center/Payload Operations Control Center, PAO=Public Affairs Officer, FD 10=Flight Day Ten, -Y=left, ET=External Tank]

E-mail informing the *Columbia* crew of debris hit, January 23, 2003 (document from *Columbia* Accident report, page 159).

8:49:53 A.M. (EI+344)	Temperature sensor on left Orbital Maneuvering system pod shows unusually low temperature. It would later show temperatures as high as 1,200 degrees Fahrenheit around twice the normal level.
8:50:53 A.M. (EI+404)	At Mach 24.1 and approximately 243,000 feet, *Columbia* enters 10-minute period of maximum heating and thermal stress on the vehicle. Penetration of the left wing by 2000-degree plasma begins. Crew and ground are unaware of any problems.
8:51:19 A.M. (EI+430)	Over Pacific Ocean, Mach 24; "remote sensors indicate off-nominal external event — earliest known event."
8:52:00 A.M. (EI+471)	300 miles west of the California coastline, *Columbia*'s wing leading-edge temperature is expected to reach 2,650 degrees Fahrenheit Main landing gear brake-line temperatures show unusual rise in left side. Shuttle is at Mach 23.5 and 236,791 feet over the Pacific.
8:52:05 A.M. (EI+476)	First computer indication of clearly "off-nominal aero increments" in yaw.
8:53:01 A.M. (EL+532)	First off-nominal roll movements as Shuttle responds to unexpected drag.
8:53:26 A.M. (EL+557)	At Mach 23, *Columbia* crosses the coast of California at 231,600 feet. Leading-wing-edge temperature is now typically 2,800 degrees Fahrenheit or more.
8:53:46 A.M. (EL+577)	Crossing California, *Columbia* appears as a bright spot of light to ground based observers. Signs of debris breaking or melting off were sighted. The *Columbia* suddenly brightens, causing a noticeable streak in the Orbiter's luminescent trail. For the next 23 seconds, four such events are witnessed by ground observers as part of the orbiter begins to break off and burn up. Hot gasses of reentry have now entered *Columbia*'s left wing and are eroding the interior of the Shuttle.

Flight	STS-7	STS-32R	STS-50	STS-52	STS-62	STS-112	STS-107
ET #	06	25	45	55	62	115	93
ET Type	SWT	LWT	LWT	LWT	LWT	SLWT	LWT
Orbiter	Challenger	Columbia	Columbia	Columbia	Columbia	Atlantis	Columbia
Inclination	28.45 deg	28.45 deg	28.45 deg	28.45 deg	39.0 deg	51.6 deg	39.0 deg
Launch Date	06/18/83	01/09/90	06/25/92	10/22/92	03/04/94	10/07/02	01/16/03
Launch Time (Local)	07:33:00 AM EDT	07:35:00 AM EST	12:12:23 PM EDT	1:09:39 PM EDT	08:53:00 AM EST	3:46:00 PM EDT	10:39:00 AM EDT

External tank bipod ramp foam losses (document from *Columbia* Accident report, page 123).

Where the left wing intersects the fuselage on the left side shows unusual temperature increase. Temperatures are now increasing inside the wing as the aluminum structure is turned to vapor and essentially becomes a fuel raising the temperature of the entire left wing.

For the next four minutes ground observers would report seeing 18 similar brightening events as parts of the Shuttle continue to break off while it flies over Utah, Arizona, New Mexico and Texas.

8:54:00 A.M. (EL+591) — Fourth brake line on left-side landing gear rises about 30–40 degrees Fahrenheit.

8:54:24 A.M. (EL+613) — In Mission Control, the reentry appeared normal until this time. Readings in four hydraulic sensors in the left wing were showing "off-scale low." Flight director (Flight) and the maintenance, mechanical and crew systems officer (MMACS) discuss the readings.

8:54:25 A.M. (EI+614) — *Columbia* crosses into Nevada airspace and a bright flash is seen by ground observers. Orbiter is now at 227,400 feet and traveling at Mach 22.5.

MMACS: Flight, MMACS.

Flight: Go ahead, MMACS.

MMACS: FYI, I've just lost four separate temperature transducers on the left side of the vehicle, hydraulic return temperatures. Two of them on system one and one in each of systems two and three.

Flight: Four [hydraulic] return temps?

MMACS: To the left outboard and left inboard elevon.

Flight: Okay, is there anything common to them? DSC (discrete signal conditioner) or MDM (multiplexer-demultiplexer) or anything? I mean, you're telling me you lost them all at exactly the same time?

MMACS: No, not exactly. They were within probably four or five seconds of each other.

Flight: Okay, where are those, where is that instrumentation located?

MMACS: All four of them are located in the aft part of the left wing, right in front of the elevons, elevon actuators. And there is no commonality.

Flight: No commonality.

8:54:33 A.M. (EI+622) — First "flash event" recorded as orbiter gas envelope suddenly brightens.

8:55:00 A.M. (EL+651) — Leading-wing-edge temperatures normally reach almost 3,000 degrees Fahrenheit. Fifth left main-gear temperature sensor indicates abnormal rise in temperature.

8:55:30 A.M. (EL+681) — Remote sensors indicate "abnormal external event."

8:55:32 A.M. (EL+683) — *Columbia* crosses Nevada/Utah border at Mach 21.8 at 223,400 feet.

8:55:53 A.M. (EL+703)	*Columbia* crosses border of Arizona into Utah.
8:55:55 A.M. (EL+705)	Mach 21 and 222,000 feet as witnesses observe the thirteenth debris-shedding episode.
8:55:57 A.M. (EL+707)	Very bright debris event departs *Columbia*.
8:56:02 A.M. (EL+713)	Flight: MMACS, tell me again which systems they're for. MMACS: That's all three hydraulic systems. It's ... two of them are to the left outboard elevon and two of them to the left inboard. Flight: Okay, I got you.
8:56:20 A.M. (EL+731)	The Flight Director continues to discuss the sensor readings with otherMission Control personnel. (GNC [Guidance, Navigation and Control Officer]) Flight: GNC, Flight. GNC: Flight, GNC.
8:56:30 A.M. (EL+741)	Flight: Everything look good to you, control and rates and everything is nominal, right? GNC: Controls been stable through the rolls that we've done so far, Flight. We have good trims. I don't see anything out of the ordinary.
8:56:45 A.M. (EL+756)	Flight: Okay. And MMACS, Flight. MMACS: Flight, MMACS. Flight: All other indications for your hydraulic system indications are good? MMACS: They're all good. We've had good quantities all the way across.
8:57:00 A.M. (EL+771)	Lower left-wing temperature sensors fail. *Columbia* is rolling to the left about 75 degrees to dissipate energy (pre-programmed maneuver).
8:57:24 A.M. (EI+795)	Flight: And other temps are normal? MMACS: The other temps are normal, yes, sir. Flight: And when you say you lost these, are you saying that they went to zero? Or, off-scale low? MMACS: All four of them are off-scale low. And they were all staggered. They were, like I said, within several seconds of each other.
8:57:59 A.M. (EI+830)	Flight: Okay.
8:58:03 A.M. (EI+834)	Leading-edge temperatures now normally down to 2,880 degrees Fahrenheit. *Columbia* elevons move "sharply" to adjust orbiter-roll axis trim as a response to increased drag on the left side of the vehicle. Shuttle now flying at Mach 20.2. (Dittemore: "That could be indicative of rough tile or missing tile but we are not sure.")
8:58:20 A.M. (EI+851)	*Columbia* now traveling at Mach 19.5 at 219,000 feet as it crosses from New Mexico into Texas airspace. At this point *Columbia* loses a thermal protection system tile which would become the most westerly piece of debris located (found in Littlefield, Texas). The tiles are beginning to peel away.
8:59:00 A.M. (EI+891)	Flight-control computers command the elevons to roll *Columbia* to the right, as left side drag pulls the vehicle to the left. Then 870-pound-thrust right yaw thrusters fire for 1.5 seconds in an attempt to maintain proper flight path. *Columbia* is fighting hard to stay on path, with sharp movements of the elevons and two right firing yaw jets. One second later the orbiter would sweep elevons in the largest motions recorded in an attempt to keep the orbiter on path.
8:59:10 A.M. (EI+901)	Commander Husband: And, uh, Hou....
8:59:15 A.M. (EI+906)	MMACS: Flight, MMACS. Flight: Go. MMACS: We just lost tire pressure on the left outboard and left inboard, both tires.
8:59:32 A.M. (EI+923)	Commander Husband: Roger. (Last verbal message received from *Columbia*.) Static indicates a possible open microphone for a moment, but no crew transmissions are heard. Hydraulic fluid pressure is zero in left wing. No signals from left wing sensors. Auxiliary power units continue to operate. Cockpit master alarm sounds.

8:59:32 A.M. (EI+923)	Flight director tells the capsule communicator (CAPCOM) to inform the crew about the data loss and relay that Houston had not understood the crew's last transmission. *Columbia* now traveling at Mach 18.1 at 200,700 feet approaching Dallas, Texas. At this point Mission Control receives its final real-time data from *Columbia* which was accepted by Mission Control computers. Tracking at Mission Control stops over Texas. (Milt Heflin, Chief of the Flight Directors Office: "I and others stared at that for a long time because the tracking ended over Texas. It just stopped. It was then that I reflected back on what I saw [in Mission Control] with *Challenger*.") CAPCOM: And *Columbia*, Houston, we see your tire pressure messages and we did not copy your last call. Flight: Is it instrumentation, MMACS? Gotta be...? MMACS: Flight, MMACS, those are also off-scale low. INCO (Instrumentation and Communication Officer): Flight, INCO. Flight: Go. INCO: Just taking a few hits here. We're right up on top of the tail. Not too bad.
8:59:36 A.M. (EI+927)	Auto pilot commands "drop left wing" in an attempt to fight increasing and overwhelming drag and vehicle movements off pre-programmed track.
9:00:03 A.M. (EI+954)	Data indicated a possible rotational hand controller activation. Possible attempt by the pilots to go to manual control. Vehicle vibration alone at this point would have indicated to the crew that something was going wrong with the reentry.
9:00:18 A.M. (EI+969)	Amateur videos taken from various locations on the ground reveal that the *Columbia* was starting to disintegrate. Best guess is that the *Columbia* is now in a flat counterclockwise spin. The crew attempting a last-ditch effort for control.
9:00:21 A.M. (EI+972)	Main body breakup of the spacecraft begins.

Bright flashes are visible (and later a dramatic change in the vapor trail seen on the videos). The main body of Shuttle *Columbia* is breaking up as the left wing fails and is ripped off at around Mach 18 and a little over 200,000 feet. Tail and right wing are ripped off. Shuttle main body tumbles and is torn apart by reentry forces as the payload is torn loose and destroyed. Reinforced crew cabin continues downrange and is probably spinning wildly, incapacitating the crew. It is probable that the crew survived for at least the next 30 seconds before their reinforced cabin failed, killing all seven instantly. The *Columbia* Accident Board would report, "It appears that the destruction of the crew module took place over a period of 24 seconds." Mission Control in Houston is unaware that they are losing *Columbia* over north-central Texas, as thousands on the ground are able to observe the breakup.

	Flight: MMACS, Flight. MMACS: Flight, MMACS. Flight: And there's no commonality between all these tire pressure instrumentations and the hydraulic return instrumentations. MMACS: No, sir, there's not. We've also lost the nose gear down talkback and the right main gear down talkback. Flight: Nose gear and right main gear down talkbacks? MMACS: Yes, sir. INCO: Flight, INCO, I didn't expect, uh, this bad of a hit on comm[unications]. Flight: GC [Ground Control Officer], how far are we from UHF [Ultra High Frequency]? Is that two-minute clock good? GC: Affirmative, Flight.
9:00:50 A.M. (EI+1001)	**Shuttle *Columbia* and her crew of seven have been destroyed.** GNC: Flight, GNC. Flight: Go. GNC: If we have any reason to suspect any sort of controllability issue, I would keep the control cards handy on page 4-dash-13.

9:02:21 A.M. (EI+1092)   Mission Control: Fourteen minutes to touchdown for *Columbia* at the Kennedy Space Center. Flight controllers are continuing to stand by to regain communications with the spacecraft.
Flight: INCO, we were rolling left last data we had and you were expecting a little bit of ratty comm, but not this long?
INCO: That's correct, Flight. I expected it to be a little intermittent. And this is pretty solid right here.
Flight: No onboard system config[uration] changes right before we lost data?
INCO: That is correct, Flight. All looked good.
Flight: Still on string two [secondary avionics link] and everything looked good?
INCO: String two looking good.
GC: Flight, GC.
Flight: Go.
GC: Flight, two minutes to MILA [Merrit Island Launch Area].
Flight: Okay.

With that call, the ground control officer had informed the flight director that on a normal reentry the *Columbia* would be able to acquire Kennedy Space Center ground station for direct communication.

CAPCOM: *Columbia*, Houston. Comm check.
CAPCOM: *Columbia*, Houston. UHF comm check.
9:03:45 A.M.   Mission Control: CAPCOM Charlie Hobaugh calling *Columbia* on a UHF frequency as it approaches the Merritt Island tracking station in Florida. Twelve and a half minutes to touchdown, according to clocks in Mission Control.
MMACS: Flight, MMACS.
Flight: MMACS.
MMACS: On the tire pressures, we did see them go erratic for a little bit before they went away, so I do believe it's instrumentation.
Flight: Okay.

No one at Mission Control knows that the Shuttle has been lost. All controllers knew for sure was that they were in a period of extended loss of signal and that data had been very spotty. The only option was to attempt to regain communication with the already doomed craft.

CAPCOM: *Columbia*, Houston. UHF comm check.
CAPCOM: *Columbia*, Houston. UHF comm check.
GC: Flight, GC.
Flight: Go.
GC: MILA not reporting any RF [Radio Frequency] at this time.
INCO: Flight, INCO, SPC [Stored Program Command] just should have taken us to STDN [Space Tracking and Data Network] now.
Flight: Okay.
Flight: FDO [Flight Dynamics Officer], when are you expecting tracking?
FDO: One minute ago, Flight.
GC: And Flight, GC, no C-band yet.
Flight: Copy.
CAPCOM: *Columbia*, Houston. UHF comm check.
INCO: Flight, INCO.
Flight: Go.
INCO: I could swap strings in the blind.
Flight: Okay, command us over.
INCO: In work, Flight.
09:08:25 A.M. (EI+1456)   INCO: Flight, INCO, I've commanded string one in the blind. [That signal commanded *Columbia*'s S-band communications to a back-up system.]
GC: And Flight, GC.

Flight: Go.

GC: MILA's taking one of their antennas off into a search mode [to try to find *Columbia*].

Flight: Copy. FDO, Flight.

FDO: Go ahead, Flight.

Flight: Did we get, have we gotten any tracking data?

FDO: We got a blip of tracking data, it was a bad data point, Flight. We do not believe that was the Orbiter. We're entering a search pattern with our C-bands at this time. We do not have any valid data at this time.

Flight: Okay. Any other trackers that we can go to?

FDO: Let me start talking, Flight, to my navigator.

Flight: Okay.

9:12:39 A.M. (EI+1710) A member of Mission Control receives a phone call on his cell phone stating that TV coverage of the Shuttle landing showed that the orbiter had disintegrated during its reentry attempt.

Flight: GC, Flight. GC, Flight.

GC: Flight, GC.

Flight: Lock the doors.

A Space Shuttle contingency was then declared by Mission Control at Houston.

Could NASA have saved the crew of *Columbia* if they had known about the massive damage done to the Shuttle's left wing? The short answer is maybe so. The *Columbia* report would say yes.[13] If the space agency had known within the first two days of orbital operations, they would have shut down all unnecessary spacecraft systems and done as much as possible to stretch out supplies, possibly as long as three weeks. It is really unknown how fast the people on the ground could prepare a second Shuttle for a rescue, but it does not take a genius to know that they would have used every trick in the book and invented a few new ones to get a bird up as fast as possible. (Shuttle *Atlantis* was already stacked and in the Vehicle Assembly Building.) "Given the history of this agency, there is positively nothing that would have been spared in our efforts to try to find out what to do to avoid catastrophe,"[14] NASA Administrator Sean O'Keefe said.

Recovery

With live coverage of *Columbia*'s descent and destruction, coupled with the fact that the Shuttle came apart over land, it was not surprising to learn that pieces of the lost craft and crew remains would soon be found. Within minutes, hundreds of pieces were located in small towns, yards, fields and country roads. In fact, many pieces were spotted in the air and visually tracked by people close to where they hit the ground. The news reports of human remains soon followed, as most networks broke into regular programming to cover the unfolding story of another lost Shuttle.

On February 2, the *Fort Worth Star-Telegram*, whose coverage was typical, reported on some of the early discoveries. "Human remains and bits of uniform were discovered with other Shuttle debris scattered across dozens of sites in East Texas on Saturday. Three brothers on all-terrain vehicles found a leg in the family's pasture along Farm Road 2024 near Hemphill close to the Louisiana border.[15] Just a few miles away, a hospital worker found a charred torso,[16] thighbone and skull with front teeth intact in the middle of Farm Road 2971. In Nacogdoches County, a shoe that appeared to be an astronaut's was found, and there were several unconfirmed reports of human remains."[17] "Resident Bob White said the leg [found in his pasture] might have been a woman's because of its small size and small foot. The leg, intact from the hip down, was scorched except for one strip."[18]

First reports indicated that human remains and Shuttle debris were scattered across some fifty or more sites in east Texas near the Louisiana border. Reports also stated that the heat of reentry, which blackened many of the remains, was not surprising. What *was* surprising was how much material, including a mission patch and an almost intact crew member's helmet, came through relatively undamaged.[19] When questioned about the finding of remains, Sabine County (Texas) Sheriff Tom Maddox reported, "Most sites contain debris, only a small number have human remains."

The day after the disaster, news reports had counted ten sites where human remains had been found, most in relation to at least some other Shuttle debris. The remains were soon confirmed to have been from the lost crew, but NASA had not been able to determine whose remains had been found at that early stage of the recovery operation. Amber Welch, who found a torso near her home, would recall how NASA employees reacted during the recovery of their fellow astronauts. "There were three astronauts that came out here. You could tell they were pretty upset, and that they had been close to this person. They weren't crying, but they were very sad. You could see it in their faces. I felt bad for those poor NASA people. This was family to them." To make matters worse, the searchers knew that they were up against a pretty grim deadline. They needed to find the remains before any wild animals did.

By February 2, enough crew member remains had been found for the recovery teams to place them into two caskets, which were then taken to Barksdale Air Force Base in Louisiana. When they arrived, the caskets, one covered with an American flag and the other with an Israeli flag, were met by an honor guard. They would later be sent to Dover Air Force Base in Delaware, where they would be identified through DNA and dental records, as well as footprints and fingerprints.[20] The medical examiners would also be looking for clues to determine exactly how they had died.

Evelyn Husband, wife of Shuttle Commander Rick Husband, released a statement that day. "Although we grieve deeply, as do the families of Apollo 1 and *Challenger* before us, the bold adventure of space must go on."[21] As the statement was being released it was announced that local residents were placing simple crosses with flowers at the sites where remains had been found. It was also announced that local churches would leave their lights on day and night in honor of the fallen crew.

By February 3, director of flight crew operations Bob Cabana would report that "all crew members' remains have been located," but that statement was later recalled. However, by the next day the remains of Ilan Ramon had been found "in the wreckage of Shuttle *Columbia*." Brigadier General Rani Falk, assigned to Washington, D.C., as Israeli Air Force attaché, stated, "NASA informed us officially that Ilan Ramon, may his name be blessed, was identified, and we can bring him for burial in Israel in the coming days."[22] Before long, medical experts at Dover Air Force Base had identified remains of all seven crew members, as plans were made to bury the final crew of *Columbia*. "We're confident that we have seven sets of remains," came the word from NASA official Bob Jacobs.[23] It was now time to locate and recover as much of *Columbia* as possible and move on to find out exactly what had brought down a second Shuttle. That task would fall to the Space Shuttle Mishap Interagency Investigation Board, turned into the *Columbia* Accident Investigation Board (CAIB) soon after the disaster. Board members included:

Retired admiral Harold W. Gehman, Jr. — Chair
Rear Admiral Stephen Turcotte — Commander of the Navy Safety Center at Norfolk, Virginia
Roger E. Tetrault — Retired president of General Dynamics
Brigadier General Duane Deal — Commander of the 21st Space Wing at Peterson Air Force Base, Ohio
Major General John L. Barry — Director of Plans and Programs for USAF Material Command
Major General Kenneth W. Hess — Commander, Air Force Safety Center, Kirtland Air Force Base, New Mexico
Scott Hubbard — Director of NASA's Ames Research Center, Moffett Field, California

Steven B. Wallace — Director of Accident Investigations, Federal Aviation Administration
Sheila Widnall — MIT professor and former Secretary of the Air Force
James Hallock — Transportation Department Aviation Safety Division chief, Cambridge, Massachusetts

Members added later included:

Dr. Sally K. Ride — Former astronaut and former member of the *Challenger* Commission
Douglas Osheroff — Professor of physics at Stanford University
John Logsdon — Director of the Space Policy Institute at George Washington University

As part of their report, the members commented, "Based on NASA's history of ignoring external recommendations, or making improvements that atrophy with time, the Board has no confidence that the Space Shuttle can be safely operated for more than a few years based solely on renewed post-accident vigilance."[24]

Spare Parts Again

As with the *Challenger* disaster, many areas would be closely examined to understand what event or events transpired to bring down *Columbia*. Investigators also sought to find out what was learned and acted upon between these twin disastrous flights. The spare parts problem, which became one focus of Congressional inquiry after *Challenger*, would again come under close scrutiny. Cannibalization (a term which is hated by NASA managers), the taking of parts from one Shuttle in order to fly another, is not something envisioned by early mission planners, but the practice continues to this day despite criticism of NASA from many sources.[25] To make matters worse, the space agency knows that with an aging fleet flying well past projected usage periods, manufacturers for replacement parts continue to fall by the wayside. Despite public denials and private assurances, Shuttle managers know that there is a real possibility that one of the three remaining Shuttles could very well end up a "hangar queen,"[26] used only for spare parts in an ever decreasing flight schedule. At some point in the near future, the Shuttle game could be over, as parts become impossible to find or repair.

Reports soon surfaced about engineers at United Space Alliance, the Houston-based contractor (a joint venture between Lockheed Martin and Boeing) responsible for the Shuttle fleet processing. It was said that the engineers had searched as recently as May 2002 on Internet sites such as Yahoo and EBay for aging computer parts, as well as used Shuttle hardware.[27] In fact, there were more than procurement problems with United Space Alliance, as seen in the audit conducted by NASA in 2001. In that report, Johnson Space Center came under criticism because it "failed to keep adequate watch on safety operations [and] United Space Alliance's safety procedures." This was not the first time however, that safety has been an issue with United Space Alliance.

In 2000, the 13-member Space Shuttle Independent Assessment Team reported, "In spite of NASA's clear mandate on the priority of safety, the nature of the contractual relationship promotes conflicting goals for the contractor (e.g., cost vs. safety). NASA must minimize such conflicts." "We haven't been able to communicate fully that safety is the one determinant for financial awards," stated NASA's chief of the Space Flight Office, Joseph Rothenberg. Yet that is exactly the job that is expected of the Johnson Space Center.

When the members of the *Challenger* Commission were told that NASA routinely traded parts from one Shuttle to another in order to keep flying their tight and underbudgeted schedule, they were shocked. At the same time, members of Congress were dumbfounded to learn that despite continued funding for spare parts NASA never seemed to spend the money. It should also be noted that many of the original manufacturers who had participated for years in the design and manufacturing of Shuttle replacement parts are long gone. Companies have bottom lines, and if the space agency does not purchase enough of their products they must move on to other areas if they wish to stay in business. One need only look at the 13 contractors who produced Shuttle tiles at the peak of the program in the early 1980s to understand the depth of the problem. After more than two decades, the tile operations have moved away from California to the Kennedy Space Flight Center in Florida and are now being produced by a single contractor. And for the most part the job involves the repair of damaged tiles after the Shuttles return to earth, not making new ones, which cost on average $2000 to $3000 each.

The problem can only get worse, as fewer companies participate in the spare parts program for maintaining the three remaining Shuttle's, *Discovery, Atlantis* and *Endeavour*. NASA will soon come to the conclusion that its Shuttle fleet is too small and too old to get the job done, and will need to look to a new method of placing crews into orbit in order to have a truly viable human space flight program. If the program gets down to only two operational orbiters, the game could very well be over.

The Shuttle Tiles

One problem which always faces designers and engineers who work on spacecraft is the problem of weight. The more a spacecraft weighs, the more fuel it needs to carry (also weight) and the more power it needs to get into space. Also, the less a spacecraft weighs per lift capability, the greater the payload it can place into earth orbit. Facing that problem, NASA engineers devised a very lightweight thermal protection system of some 24,000 ceramics tiles to protect the Shuttles when they reentered the earth's atmosphere. They were a major technical achievement yet they have been very subject to debris impact damage, a concern which has never left the minds of Shuttle managers or the crews who fly them.

One of the most often quoted concerns was the so-called "zipper effect" involving the loss of a single tile in a critical area which then exposes nearby tiles to super-heated gases of reentry, pulling off many more tiles. At the same time the missing tiles would create a very turbulent airflow across the wing's surface, which would increase the drag at that point when entering the atmosphere. This increased drag would tend to pull the Shuttle to one side in a yaw maneuver, which could cause the Shuttle to spin out of control and disintegrate.

Very early into the investigation of the *Columbia* accident, it became clear that some kind of unusual event had caused a breach in the tile system, evident by temperature increases relayed to Mission Control during reentry. It would take a long time to pinpoint the cause of the breach, but foam insulation breaking away from the external tank during launch was the prime suspect, even though the possibility of impact by a micrometeorite was investigated and discounted.

Even if the crew of *Columbia* had known about damage to the protective tiles, was there anything they could have done in orbit? The short answer is no. The crew not only did not have any means to repair tile damage, they had no way of getting to the area of possible damage. There were no EVA (Extra Vehicular Activity) space walk suits on board the Shuttle and no Canada arm to place a space-walking astronaut close enough to do any repairs. The robotic arm was not required for this mission, so it was not brought along. There is also the possibility that the arm would not have been able to bring anyone close enough to repair any damage even if it was on board. The only way to save the crew would have been to send up a second Shuttle to rendezvous with the crippled spacecraft and somehow transfer the crew from one Shuttle to the other. That has never been attempted before and time would not have been on the side of NASA. The real question would be: could NASA launch another Shuttle in time to reach the *Columbia* before its air, water and other critical expendables ran out?

Certainly NASA managers will look for some new method to repair possible tile damage in the future, but it would not be the first time that issue has been addressed. Even before the space agency launched its first Shuttle mission in April 1981, NASA developed a tile repair kit designed to be used by a space walking astronaut. The system called for stowing on board the Shuttle 162 pre-formed tiles of many shapes and sizes. Depending on the amount of damage and where the damage had occurred, an astronaut could select a replacement tile based on size, shape, density and color. However, former astronaut Charles Bolden, who had worked on the on-orbit tile repair project,[28] stated that the space agency "could not find fillers and adhesives that would cure in a vacuum." At that point NASA dropped its work to develop an on-orbit repair kit, and for twenty years nothing has been done to solve the problem.

In hindsight it seems to be a very good idea to have on-board repair capabilities and the space agency is now giving the repair kit another close look, announcing on February 24, 2003, that "NASA will study whether it can give astronauts the ability to inspect and repair heat-resistant tiles while the space shuttle is in orbit." This represented a major shift in policy, but will it end with a workable solution to the tile problem? Perhaps NASA need not look at replacing a tile with another tile. The problem is to repair the damaged area with a protective covering that will last for only one reentry. That covering could be made of much stronger stuff, such as the carbon/carbon material used on high-heat areas and simply bolted into place through the damaged area right onto the Shuttle's aluminum skin. It does not have to be pretty; it only has to be tough and last one time. Any needed corrections to the on-orbit repair area could then be done on the ground where there would be plenty of time to remove the patch and replace it with tiles before the Shuttle was ready to fly again.

At the same time, NASA announced that for the time being, Shuttles launched would be sent exclusively to the space station. This would allow the station's 58-foot long robot arm the opportunity to examine the Shuttle for damage and, if needed, a space-walking astronaut could make repairs from the station. This method would also give the astronauts the added safety of staying on board the station if the docked Shuttle was too heavily damaged to be repaired. The damaged Shuttle would then be undocked for a reentry without a crew to whatever fate awaited, while her original crew would wait for a rescue Shuttle to launch and bring them home.

Columbia's External Tank

Two and a half years before *Columbia*'s final flight, John Ehlers, a product assurance engineer at Lockheed Martin's Michoud Assembly Facility, had warned Michoud management that there were grave problems with how the epoxy primer was being applied to the massive external tanks. In fact, one of the tanks he was concerned about was used on mission STS-107 — the final *Columbia* mission. "If the primer is not given sufficient time to dry, moisture can form and its seal can be weakened." "In any case it wasn't dry. Obviously, there was a problem." Ehlers was laid off in 2002 and did not press the issue further outside of the company until after the loss of *Columbia*.

For more than twenty years, NASA managers struggled with the problem of insulation debris breaking off of the external tank during launch, and it was well documented that the insulation could easily damage the Shuttle thermal tiles. It was called "debonding" and no fewer than five studies (done in 1994, 1996, 1997, 1999 and most recently 2001[29]) reported on the problem. On August 8, 2005, *Newsweek* would report, "In what now seems like a premonition, a NASA engineer reported in 1997 that the newly formulated "environmentally friendly" foam could easily pop off the tank and damage the Shuttle's heat-resistant tiles. The newly formulated foam did not directly affect the lost bipod section which was the cause of *Columbia*'s loss, but it continues to be a flight-safety issue.

In the 1999 report, Christopher K. Davis, who works as an engineer at the Kennedy Space

Center, reported on "the major damage flying insulation had done to the Shuttle body."[30] "Debonding [insulation] causes most of the damage to the orbiter belly tile and exposes ET [external tank] to the point of thermal loading." The safety panel in 1999 also highlighted the loss of skilled people working on the Shuttle program. As reported in the *Los Angeles Times* on February 5, 1999, "[NASA's] independent safety panel warned [that] the shortage of trained technicians at the National Aeronautics and Space Administration's manned space flight centers 'can jeopardize otherwise safe operations.'" Yet NASA managers discounted that concern. "William Readdy, head of NASA's space shuttle program, acknowledged Thursday that the agency has not been able to hire the new engineers and scientists it would like, but said that budget cuts and hiring freezes have not eroded NASA's commitment to safety," the article continued. This from an agency that had lost two Shuttles—and 14 lives.

In the 2001 report, Davis wrote of continuing problems which he felt were not being given serious consideration within NASA. Using data from the Johnson Space Center, Kennedy Space Center and Marshall Space Flight Center (which was responsible for the external tank), Davis concluded that the insulation "has a history of debonding, sometimes striking the orbital tile and causing damage."[31] This data coincided with information gathered from United Space Alliance, the company responsible for Shuttle maintenance. NASA managers knew or suspected that insulation had come off the external tank on every single flight. That is a 100 percent failure of design and yet Shuttles were allowed to fly.

These reports also focused on why the spray-on foam insulation could separate from the metal used to construct the huge tanks. Dirt and the method used to spray the tanks were a concern, as was the fact that the external tank would slightly contract when super-cooled, cryogenic liquid hydrogen was loaded. The liquid caused the aluminum tank to shrink, which could pull small areas away from the skin, something that could not be seen during visible inspections. Given the knowledge that this debonding could not be seen, NASA did not use a known method of detection called shearography which had been recommended in several studies. While the studies had recommended using laser shearography to detect flaws on the external tank, it had yet to be approved. It would also be very difficult to test the debonding theory on the external tank used on *Columbia*'s final mission as it is now resting, probably in several pieces, at the bottom of the ocean.

When NASA switched to a different type of foam insulation which was "developed for environmental reasons," the new insulation created more problems than it supposedly solved. The 2001 NASA report on debonding reported "greater damage" to Shuttle tiles after NASA went to the new "environmentally correct" material. A similar problem had occurred after NASA went to more environmentally correct putty on the Shuttle's O-Rings. The new putty did not do its job very well, as heat continued to reach O-Rings, and the result of that was deadly.

The *Columbia* report would focus on management failures in this area as well. "Foam loss may have occurred on all missions, and left bipod ramp foam loss occurred on 10 percent of the flights for which visible evidence exists. The Board had a hard time understanding how, after the bitter lessons of *Challenger*, NASA could have failed to identify a similar trend."[32]

The E-Mails

As the Shuttle *Columbia* went about the business of discovery in low earth orbit, several engineers on the ground at NASA's Langley Research Center in Hampton, Virginia, were debating the possible effects of foam insulation hitting the *Columbia* during launch. It became clear to several of the engineers that a good deal of damage could have been done and they were not fully convinced that the Shuttle could make a safe reentry and landing. It was also clear that these engineers were running into a problem getting top NASA officials to pay much attention to their concerns. One engineer, Robert L. Daugherty, would eventually complain that his concerns were "being treated like the plague."[33] He would also write that NASA was "unwilling to test their theories" and that engineers might have to work on their own time at night to do the tests.[34]

Another engineer e-mailed that he did not believe the published statements that a piece of insulation was the potential problem, and reported that the most likely object was a solid piece of ice. The launch videos did indeed show a bright white object making contact with the Shuttle and disintegrating. If this were correct, the engineer wrote, "the impact of a large chunk of ice would be equivalent to a 500-pound safe hitting the wing at 365 miles per hour."

These engineers, reported as "12 mid-level NASA engineers," were for the most part concerned with the dynamics of the landing gear, which is where many of them felt the impact had done most of its damage. If they were correct, one engineer stated, flight controllers "would sure as hell want to know whether they should land gear up, try to deploy the gear or go bailout."[35] "We can't imagine why getting information is being treated like the plague."[36] NASA Administrator Sean O'Keefe would later describe some of the e-mails as "pleading."[37] Looking at the need for a close examination of possible damage to the spacecraft, Alan R. Rocha from the Johnson Space Center and chief engineer for the Space Shuttle's Structural Engineering Division wrote, "In my humble technical opinion, this is the wrong (and bordering on irresponsible) answer. Not to request additional imaging help from any outside source. Any damage to the protective tiles could lead to heating and damage to the underlying structure of the Shuttle, particularly near critical spots like the main landing gear door. That could present potentially grave hazards."[38] "Remember the NASA safety posters everywhere around site stating, 'If it's not safe, say so'? Yes, it's that serious."[39] It was an e-mail that he never sent, but he did forward a copy to his upper management. The bottom line was that photos should have been taken and they were not, and that was a deadly error.

On flight day 8, January 23, Mission Control sent an e-mail to Commander Husband and Pilot McCool showing that debris from the external tank had struck the orbiter's left wing during ascent. They also informed the crew that there was "no concern for RCC or tile damage" and that Mission Control had "absolutely no concern for entry."

After the destruction of the Shuttle, NASA official John Petty would state that the engineers were "playing devil's advocate." "These are engineers ... looking at worst-case scenarios. They are postulating. That's what these guys get paid for."[40] What he did not state was why this particular "postulating" did not reach NASA managers responsible for bringing the Shuttle in for a safe landing. Only later would senior engineer Robert L. Daugherty report that his e-mails were "misinterpreted" and that he did not feel that the Shuttle crew was in any "real" danger. "I had been absorbed in what-ifing all week, so of course there was some natural uneasiness on my part. I certainly believed that everything was going to be perfectly fine."[41] Yet he would also write, "Apparently the thermal folks have used words like 'survivable' [and] 'marginal.'"[42] Events would prove that the reentry was not survivable that day.

The day before the landing attempt, NASA engineer William C. Anderson would note that the discussion was a bit late in the process. "Why are we talking about this on the day before landing?"[43]

Columbia's Long History of Problems

When you go into space you risk your life.
— NASA Administrator Dan Goldin

After a test series of only four space flights flown by *Columbia*, President Reagan declared that the Space Shuttle program was "operational." Yet, the reality was that Space Shuttle *Columbia* was a test model. It was the first manned spacecraft designed and built to be reusable in the harsh environment of earth orbit, and would never move beyond being a test bed. It was perhaps best described by space analyst Howard McCurdy of American University as having plenty of problems needing to be worked out. "I don't remember any astronaut saying it was a dog, but the reality is that it wasn't meant to be operational. It was a test model. It still had a lot of bugs to work out." And from the very beginning, technical problems would crop up with *Columbia* even before its first launch. As the first reusable manned spacecraft, *Columbia* was the craft on which many new systems were first tried, including the reusable tile system. The tiles would prove to be a major problem, as 200 permanent and some 4,800 temporary tiles were lost as *Columbia* was being transported across the country to Cape Canaveral. When it finally began operations, more than 70 percent of its launches were delayed by technical problems. During its 26 missions it recorded the worst launch-delay record in the Shuttle fleet.

The problems did not end once *Columbia* reached orbit, as on-orbit systems failed at an alarming rate. One of the most problem-plagued flights was launched on January 9, 1990, as mission STS-32 with a five-member crew. That flight would be required to delay its return due to a failure in one of its main flight computers. The mission also experienced several false alarms in its smoke-detector system, as well as an under-reported incident in which the Shuttle tumbled out of control for twenty minutes while the crew was asleep.[44] The orbiter had been sent an incorrect navigation command by the ground. Controllers at Mission Control in Houston woke the crew and informed Shuttle Commander Daniel Brandenstein of the problem. In short order Brandenstein and pilot James Wetherbee were able to manually fly the orbiter into its proper orientation. As for the smoke-alarm problem, it should be noted that NASA managers had documented a "concern" with the fire-alarm system before launch. "MLP-3 Fire Alarm Signal Circuit. Potential exists for failure of MLP signal circuits without indication of failure. Bells may not ring when alarm activates." There was also a problem with the dehumidifier when two gallons of water leaked into the cabin. Finally, a defect on one of the Shuttle main engines was discovered *after* the mission successfully landed.

In 1983, a major disaster was narrowly averted when an auxiliary power unit (APU) began to leak highly volatile hydrazine fuel. The leak occurred while *Columbia* was in orbit for mission STS-9, commanded by John W. Young, and was not detected until well after *Columbia* came to a halt on the Shuttle runway. As it came to a stop, smoke could be seen coming out of the rear of the spacecraft and heat waves were easy to see on live NASA video. When safety inspectors examined the rear of the spacecraft they were shocked to find it charred and blackened.[45] Later study would show that the hydrazine had caught fire when enough oxygen in the earth's atmosphere allowed it to ignite. There were also two main Shuttle computer failures on that flight. "Young had radioed that two of *Columbia*'s five computers had stopped working after an unexplained jolt hit the ship as a thruster jet in its nose fired."[46] Although one of the two failed computers came back on during reentry, it failed again just as the *Columbia* was landing. It would be the only mission *Columbia* would fly in 1983 and it nearly ended in disaster.

On what was to be *Columbia*'s only flight in 1985, the Shuttle was moved to launch pad 39A for a scheduled December 20 launch. However, delays would keep the Shuttle on the pad for 25 days as *Challenger* was bolted to launch pad 39B, marking the first time that two Shuttles had been on launch pads at the same time. Delays in *Columbia*'s launch involved an electronic malfunction only 14 seconds before launch, as well as a halt to the countdown on January 6, 1986, when engine sensors indicated that the Shuttle main engines were too cold.[47] The problem was soon traced to liquid-oxygen tanks which had been mistakenly drained of thousands of pounds of cryogenic oxygen. If launched, the Shuttle would not have had enough to reach orbit.[48] The Shuttle could easily have crashed when it had no place to land. When the Shuttle finally launched on January 12, it began to leak helium. Commander Robert "Hoot" Gibson would recall, "We had one of the more interesting ascents. We had a helium tank leaking, and in the midst of working that problem, we had failures of several maneuvering thrusters." Only days after *Columbia*'s troubled flight, NASA would launch *Challenger* on her doomed mission.

On July 23, 1999, *Columbia* (STS-93) was launched with a crew of five to deploy the Chandra X-Ray Telescope. By all accounts, mostly postflight, *Columbia* was in no condition to fly that mission, and the crew was very lucky to have made it back on the aging Shuttle. During the launch, one of the Shuttle's main engines leaked explosive hydrogen fuel "all the way into orbit."[49] The day before, controllers had stopped the launch due to a leak, yet they failed to scrub the mission. Flight director Ralph Roe stated, "We are convinced this is not a real leak and we have no serious concern with launch."[50] Because the engine did not receive enough fuel it consumed liquid oxygen at a much higher rate, which depleted the oxygen, which in turn caused all three engines to shut down prematurely. This shutdown placed the Shuttle into a lower than planned orbit and could have caused the mission to be aborted before it began orbiting. Additionally, damaged wiring, which had not been spotted, short-circuited during the launch, shutting down several engine-control modules on two of the main engines. Upon return, Shuttle managers grounded *Columbia* for a major overall. The possible destruction of *Columbia* on that flight was much closer than anyone at NASA will admit, even today.

Inspectors showed that *Columbia* had a great many more wiring irregularities than the other Shuttles and there was a good deal of frayed wiring insulation in many areas of the ship. *Columbia* was sent back to Rockwell's Palmdale facility for a major overhaul, and over 3000 defects were discovered in the over 200 miles of wiring that made up the Shuttle's electrical system.

The first operational Shuttle was indeed a test bed, which may have simply flown past its operational abilities. It was the first, which meant it had much to teach those who would build future Shuttles. The newer models would be made stronger, lighter and more adaptable to the tasks ahead. Even with three major overhauls, *Columbia* was clearly showing her age, even before her final mission.

An Issue of Safety Once Again

A *Los Angeles Times* report of February 9, 2003, summed up the situation in one sentence. "Under relentless pressure to cut costs, NASA has scrapped hundreds of millions of dollars in safety upgrades for the Space Shuttle program, canceled efforts to modernize equipment and laid off thousands of workers performing maintenance or safety inspections." In February 1999 it had been reported, "At NASA's Kennedy Space Center, the Shuttle work force has been cut by 50% in the past five years, down to 3,800 contractors and 600 civil servants." John Pike, a long time space-industry observer, would state at the time, "NASA has been cutting into muscle and bone. They were taking a calculated risk they could reduce overall staff and cut the number of safety and quality assurance inspectors, without appreciably increasing the risk of another *Challenger* accident."

For four years after the *Challenger* disaster, spending on the Shuttle program increased, topping out at $5.5 billion in 1990. Much of the new funding paid for upgrades for the fleet and construction of the replacement Shuttle, *Endeavour*. However, the budget fell for the most part during the next twelve years, to $3.3 billion in 2002. This was not enough to fund a safe Shuttle program. As John Pike stated after the *Columbia* accident, "It is not like this is completely unexpected. You can't do this on the cheap. There was a hallucination that reduced spending would actually improve safety."[51] And even if funding was not a direct cause of the *Columbia* disaster, the program needs to be properly funded or not flown at all.

After the loss of *Challenger*, NASA managers proposed a major new program to develop a safer set of solid rocket boosters. That program was funded for $4 billion and was called the advanced solid rocket motor (ASRM). The idea was to develop a rocket with greater lift capability with less cost and fewer joints that could fail as they did on the final *Challenger* flight. However, halfway through the program the research was ended when Congress voted 401–30 to kill funding in 1993. The program was behind schedule, but that is not unusual in big-budget space research; it was not the pork-barrel situation reported in the newspapers.

With the loss of funding, NASA looked to other possible solutions, such as developing a new liquid-fuel rocket not unlike the ones the agency had used for years on earlier manned space projects including the powerful Saturn V rocket which had sent astronauts to the moon. Unlike the solid rocket boosters which provide 80 percent of the initial lift for Shuttle launches, liquid boosters can be throttled up or down and can be turned off. This would allow a measure of safety not available with solids if an abort were required during the initial phases of the launch. However, with NASA budgets continuing to fall and the Shuttle continuing to launch "safely," Congress was in no mood to fund any new programs, especially one expected to cost upwards of $5 billion.

The space agency also requested funding to replace its aging Shuttle computers, which were designed and built in the 1970s. As late as 2002 the problem was reviewed, but without proper funding NASA managers cancelled the upgrades. Most people would be astonished to learn that the home computers they use every day have upwards of 2000 times the capacity of the obsolete ones presently used to launch Shuttles. In fact, the computers are so old that not only is the company which designed and built them out of the computer business, the space agency has trouble locating people who understand the language used, so that needed repairs can be made. The bottom line is that the more NASA can put on its launch computers the more effective the Shuttle is at doing its job, and that relates to being able to better respond to critical situations. Better computers relate directly to a safer Shuttle during launch, orbit, and landing, keeping an eye on the systems.

In 2000, NASA also took a hard look at replacing the aging auxiliary power units. Its *Space Shuttle Program Annual Report* states, "The power units are considered [by NASA] the Shuttle system most likely to cause a safety problem." Indeed, these power units use highly explosive hydrazine. Hydrazine catches fire when it comes into contact with air, and *Columbia* caught fire as it came in for a landing at Edwards Air Force Base in 1983. The system is also dangerous to work with on the ground, and a fire broke out in 1999 as maintenance crews were working on the system. The fuel had leaked out and ignited, and had to be put out by the three people working on the system. Those same APUs are still being used, three at a time, on each remaining Shuttle in the fleet. So far they have not failed catastrophically, but NASA needs to address the problem now before the fleet is brought down to two Shuttles. For the APUs, it is only a matter of time.

Slowly but surely, NASA began to back away from Shuttle safety assessments. After the *Challenger* Commission had done its work, the space agency published an estimate that another *Challenger*-like disaster was likely in 1 of 78 flights using the Shuttle system. However, in 1996, with several years of success, and funding flat at around $3 billion, NASA upped the estimate to 1 in 248, and by 2001 had published a new level of probability: 1 Shuttle loss in 483 flights. At an average of ten flights a year, what the space agency was trying to tell the American people was that it could fly the Shuttle for twenty more years and not lose another Shuttle, since their estimate covered almost fifty years. After the *Columbia* disaster and after the *Discovery* flew STS-114 NASA would revise that estimate to 1 in 220 flights. The reality of Shuttle flights has shown a loss rate of 1 in 56.5 flights, and with an aging fleet that rate will not improve, and could very well become even more deadly. The Shuttle is simply not a safe system to fly and never will be.

Finally, NASA was instructed to increase the maintenance and safety staff responsible for the Shuttle, after the loss of *Challenger*. However, as Shuttles continued to fly successfully, NASA once again began cutting back. As the budget fell in the early 1990s, the maintenance staff responsible for the Shuttles was cut from 4000 people to around 1700. Numbers of safety personnel directly responsible for checking the work done on Shuttles have also been cut, and this could easily mean disaster to a space flight system which is the most complicated ever developed. In fact, with a "can't fail" culture within the halls of NASA, the space agency can expect to lose the next Shuttle within the next 25 flights, most probably on or near the launch pad, as a main engine or an APU fails catastrophically.

The Decision-Making Process

In 1986, Robert B. Hotz, a member of the general advisory committee to the U.S. Arms Control and Disarmament Agency, found himself on the *Challenger* Commission investigating the loss of that Space Shuttle along with twelve other distinguished members. In February 2003, the retired editor of *Aviation Week and Space Technology* was asked about the space agency he and others had investigated in 1986 after the *Columbia* accident. "There is a lot of déjà vu here. They knew they had a problem, but they lived with it. It's an old issue in flight that if you get away with it once, and you get away with it twice, it can come back and bite you."[52]

What Hotz was referring to was the heat-resistant yet very fragile tile system which protects the Space Shuttles when they reenter the atmosphere and many times have been damaged during launch. Time after time tiles had been hit by debris falling off the external tank as well as by ice, yet each time the spacecraft was able to return from space safely. In fact, there has never been a Shuttle flight which has come back to earth after a mission without at least some damage to its tiles. On the 113th mission, NASA's luck ran out and the fragile tile system bit back.

NASA managers had come to expect some type of damage to the tiles on each flight, which gave them false confidence for a safe return, even as more damage began to turn up with the new bonding material being used. Yet the tiles were never designed to take such hits, and any major damage to any critical area could have brought disaster, which is exactly what happened to *Columbia*. NASA relied on the fact that on average, over the twenty-two years Shuttles had been flying, 100 tiles per flight would have impact damage and the bird returned home. What managers did not know was how extensive the damage must have been on *Columbia*'s final mission. The bottom line, when all was said and done, is that some very bright people in our space program made an educated guess—and it turned out to be wrong. Yet there still seems to be a sense in NASA that they did not miss anything. NASA manager Michael Kostelnik stated "I am very comfortable that this team did as well as they could in trying to understand what the problem was. I believe we made the right decision at that time." How could anyone lose a Shuttle and the lives of seven astronauts and still believe they made a "right decision"? For now, at least, it seems that NASA needs to take another close look at its decision-making process if it is to continue launching astronauts into space using an aging Shuttle fleet.

From the *Columbia* report we learn: "First, despite all the post–*Challenger* changes at NASA and the agency's notable achievements since, the causes of the institutional failure responsible for *Challenger* have not been fixed. Second, the Board strongly believes that if these persistent, systemic flaws are not resolved, the scene is set for another accident."

Planning Ahead

Even before the recovery teams had left the field, and well before the *Columbia* Accident Investigation Board had made any recommendations, NASA in mid–March of 2003 was making contingency plans to launch Shuttles on a full schedule as early as six months in the future. That is exactly what happened soon after the *Challenger* disaster in 1986 when NASA began giving out contracts for new booster rockets before the president's commission told NASA managers to stop and wait for the results of their investigation. In 2003, NASA Administrator Sean O'Keefe announced, "We're not just sitting here waiting for a report. [Our aim] is to get ourselves ready and prepared to move ahead." Even some members of Congress were up-front about NASA pushing ahead. Representative Bart Gordon of Tennessee stated, "I don't think it's ever too early to make plans, but it is very, very optimistic to think there can be a return to flight this fall."[53] This time, however, no one was telling the space agency to stop and take a hard look before moving on, because it was the space agency, for the most part, investigating itself.

Clearly, top NASA management were pushing to move into a launch cycle even faster than the 32 months it took to launch after the *Challenger* was lost, but was this the best way to go? NASA did not yet know if the Shuttle system could even be made operational in the long term again, let alone be ready for missions in six months. There is the inescapable reality of the International Space Station, which is manned and is expected to stay that way for many years. NASA knows that the only system now available to continue its construction and maintenance is the Space Shuttle. It was soon announced that a crew of one American and one Russian would launch near the end of April 2003 from Russia in a Soyuz spacecraft to replace the crew of three on the space station. Yuri Malenchenko and Edward Lu would simply be onboard to maintain the station for future use. However, if the station is to continue to grow and operate above simply being maintained, then the Shuttle must fly—and relatively soon. There is no other short-term solution, because NASA placed all of its eggs in one Shuttle basket.

The *Columbia* Accident Investigation Board went beyond the accident itself, focusing on the so-called "culture at NASA." The board wanted to see if the agency had allowed a faulty or dangerous Shuttle system to continue to operate beyond what was reasonable, based on what in the old days would have been called "go fever." In short, did NASA believe its own propaganda—that it could not fail? Did managers launch when they knew or suspected a major failure could occur?

There was also the possibility that some type of criminal charge could be filed. The only comment on that matter came from Nancy Herrera, a staffer for U.S. Attorney Michael T. Shelby: "The Department of Justice does not confirm or deny the existence of an investigation, nor do we provide opinions about the possibility of criminal charges being filed."[54]

What about moving on to the next great adventure? After the *Columbia* disaster, former moon-walking astronaut Alan L. Bean was asked why we stopped after the moon and what he and the other astronauts thought about the future at the time. "When we did our moon flights, we all thought that the program would keep on, and then we would start construction of a moon base and space stations. At that time in our culture's history, we were doing the most that was possible to be done. We naively assumed that's what would continue, but it didn't. It's the normal thing for a culture, in history, that we respond to emergencies."[55] Only time will tell how this generation responds to this Shuttle disaster and whether we will choose to continue going into space or leave that to others. On August 26, 2003, the *Columbia* Accident Investigation Board released their report.

From the very time I was 4 years old, when the Mercury program first got started, I was in front of the TV for every one of the launches.
—*Columbia*'s last Commander, Colonel Rick D. Husband

Recommendations by the *Columbia* Accident Investigation Board

It is the Board's opinion that good leadership can direct a culture to adapt to new realities. NASA's culture must change, and the Board intends the following recommendations to be steps towards effecting this change.

Part One—The Accident

Thermal Protection System

1. Initiate an aggressive program to eliminate all External Tank Thermal Protection System debris-

shedding at the source with particular emphasis on the region where the bipod struts attach to the external tank. (Return-to-flight critical)

2. Initiate a program designed to increase the Orbiter's ability to sustain minor debris damage by measures such as improved impact-resistant Reinforced Carbon-Carbon and acreage tiles. This program should determine the actual impact resistance of current materials and the effect of likely debris strikes. (Return-to-flight critical)

3. Develop and implement a comprehensive inspection plan to determine the structural integrity of all Reinforced Carbon-Carbon system components. This inspection plan should take advantage of advanced non-destructive inspection technology. (Return-to-flight critical)

4. For missions to the International Space Station, develop a practicable capability to inspect and effect emergency repairs to the widest possible range of damage to the Thermal Protection System, including both tile and Reinforced Carbon-Carbon, taking advantage of the additional capabilities available when near to or docked at the International Space Station.

5. For non–Station missions, develop a comprehensive autonomous (independent of Station) inspection and repair capability to cover the widest possible range of damage scenarios.

6. Accomplish an on-orbit Thermal Protection System inspection, using appropriate assets and capabilities, early in all missions.

7. The ultimate objective should be a fully autonomous capability for all missions to address the possibility that an International Space Station mission fails to achieve the correct orbit, fails to dock successfully, or is damaged during or after undocking. (Return-to-flight critical)

8. To the extent possible, increase the Orbiter's ability to successfully re-enter Earth's atmosphere with minor leading edge structural sub-system damage.

9. In order to understand the true material characteristics of Reinforced Carbon-Carbon components, develop a comprehensive database of flown Reinforced Carbon-Carbon material characteristics by destructive testing and evaluation.

10. Improve the maintenance of launch pad structures to minimize the leaching of zinc primer onto Reinforced Carbon-Carbon components.

11. Obtain sufficient spare Reinforced Carbon-Carbon panel assemblies and associated support components to ensure that decisions on Reinforced Carbon-Carbon maintenance are made on the basis of component specifications, free of external pressures relating to schedules, costs, or other considerations.

12. Develop, validate and maintain physics-based computer models to evaluate Thermal Protection System damage from debris impacts. These tools should provide realistic and timely estimates of any impact damage from possible debris from any source that may ultimately impact the Orbiter. Establish impact damage thresholds that trigger responsive corrective action, such as on-orbit inspection and repair, when indicated.

Imaging

1. Upgrade the imaging system to be capable of providing a minimum of three useful views of the Space Shuttle from liftoff to at least Solid Rocket Booster separation, along any expected ascent azimuth. The operational status of these assets should be included in the Launch Commit Criteria for future launches. Consider using ships or aircraft to provide additional views of the Shuttle during ascent. (Return-to-flight critical)

2. Provide a capability to obtain and downlink high-resolution images of the external tank after it separates. (Return-to-flight critical)

3. Provide a capability to obtain and downlink high-resolution images of the underside of the Orbiter wing leading edge and forward section of both wings' Thermal Protection System. (Return-to-flight critical)

4. Modify the Memorandum of Agreement with the National Imagery and Mapping Agency to make the imaging of each Shuttle flight while on orbit a standard requirement. (Return-to-flight critical)

Orbiter Sensor Data

1. The Modular Auxiliary Data System instrumentation and sensor suite on each Orbiter should be maintained and updated to include current sensor and data acquisition technologies.

2. The Modular Auxiliary Data System should be redesigned to include engineering performance and vehicle health information, and have the ability to be reconfigured during flight in order to allow certain data to be recorded, telemetered, or both as needs change.

Wiring
As part of the Shuttle Service Life Extension Program and potential 40-year service life, develop a state-of-the-art means to inspect all Orbiter wiring, including that which is inaccessible.

Bolt Catchers
Test and qualify the flight hardware bolt catchers. (Return-to-flight critical)

Closeouts
Require that at least two employees attend all final closeouts and intertank area hand-spraying procedures. (Return-to-flight critical)

Micrometeoroid and Orbital Debris
Require the Space Shuttle to be operated with the same degree of safety for micrometeoroid and orbital debris as the degree of safety calculated for the International Space Station. Change the micrometeoroid and orbital debris safety criteria from guidelines to requirements.

Foreign Object Debris
Kennedy Space Center Quality Assurance and United Space Alliance must return to the straightforward, industry-standard definition of "Foreign Object Debris" and eliminate any alternate or statistically deceptive definitions like "processing debris." (Return-to-flight critical)

Part Two—Why The Accident Occurred

Scheduling
Adopt and maintain a Shuttle flight schedule that is consistent with available resources. Although schedule deadlines are an important management tool, those deadlines must be regularly evaluated to ensure that any additional risk incurred to meet the schedule is recognized, understood, and acceptable. (Return-to-flight critical)

Training
Implement an expanded training program in which the Mission Management Team faces potential crew and vehicle safety contingencies beyond launch and ascent. These contingencies should involve potential loss of Shuttle or crew, contain numerous uncertainties and unknowns, and require the Mission Management Team to assemble and interact with support organizations across NASA/Contractor lines and in various locations. (Return-to-flight critical)

Organization
1. Establish an independent Technical Engineering Authority that is responsible for technical requirements and all waivers to them, and will build a disciplined, systematic approach to identifying, analyzing, and controlling hazards throughout the life cycle of the Shuttle System. The independent technical authority does the following as a minimum:
• Develop and maintain technical standards for all Space Shuttle Program projects and elements.
• Be the sole waiver-granting authority for all technical standards.
• Conduct trend and risk analysis at the subsystem, system, and enterprise levels.
• Own the failure mode, effects analysis and hazard reporting systems.
• Conduct integrated hazard analysis.
• Decide what is and is not an anomalous event.
• Independently verify launch readiness.
• Approve the provisions of the recertification program called for in this report.
2. The Technical Engineering Authority should be funded directly from NASA Headquarters, and should have no connection to or responsibility for schedule or program cost.
3. NASA Headquarters Office of Safety and Mission Assurance should have direct line authority over the entire Space Shuttle Program safety organization and should be independently resourced.
4. Reorganize the Space Shuttle Integration Office to make it capable of integrating all elements of the Space Shuttle Program, including the Orbiter.

Part Three—A Look Ahead

Organization
Prepare a detailed plan for defining, establishing, transitioning, and implementing an independent Technical Engineering Authority, independent safety program, and a reorganized Space Shuttle Inte-

gration Office. In addition, NASA should submit annual reports to Congress, as part of the budget review process, on its implementation activities. (Return-to-flight critical)

Recertification
Prior to operating the Shuttle beyond 2010, develop and conduct a vehicle recertification at the material, component, subsystem, and system levels. Recertification requirements should be included in the Service Life Extension Program.

Closeout Photos/Drawing System
1. Develop an interim program of closeout photographs for all critical sub-systems that differ from engineering drawings. Digitize the closeout photograph system so that images are immediately available for on-orbit troubleshooting. (Return-to-flight critical)
2. Provide adequate resources for a long-term program to upgrade the Shuttle engineering drawing system.
• Reviewing drawings for accuracy.
• Converting all drawings to a computer aided drafting system.
• Incorporating engineering changes.

Final Resting Places of *Columbia*'s Last Crew

RICK HUSBAND	Buried in his hometown of Clear Lake, Texas, with full military honors.
WILLIAM C. MCCOOL	Buried at Annapolis Naval Academy with full military honors.
MICHAEL ANDERSON	Buried at Arlington National Cemetery with full military honors, near crew members of the *Challenger*.
DAVID M. BROWN	Buried at Arlington National Cemetery with full military honors, near crew members of the *Challenger*.
KALPANA CHAWLA	Cremated, with her ashes given to her family in India.
LAUREL B. CLARK	Buried at Arlington National Cemetery near crew members of the *Challenger*.
ILAN RAMON	Buried during a private, traditional Jewish funeral in a northern Israeli town, in the same graveyard as Moshe Dayan.

Chapter Fourteen

On the Road Again

Not only could the Columbia *accident have been prevented, but ... the* Challenger *management findings made years ago provided ample direction on how to avoid the* Columbia *tragedy.*
—Aviation Week & Space Technology, *April 28, 2003*

Press to Launch

The Russians could keep the International Space Station operational with a crew of two for many months and even years, but they would be little more than custodians and would be able to do little or no real science. Long-term plans for the space station required visits from a fully operational Space Shuttle, and NASA knew that they would need to get Shuttles flying as soon as possible to have any hope of saving the station. For the public, however, William F. Readdy, associate NASA administrator for space flight, would state, "We're going to take however long it takes to get to the bottom of it, to find the root cause, fix it and get back to flying. Whether it takes three months or three years, so be it. Our primary objective is to get back to flying safely." By March 2003, NASA was already gearing up to get back into space with Shuttles, even though the investigation had yet to pinpoint the cause of the disaster and the report was months away. NASA appointed astronaut James D. Hassell, Jr., to head the new return-to-flight plan. It was beginning to look like the old days just after the *Challenger* disaster.

It did not take long for *Columbia* Accident Investigation Board members to speak out on three critical areas that needed to be fully addressed before the space agency could fly again. Unless they were addressed, the report stated, "the scene is set for another accident." First, the space agency would need to fix the external tanks to prevent debris from falling off and striking the vulnerable Shuttle tiles. Second, as with the *Challenger* report, NASA was instructed "to change the way it thinks about safety and the way managers communicate possible safety problems and appraise them." Finally, the board said that NASA and national political leaders must properly fund the program while agreeing to replace the aging Shuttle fleet with a new series of vehicles as soon as possible.

Board chair Harold Gehman, a retired navy admiral, stated, "We believe another vehicle, either as a replacement or a complement, is a very, very high priority." In the twenty years leading up to the *Columbia* disaster the space agency had proposed at least five Shuttle replacements/follow-up spacecraft, only to have all of them quashed due to a lack of political support or the projected high cost of replacement. American University public policy professor Howard McCurdy wondered about the timing and if momentum could be sustained. "[NASA] has a window of opportunity [of 6 months]; if not then, it will never happen because the accident will fall off the agenda. Then you can almost guarantee that there will be another accident." This was an echo of what had happened after *Challenger*.

By September 2003, the space agency decided that it had had enough time to absorb the *Columbia* Accident Investigation Board report to develop and publish its own return-to-flight document. The 158-page NASA report focused on safety, but indicated that Shuttles could fly as early as March 2004, giving the space agency only six more months to solve all of the problems exposed by the accident board. Most who read the report believed that NASA was being overly optimistic for a flight return, as many tests had yet to be conducted and no one could guess what the results would be. However, space observers could point to at least one area of concern corrected, as NASA began to clean house. Shuttle director Ron Dittemore was replaced, along with the directors of Marshall, Johnson and Kennedy Space Centers. Marshall Space Flight Center's manager for the external tank project had retired and Linda Ham, head of the flight director group that ran the *Columbia* mission, was reassigned.[1]

The NASA report spoke of the space station as "an emergency shelter" to be used if a Shuttle were damaged during launch, allowing a full inspection and possible repair. NASA would also press to have a backup Shuttle ready to go to the pad and launch for a rescue mission within the flight time of the original mission. Within days the space agency would back away from the projected March 2004 launch target, reporting that even a mid-summer launch could be too soon.[2] NASA also decided to embed 88 high-tech sensors on the critical wing edges of the three remaining Shuttles to detect debris hits in real time. They would be able to record 20,000 readings per second during launch, although they would not be able to send data on the extent of damage any debris hits had caused.[3]

On September 24, the *Los Angeles Times* reported that all eleven members of NASA's safety board had resigned. They resigned so that the NASA administrator could appoint a new group of people to take a fresh unbiased look at Shuttle safety. The previous group were also not happy with the way safety was viewed.

On October 4, 2003, the forty-sixth anniversary of the launch of Sputnik 1, NASA announced that they needed more time to make necessary changes to the Shuttle and that they could launch no earlier than September 2004. A launch window of September 12 to October 10 was announced. However, by February 2004 it was again clear that NASA had been far too optimistic, as NASA chief Sean O'Keefe told a congressional panel that November or later was a more realistic target date. The next time O'Keefe would testify would be to tell the Senate Commerce Science and Transportation Committee that fixing all of the problems uncovered by the *Columbia* Accident Investigation Board would cost upwards of $2.2 billion,[4] nearly double the earlier projected costs. Safety and reliability was not going to be cheap. A May 2005 launch date was now on the table and it was a date that the space agency planned to keep.

In a surprise development, Russian cosmonaut Sergei Krikolev questioned using the space station as a rescue platform, due to the station's limited supplies. In February 2004, Krikolev stated, "I think now people are aware of this and that's why I got assurance from station managers that everything ... needs to be done to be sure that this scenario would be avoided."[5] Later, the new Shuttle crew would make it clear that they would prefer a pickup by a rescue Shuttle as they waited at the space station than to reenter with a Shuttle repaired in orbit.

The Shuttle's Return — STS-114

By February 2005, the crew of *Discovery* were well on their way to completing their training for a May launch. Safety was on the minds of many as Commander Eileen Collins answered questions from the press. "Clearly, I'm not going to fly on something that's unsafe," she said.[6] However, as these words were being spoken, NASA officials were privately admitting that not all of the changes recommended by the *Columbia* Accident Investigation Board had been made and would perhaps not be able to be made by the May launch date. The space agency had not solved the external tank debris problem and had not yet developed a satisfactory method to repair damaged tiles while in orbit. This next mission would be a risky test flight, despite the usual assurances by NASA. "Go fever" had begun on what would be *Discovery*'s thirtieth mission in a series begun on August 30, 1984. On March 1, NASA issued requests for proposals for the CEV (Crew Exploration Vehicle), the earth-orbital and space-station docking vehicle slated to replace the Shuttle.

On March 31, Shuttle *Discovery* began its slow crawl to launch pad 39B for mission STS-114, but not before a brief delay caused by the discovery of a small crack in the external tank's worrisome foam insulation. What may have been even more worrisome was the expressed opinion by NASA management that the crack was "no reason for concern."[7] In the meantime, Shuttle *Atlantis* was being prepared for a possible rescue mission designated STS-300.

The new crew would be on hand to escort their craft to the pad. The crew consisted of Commander Eileen Collins, Pilot James Kelly, and mission specialists Charles Camarda, Wendy Lawrence, Stephen Robinson, Andrew Thomas, and Soichi Noguchi, the latter from Japan's space program. They were ready to go, but was their spacecraft truly "Go for Launch?"

By April 29, the Shuttle launch was again delayed due to new safety concerns. "After a great deal of testing and analysis, we have been able to cross about 175 potential debris sources off our concerned list. There are still about three or four more items to work on.... We'll take a few more weeks to deal with them,"[8] stated N. Wayne Hale, Jr., deputy manager of the Space Shuttle program.

Ice buildup on fuel lines on the external tank and malfunctions of hydrogen sensors in the tank were also causing concerns. These caused the space agency to further delay the return-to-flight launch of *Discovery* by two months. A new target date was set for no earlier than July 13. Of the latest delay, the new NASA chief Michael D. Griffin said, backing his team in a prepared statement, "It is prudent to have additional verification and validation of our extensive engineering work to ensure a safe flight."[9]

With two additional months before launch, NASA decided to bring *Discovery* back to the hangar to have a new, supposedly safer external tank attached, along with new, matched solid rocket boosters which had been earmarked for *Atlantis*.[10] This new tank had additional heaters installed to prevent dangerous ice buildup. Why did NASA place *Discovery* on the pad with an old, less-safe tank in the first place? At the same time, Michael Griffin began to push the space agency to seek designs for the Shuttle's replacement as soon as possible. NASA planned to retire the aging Shuttle fleet in 2010, with the new vehicle coming online in 2014, leaving the agency out of the human space flight business for at least four years.[11] With a space station on orbit, Griffin wanted this gap closed as much as possible.

In the meantime, NASA was also directed to focus on President Bush's goal of returning Americans to the moon for much longer stays than those of the old Apollo missions, and then on to Mars. NASA had a tough road ahead, but if the spirit of Apollo could once again capture the public's imagination, it would be a road worth traveling. On June 15, 2005, *Discovery* returned to the launch pad with its new boosters and fuel tank. As reported in the *Los Angeles Times* from June 28, 2005, "NASA has failed to fulfill the three most critical criteria for improving Shuttle safety recommended by the board that investigated the *Columbia* Shuttle accident, an independent advisory group said Monday. Despite 2½ years of intense effort the space agency has not eliminated the risk of debris striking the Shuttle during liftoff or sufficiently strengthened the orbiter to resist such impacts, the Stafford-Covey Return-to-Flight Task Group said. NASA also has failed, so far, to develop the ability to repair such damage in orbit, the group said."

In response to reports that the launch team was not ready to return to space, NASA Chief Griffin stated, "We are go for launch. We still have a lot of work to do, but we are ready for a return to flight."[12] NASA had laid out a flight schedule which included numerous Shuttle missions to be flown over the next five years, most focused on finishing the construction of the International Space Station, with one possible flight to upgrade the aging Hubble Space Telescope. NASA had spent over two years, employing over 20,000 workers, and spent $1.4 billion on Shuttle upgrades and safety improvements. There was a great deal riding on this flight. One more Shuttle disaster would surely end the program, cripple the space station, end any possible Hubble repair, and could very well prohibit return to the moon and a chance to go on to Mars. The American people might have had enough of human space flight at that point, at least for the foreseeable future. NASA's future was on the line.

July 13 would not prove to be the day. Two and a half hours before its scheduled liftoff, the launch of *Discovery* was scrubbed due to a faulty fuel gauge in the seemingly trouble-plagued external tank.[13] In April, NASA engineers had noted during a test that two other fuel sensors had failed, but technicians could not discover the reason. The fix had been to simply change out the tank for a new one without any real idea of why the sensors had failed in the first place. These sensors are critical to understanding the fuel levels. Without them working properly the faulty readings could cause the Shuttle's main engines to shut down at the wrong time. Depending on when that happened, such an event could be critical to crew survival. It is the job of the ECO (Engine Cut-Off) sensors to signal a cut-off of the Shuttle's main engines before the super-cooled liquid-hydrogen tank goes empty, something which could cause a "catastrophic engine failure." In non–NASA terms, it could blow up.

During the delay, the space agency was taking its usual hits from the media. In an Ed Stein cartoon published across the nation, a reporter stands in front of the Shuttle on the pad, addressing the reader. "The Space Shuttle remains grounded while NASA engineers feverishly search for a reason to risk launching this turkey again."

NASA would later trace the sensor failure to a possible electrical grounding problem with the point sensor box. Shuttle managers would decide to launch even if only three of four sensors were working and even if they did not fully understand the problem.[14]

Go for Liftoff — July 26, 2005

It had been 32 months since the *Columbia* disaster and NASA was once again on the pad with a crew of seven ready to "put up a bird." Needless to say, *Columbia* and *Challenger* were on everyone's mind as the clock began to countdown. N. Wayne Hale, Jr., would express what most of NASA was feeling. "I wake up every day and ask myself, 'Are we pushing too hard?' We are still struggling with the ghosts of *Columbia*."[15]

Launch was scheduled for 10:39 A.M. EDT. Ten hours before launch, fuel began to be loaded, along with super-cooled liquid oxygen. At minus 423 degrees Fahrenheit, the liquid hydrogen fuel made contact with the ECO sensors, and all appeared ready to go. For the most part, the countdown went very smoothly, as tens of thousands of people stopped along highways to watch astronauts return to space. After *Discovery* (STS-114) was in orbit, NASA Administrator Griffin held a news conference. "Take note of what you saw here today," he said. "The power and majesty of the launch, but also the competence and professionalism, the sheer gall and pluckiness of this team that pulled this program out of the depths of despair two and a half years ago and made it fly."[16] The space agency had yet to explain why "an unidentified orange object" had been seen to fly off of the "repaired" external tank, barely missing the Shuttle's wing during launch. As newspapers across the county reported on the launch, it became clear that at least three separate incidents had occurred of debris falling away from the external tank. Some of the debris was large enough to have caused the same damage to *Discovery* that had destroyed *Columbia*. The ghosts of *Columbia* were indeed still around for NASA.

The day after the launch, all Shuttles were grounded after the space agency confirmed that a piece of insulating foam had separated from the external tank, narrowly missing the spacecraft, grounded even before *Discovery* could dock with the space station for a complete inspection. Images of the external tank clearly showed where a large piece of insulation had fallen off. Shuttle manager Bill Parsons stated, "We have to take a step back. Until we're ready, we won't go fly again."[17] As for a possible rescue of the crew from the space station if needed, NASA managers revealed that *Atlantis* would not be ready to launch for at least 25–30 days. That would be a long time to wait at the space station for rescue.

Harold Gehman, *Columbia* Accident Investigation Board chair, was not surprised that debris had flown off the tank. "We had precious little faith that they [NASA] could stop this stuff from coming off and lo and behold, they couldn't."[18] Speaking of the Board, he said, "At the time, we got mixed and inconsistent explanations why foam fell off. When we went into the body of research, it was inconsistent and unpersuasive." It would be back to the drawing board for NASA.

Back to the Drawing Board — Again

I have had a lot of thoughts about Columbia *and I will have thoughts after the landing.*
— Colonel Eileen Collins

As *Discovery* moved toward a docking with the space station to deliver supplies and upgrade station systems, NASA engineers and managers struggled with several problems on the aging spacecraft. Due to the unprecedented ability of astronauts to inspect their craft, NASA was able to spot two protruding gap fillers on *Discovery*'s underbelly near the nose. These gap fillers, which are glued between Shuttle thermal tiles and are as thin as a credit card, could cause heating problems downstream of where they are located as reentry heat is focused on a compressed area. That extra heat could be as much as 600 degrees Fahrenheit which could be enough to compromise the thermal system. Managers would need to decide whether or not to send the astronauts out to remove or cut away the ceramic-coated gap fillers or simply allow them to stay as they were. This problem had been noted on earlier flights, but only after the spacecraft had landed, and only one mission, STS-73, flown in 1995, had a similar situation, with a gap filler pushed out as far as 1.4 inches. On that *Columbia* mission, "some noticeable additional heating" was noted but the thermal system held and the crew came home safely.[19]

In the meantime, *Discovery*'s astronauts had work to do. Astronauts Stephen Robinson and Soichi Noguchi stepped outside to test procedures to repair damaged thermal tiles and reinforced carbon panels used on the nose and leading edges of the spacecraft. A space-age version of a caulking gun was used to spread a "goo" called NOAX on the carbon panels as Noguchi spread a thin "emittance wash" over the pre-damaged tiles set up in the Shuttle's cargo bay.[20] NASA reported that the work had gone well; however, most of the astronauts would prefer to be picked up by a rescue Shuttle than return to earth on a Shuttle which had gone through on-orbit repairs. In short, there was no confidence in the "goo" which was being used for the repairs.

Before long, managers gave the go-ahead for the first-ever on-orbit repair by authorizing an extra space walk (EVA) to the critical underbelly of the Shuttle. The astronauts would be held at the end of a long remote-arm extension and remove the two gap fillers which were causing concern. To the relief of all on board, not to mention those on the ground, the repair proved to be an easy one as the space-walking astronauts simply pulled out both gap fillers with tongs. Astronaut Stephen Robinson was having a good day. Earlier he stated, "We predict it won't be too complex. It's very simple, but it has to be done very, very carefully." He was right.

Next on the list of problems for NASA managers was a 7.7-inch gash in one of the Shuttle's thermal blankets, this one located near the commander's side cockpit window. The thermal blankets are used to cover areas not expected to encounter heat above 1,500 degrees Fahrenheit. Along with the gash was a section of the blanket sticking out and measuring four inches wide and twenty inches long. Clearly the Shuttle had taken a debris hit, but NASA could not discover what had caused the gash. This had been the first significant damage to a thermal blanket during flight and NASA managers were going to have a close look at what, if anything, needed to be done.

Before long, a decision was made to not perform an EVA, as any repairs done on-orbit could very well make things worse. Ground testing indicated that the blanket would hold during reentry with a small, 1.5 percent chance of anything tearing off and causing damage to the spacecraft. N. Wayne Hale, Jr., informed the public, "If anything does come off it will be very small. We feel we are good to fly as is." It turned out that NASA managers were correct, although Hale had further cautioned that a piece as light as 0.8 ounces from the ceramic-coated blanket could flake off and "tear a six-foot hole in *Discovery*'s rudder area."[21] Hale further stated that *Discovery*'s astronauts would be able to land even with that extensive damage to the Space Shuttle's aero surfaces. That confidence has yet to be tested. On Tuesday, August 9, 2005, *Discovery* sailed to a perfect landing at Edwards Air Force base in California. No one could guess how long it would be grounded.

Grounded Again

NASA managers were more than a bit embarrassed when a one-pound piece of foam, large enough to do significant damage to the Shuttle, broke off *Discovery* during its latest mission. It was one of 16 that fell off. After 32 months of work and $1.4 billion, the space agency had essentially failed to solve the problem. After *Discovery* landed, it was reported that the external tank used for this mission had been repaired in February, a week before the tank was shipped to the Kennedy launch center. Workers at the Louisiana manufacturing plant had found what was called "a small crack in the foam"[22] and had effected repairs. This was the same area where the largest piece of insulation foam had broken off during *Discovery*'s launch. The space agency further reported that "non-destructive examination of that section of the foam, before the flight, showed what appeared to be thin lines or pockets of low-density areas."[23] In other words, part of the foam insulation was not in full contact with the fuel tank. Nevertheless, NASA launched.

The space agency soon began to report that no matter what they did, foam was going to come off during every mission. Their job would now be to make sure that the foam that came off was small and came off late in the launch sequence. Bill Gerstenmaier, the NASA official responsible for leading the investigation into the foam loss problem, addressed the new reality. "Frankly, even the next time we fly the tank, I would expect to see a little bit of foam loss somewhere in the tank. I think it's an extremely difficult engineering problem to solve. There's no immediate answer or problem that jumps out at us. We're going to have to really understand why this foam came off."[24] NASA's new reality went beyond the original design which stated that no foam was to be allowed to break off and make contact with any of the Shuttle's thermal protection system.

In August 2005, the Return-to-Flight Task Group, headed by former astronaut Tom Stafford from the Apollo program and Shuttle astronaut Richard Covey (who had served as capsule communicator on *Challenger*'s final flight), released their 216-page report. Their report stated that NASA had made a good deal of headway in correcting the conditions which had led to the loss of *Columbia* and her crew, but that lingering problems continued to plague the space agency. A 30-page section of the report blamed a continued "lack of focused, consistent leadership and management within NASA" for an ongoing series of problems. "The agency continues to put schedule ahead of safety, follows lax engineering practices and exhibits an unwillingness to learn from its mistakes."[25] That said, co-chair Tom Stafford felt that NASA had made launching the Shuttle safer than at the time of the loss of *Columbia*, speaking of "this false schedule pressure."

Former NASA Administrator Sean O'Keefe, who had formed the group in 2003, made his thoughts known. "You've got to be pretty thick skinned with this stuff. You don't learn anything by everybody saying you did a great job. If continuous improvement is part of the objective, you won't get there if you just sit back and rest on your achievements."[26]

NASA Administrator Michael Griffin was quick to state that he felt good that the task group members were able to "speak their minds. We do not shrink in NASA from criticism of our engineering processes, our decisions or anything else. We will listen to it, we'll evaluate and we'll make a decision and we'll move on." Griffin's next decision was to push the next launch back to no earlier than March 2006 to allow a complete review of the main tank's safety issues when it came to foam.

Discovery on the Pad — STS 121

In February 2006, *Discovery*'s new commander Steve W. Lindsey said at a news conference that he was ready to fly, even though foam was expected to still pose a problem. "The program has never advertised that we would never lose any foam. We will lose foam on this flight, just like every other. The key is that the foam we do lose is small enough ... that it can't hurt us if it hits the vehicle."[27] Within days, NASA announced it would attempt to launch three Shuttle missions in 2006. The agency soon postponed the planned May launch to July so that engineers could replace sensors which had failed on the hydrogen fuel tank. These were the same sensors that had caused so much concern on the last mission. The Shuttles were showing their age, as the 10-year-old sensors were replaced during a three-week process. The delay would also allow the space agency the time to send the damaged robotic "Canada Arm" back to its manufacturer for repairs. The arm had been cracked a week earlier when workers hit it with a platform as they were working in the cargo bay.

By May 19, all of the repairs and flight preparations had been completed, as NASA rolled *Discovery* out of the Vehicle Assembly Building on top of the crawler, attached to its two solid rocket boosters and huge external tank. It took nearly eight hours to make the four-mile trip to launch pad 39B. This mission to the International Space Station would be the fifteenth for the space agency's aging Shuttle fleet. Shuttle manager N. Wayne Hale, Jr., reported, "Based on what we know today, there is no reason not to launch on July 1." However, referring to the fact that flying Shuttles is simply not safe, he stated, "What we've done is eliminated the largest hazards. This is a 1-in-a-100 vehicle. It is a risky vehicle to fly."[28] During the grounding, NASA engineers replaced some 5000 ceramic spacers held between *Discovery*'s heat-resistant tiles, replaced 242 tiles and changed out 44 insulating thermal blankets, most of them located near the cockpit. The agency had also removed around 34 pounds of foam insulation on the huge external tank which engineers felt was not needed, mostly around outside cables which run the length of the tank. NASA now said that no foam piece larger than one-fifth of a pound would come off of the external tank during launch. Some engineers at the space agency were not so sure.

Two Say "No Go"

The schedule said July 1, but two top Shuttle officials had grave doubts that any Shuttle should launch until problems with the foam on 34 brackets attached to the external tank were solved. NASA's chief engineer Chris Scolese and former Shuttle astronaut Bryan D. O'Connor, now head of the office of Safety and Mission Assurance, gave a "no go" to the Shuttle launch. Twenty-five

Shuttle management-team members had met for a two-day flight-readiness review. Twenty-three managers had given a go, but two had "recommended that we not fly, but they did not object to us flying" according to Bill Gerstennaies, NASA's associate administrator for space operations. The no go/approval was contradictory and NASA officials knew it.[29] N. Wayne Hale, Jr., stated, "People concerned about culture change ought to take heart. The agency has really changed. I think it is acceptable for a number of reasons to go fly for a limited number of flights until we come up with a new design." It would not take long for the nation's press to be all over this story. With two Shuttles lost, the American people wanted to know just what the problem was and why NASA was willing to risk the lives of seven more astronauts when two top managers in critical safety positions put it on record that they were not satisfied with the Shuttle's safety and its ability to get into orbit in one returnable piece.

O'Connor and Scolese soon found themselves clarifying their "no go" votes. As published in the June 21, 2006 issue of the Orlando *Sentinel*, "Possible debris shedding from foam-insulation ramps on the ship's External Fuel Tank posed an unacceptable risk to *Discovery*. However, both said Wednesday that proceeding with launch would be acceptable because NASA could provide astronauts refuge on the International Space Station and launch a rescue flight to bring them home."

O'Conner stated, "I thought when I sized all of this up that if we were in the red area — in other words, the unacceptable risk area for loss of the vehicle — I did not consider us to be there for the loss of the crew. Even if I disagreed with some on loss of the vehicle, I think everybody in that room agreed that the loss-of-crew risk for this mission is acceptable. This is not a new topic. We've been losing pieces of foam off of these things for quite some time. As we looked at it, the teams realized that the potential here was higher than we thought before for damage to the orbiter." To be sure, this was the first time in the Shuttle program's history that NASA was fully prepared to launch a vehicle despite two top officials giving the launch a "no go."

Scolese would also clarify the reasons for his "no go" vote. "Everybody was free to express their views, something that people have been concerned about in the past. We had a very good, productive dialogue, and I think that's an indication that we've come a long way. I remain 'no go' upon potential loss of the vehicle; however, for this mission I have no intention to appeal the decisions based upon capability to provide [safe haven at the space station]."

On the day before *Discovery* was scheduled to launch, a three-inch, triangular piece of insulation foam was discovered to have fallen away from an area of the external tank near a five-inch-long crack in the foam. This completely unexpected event prompted an immediate look at the expandable bracket which housed a liquid-oxygen fuel line.[30] The external tank had expanded when the super-cooled fuel had been drained after the Sunday launch had been scrubbed due to weather. Was this the only area that had been affected by the expansion of the tank? Once again it would be a long night of "what ifs." After a day-long consultation with top NASA engineers, it was decided that the loss of such a small piece of insulation would not adversely affect tank temperatures. There would not be too much extra heating of the tank during the launch and the cooling effect while on the pad would not be great enough to allow extra ice to form on the tank. The second area of concern was answered when a jerry-rigged camera held at the end of a flexible pipe took close-up photos of the damaged area. Engineers quickly realized that there was no additional danger of more foam flaking off, no more than usual that is. Once again NASA was go for launch.

At 2:38 P.M. EDT on July 4, 2006, *Discovery* lifted off from pad 39B on a mission to the International Space Station. As reported by the Associated Press, "About three minutes later, as many as five pieces of debris were seen flying off the 154-foot tank, and another piece of foam popped off a bit later, Mission Control told the crew. The latter piece seemed to strike the belly of *Discovery*, but NASA assured the seven astronauts it was no concern because of the timing." N. Wayne Hale, Jr., was quick to report that the foam had come off at such a high altitude that "there wasn't enough air to accelerate the foam into the Shuttle and cause damage."

In an interview with the Associated Press, NASA Administrator Griffin addressed the foam issue, echoing astronaut Steve Lindsey: "If foam hits the orbiter and doesn't damage it, I'm going to say ho-hum because I know we're going to release foam. The goal is to make sure that the foam is of a small enough size that I know were not going to hurt anyone. It's hardly the only thing that poses a risk to a Space Shuttle mission."[31] The next day, NASA managers reported that at least six pieces of foam had come off the external tank during the launch, with one photo of the tank showing an 8 by 10-inch gap near the fuel lines. However, most of the pieces had come off well over 200,000 feet and would not have caused any damage if they had struck the orbiter. NASA had once again dodged a foam bullet.

After docking with the space station, the Shuttle crew went about the business of checking *Discovery*'s thermal protection system, finding only one gap filler sticking out about one inch and showing a crack. It was soon decided that the gap filler would pose no problem to a safe return and would not be removed by a space-walking crew member. Within days the crew was told that they were "100 percent cleared for entry." After three space walks, maintenance on the space station, and the delivery of a station crew member, *Discovery* was ready to come home. The crew had also performed more tests on NASA's NOAX material used to repair thermal tiles in orbit. NOAX (non-oxide adhesive experimental)—"goo" as the astronauts refer to it—performed well, but NASA was not yet ready to certify it for actual Shuttle repairs. Neither NASA managers nor any of the Shuttle astronauts were willing to risk their lives on reentry using the "goo" for repairs.

On July 17, *Discovery* landed at the Kennedy launch center, having met all of the mission's test and space-station objectives. The only problem encountered was a worrisome leak in one of the auxiliary power units. The fix was to turn the unit off and land using only two operational APUs. As of this writing, a failed APU has yet to destroy a Shuttle—but the problem persists.

Howard McCurdy gave NASA a thumbs-up after the landing. "What's important is that they changed their approaches to space flight considerably; it was an organizational test. I don't give many A's. They're clearly back to where they want to be. A B-plus."[32] With perhaps a dozen Shuttle flights in NASA's future at this writing, only time will tell if B-plus will be good enough.

> *Based on NASA's history of ignoring external recommendations, or making improvements that atrophy with time, the Board has no confidence that the Space Shuttle can be safely operated for more than a few years based solely on renewed post-accident vigilance.*
> —*Columbia* Accident Investigation Board

Section VI

The Future of Manned Spaceflight

Chapter Fifteen

A Look to the Future

You accept risks but you feel that the risks are worth it for the county and the importance of what you are doing. And you are willing to take those risks.
— Senator John Glenn, January 28, 1986

Why Space?

We do not send men and women into space because we need to; we send them into space because we choose to do so. It is a leap of human faith and a personal challenge that we see as worth the risk and the effort. All who accept the challenge of space know these risks, and there are many. However, robotic spacecraft, programmed to take the risks now being accepted by our Shuttle crews, could accomplish many of the tasks we ask our astronauts to perform. Why then do we risk our nation's resources and the lives of those who venture into the unknown? The answer is surprisingly simple. We explore and take the risks because it is a profoundly personal adventure and one we are unwilling to leave to unemotional robotic scouts. Perhaps President Teddy Roosevelt said it best when he spoke of others who took risks and explored: "The credit belongs to the man who ... spends himself in a worthy cause; and if he fails, at least he'll never be with those cold and timid souls who know neither victory nor its pursuit."

Those who say that space is explored more cheaply and safely with unmanned craft are correct, but I suspect that the exploration is not being accomplished as well. Robots such as the Soviet Lunakod roamed the lunar surface and sent back many interesting images, and others even returned with a small amount of lunar dust. But it took a human walking on the lunar surface to look over that soft, gray landscape and declare it to be, in amazed admiration, "Magnificent desolation." It took on-site inspection of the lunar surface to spot orange soil and select the best group of rock samples to be returned and studied by many scientists around the world. And it took a human in an Apollo 8 spacecraft orbiting the moon for the first time to look away from his pre-programmed list of tasks and take an unauthorized photograph we now call "Earthrise." Yes, it is cheaper and much easier to explore without sending humans into outer space, but it certainly would not convey the excitement and grandeur involved with mankind's personal exploration of our solar system. It is part of who we are as human beings, and it always will be as long as we have explorers among us with the courage and curiosity to look beyond the horizon and go explore. And we have only just begun the grand adventure of mankind.

Having now accepted the risks, we must decide with a certain amount of clarity and managed direction where we will go from here.[1] The United States, through our National Aeronautics and Space Administration, needs to develop a detailed, well-publicized, step-by-step plan to expand human presence in space. Within that plan should be an explanation to the American people of exactly what will be gained economically, culturally, scientifically and educationally. By not presenting the manned lunar flights as opportunities to develop this nation's infrastructure, the space agency and the country failed to fully take advantage of the science and technology which had been developed by going to the moon. When the United States put the first human on the moon, we did more than gather a few rocks; we changed a nation and the future direction of our world. That opportunity does not come around very often. We need to make such an opportunity again and clearly explain the reasoning behind it to the American people.

Over thirty years ago, the United States ended its first reconnaissance of the moon, but to this date has never looked forward from those accomplishments. Even though the original reason we decided to go was political, it was discovered that much more than political gain was at stake. As President John Kennedy said, "We choose to go to the moon in this decade and do the other things not because they are easy but because they are hard." We need to do hard things again. This nation needs to begin a new process of space exploration which will not only challenge our nation's technological abilities and managerial skills, but amaze the world. We need to continue the journey if we are to stay the most powerful, technologically advanced nation in the history of the world. The world is looking to the United States for leadership.

The Space Shuttle's Future

Our most expensive and risky manned vehicle presently used to access near-earth orbital space is the Shuttle, and the United States needs to use that resource sparingly. If it is to fly at all, the system must be dedicated to shuttle-specific payloads which cannot achieve orbit by any other means. At the same time, America's space agency should be hard at work designing the Shuttle's replacement as well as developing ways to extend the lifetime of the present fleet of three to be used unmanned as carriers in fully automatic mode. That method has already been tested and flown by the Soviet Shuttle, and although no actual payload was deployed, the flight was generally successful. (It came down almost in one piece, despite a bumpy, off-runway landing.) Indeed, when the *Buran* flew its only—and highly propagandized—mission, there were no life-support systems on board to support a crew, so it had to be flown unmanned. The costs of stripping their Shuttle down and building life-support systems into *Buran* was given as one reason it never flew again, but it is doubtful that that was the only, or main reason; after all, their government was falling around them.

After the *Challenger* disaster, NASA began to study several internal proposals to fly the Shuttles unmanned. These plans were developed for so-called short missions, which could automatically deploy a Defense Department satellite in the event an emergency occurred and could be landed after only a few hours. The Shuttle *Columbia* was targeted as the vehicle of choice because of its age even then, but there was no consensus of opinion within the space agency for this project. It was supported by Marshall Space Flight Center who saw it as a less risky proposition, which would cut from $30 to $40 million from the cost of each mission. However, managers at Johnson Manned Space Flight Center opposed the idea based on the costs involved in modifying the orbiter. But there was the unmanned aspect of the proposal to consider, plus the year or more it would have taken to do the modifications.

After the final *Challenger* flight, NASA began plans for a Shuttle-derived heavy-lift vehicle called Shuttle-C ("C" meaning "cargo") The system would use Shuttle-derived hardware such as the SRBs and external tank with a new cargo pod shaped like a Shuttle without wings. The cargo pod would carry the payload into orbit and then reenter to burn up in the atmosphere. The problem with this system was that each launch meant losing the pod and the two main Shuttle engines attached to it. Other than the solids, most of the reusable capabilities would be lost and the costs would have continued to climb.

As a replacement for the Shuttle system comes online, the space agency should in turn rework the existing Shuttles to eventually fly all of them unmanned. This system could either use the Shuttles in automatic mode, or with the Shuttle-C configuration. Either way, the method would extend Shuttle technology and usefulness well into the twenty-first century, perhaps as far as the year 2025. By that time, the Aerospace plane, X-33, or some other advanced manned spacecraft should be well into its operation. Flying a 50-year-old manned Shuttle, no matter what propaganda might be forthcoming from NASA, is simply not an option.

America's Space Station

Without the United States there would be no International Space Station. The Space Station's name should be *Freedom* or perhaps *Liberty Station*. The bottom line is that nothing would fly without U.S. technological know-how and U.S. taxpayer funding. Some would have looked to Russia's *Mir* Space Station as the future, but in reality it was a technological dead end, long past its orbital prime when it was finally allowed to crash into the Pacific Ocean. It was extremely dangerous to the crews who manned it. It was a small, very noisy, earth-orbital outpost held together more by the tenacity and raw courage of its mostly Russian crews, than by the technological prowess of Russian science and engineering, which are decades behind the curve.

Russia's *Mir* station was more an example of what a dictatorial nation can accomplish when it puts the welfare of its own people in the waste can as it directs its limited resources towards international propaganda projects. It is an example which demonstrates that long orbital stays which break records are not necessarily translated into good or even usable science. It was an excellent test bed on how to survive in the unforgiving realm that is space; however, *Mir* will not be missed.

In order to succeed in this area, the United States must continue to build a space station far beyond the limited capabilities of *Mir*, and U.S. citizens must be shown that it is their nation leading the way, not a group from Europe or anywhere else. In this way, support for the station and other future programs could be sustained. Taxpayers need to be shown what they will receive by supporting such programs. In an era of tight budgets, perhaps the best mix for the station would be one of pure science combined with commercial applications. After thirty years of planning, talking, designing, redesigning and spending taxpayers' money, it's time to put our station into full-scale operation so that the Shuttle will have something completed to shuttle to. Experience gained on the space station would also directly relate to a manned Mars flight, which is an adventure mankind will experience in the not-too-distant future. Indeed, the first human who will walk on the planet Mars is at this moment walking on planet Earth. The only question is: which country is this individual living in today and who will lead the way? If the United States does not take the lead, then someone else will and it will gain the technological advantage and may not be all that friendly to the United States. It is time to set aside political correctness and get on with the business of discovery and technological advance for ourselves and the rest of the free world.

A Return to the Moon

Learning to live in space can be accomplished by operating an earth-orbiting space station, but where does mankind go to test the technologies required to explore the planets? The answer, of course, is the moon. If we are to someday explore our solar system with human-crewed spacecraft, the earth's moon is a perfect test bed. It is much better to test the techniques and equipment of exploration three days from earth, rather than months away on Mars. It is time to plant new American flags on the moon and make certain the world knows who is leading the way. However, beyond presenting an opportunity to test Mars equipment, the moon offers a laboratory of its own toward the understanding of how our planetary system developed and opens a window on earth's early history. To understand how the earth developed, we must learn more about our natural satellite, and the best way to accomplish this is to establish a manned lunar outpost. The reconnaissance of the Apollo days is complete. It is now time to get down to some real long-term exploration. We need to spend months on the moon, not simply days.

In 1992, NASA's newly formed Space Exploration Office developed plans for returning to the moon. The plan, named FLO for "First Lunar Outpost," would continue the work left undone by the Apollo reconnaissance missions. Beginning with an inexpensive series of unmanned lunar orbiters designed to fully map the moon and explore its resources from orbit, the program would eventually land teams of four astronauts on the lunar surface. These teams would arrive after an unmanned lander had pre-positioned a lunar-habitat module. It would allow stays up to 45 days at a time as the outpost expanded. This system is based on the step-by-step build up of laboratories developed at the north and south poles of the Earth.

Beyond the pure science of lunar geology, these new missions would focus on developing the techniques needed for permanent occupation of the moon, such as extracting oxygen from the lunar regolith (soil) and creating lunar building materials. NASA planners designated Mara Smythii near the equator as the primary landing site for the FLO, and have developed nine long traverses to cover an area as much as thirty square miles. An area greater than the combined Apollo program visits would be explored during the first mission. But once again the space agency found little support in Congress for this program. The interest in exploration was strong, but the costs were higher than the Congress was willing to expend. They should have asked the American people.

Perhaps NASA needs to look more closely at developing the moon's resources as a way of reducing some of the costs of exploration. Certainly the act of returning to the moon would create whole new technologies, which would very quickly be translated to products applicable on earth, but that is a circular route. Most Americans would be looking for more direct contributions from space exploration. One method, which appears to fit the bill, would be to develop energy resources. With oil resources being consumed at an ever-increasing rate, any new energy source would have a direct impact on everyone. With many of the world's oil fields in generally unstable areas, this is also clearly a national defense issue.

The answer could be Helium-3. On earth this element is extremely rare, but analysis of lunar regolith returned by the Apollo astronauts showed the element to be plentiful on the moon. Charged particles from the sun have been depositing Helium-3 directly onto the surface for four and a half billion years. Calculations by University of Wisconsin professor Jerry Kulcinski showed that there could be enough Helium-3 in lunar surface materials to supply the entire earth's needs for centuries. As for the price, calculations show that one metric ton of Helium-3 would supply an equal amount of energy to $3 billion worth of coal. If oil companies were to be included as funders, they could very well reap the rewards of any energy revenues returned from such explorations. And only Americans have the proven ability to do the work.

Costs for processing Helium-3 for that metric ton have been estimated at $20–40 million. It would be a very nice return on investment even if it costs ten times that amount. The profits would be astronomical, and the processing would not be all that difficult. First, an unmanned test rover would be sent to the surface to test the systems. Then, a simple rover the size of a small bus could be programmed to roam the surface, scooping up the regolith, and heating the material to release the Helium-3. The gas would then be cooled to a liquid and then stored in tanks as the rest of the material was pushed out the back. When the tank was full, the super rover would signal the astronauts who would go out and replace the tank with an empty one. Before long, mining Helium-3 would become fully self-sufficient, at least as far as further monetary commitments from earth would be concerned.

There are many other valuable resources which may be available from lunar mining operations which will need to be extracted in order to sustain lunar operations. Methane, hydrogen and even molecular water could be removed from the regolith by similar methods used to obtain Helium-3, and the roving processors mining them could be fully powered by solar energy, with storage batteries for those long and cold lunar nights. And developing new solar cells would directly translate to new uses on earth.

It is no longer a question whether mankind will return to the moon. The questions are when and who will lead? The nation who develops the lunar environment first and stays for the long haul will become (and stay) the undisputed technological and economic leader on earth for many decades, and perhaps centuries to come. That nation will use the experience, technology and resources gained from lunar exploration to begin the manned exploration of Mars and beyond. The question is whether that country is Japan, communist China, socialist Russia, a European group or the United States of America.

The Red Planet

Ever since the turn of the twentieth century, mankind has been on Mars. No, there are no footprints of any "small steps for man," but we have been there all the same. Our imaginations saw canals and ancient civilizations struggling to survive on a difficult world millions of miles away, yet as close as the planet next door. We even see a face on its ancient, wind-blown surface, one not too unlike our own.

Some of those who fought so hard to place men on the moon over three decades ago view Mars as a goal unlikely to ever be accomplished. In an interview in September 1988, Christopher Columbus Kraft, the dean of Mission Control for the Apollo program has stated, "Mars is bull. Trying to sell a Mars mission is the way to kill the space program. It would cost billions and billions and no damn fool in Congress is going to sign the check."[2] Perhaps so, but that was one of the reasons argued for not going to the moon, but we went anyway, and a nation was forever changed and enriched by the act of going.

When humans last went to the moon, the costs were all up front with payoffs somewhere down the road. These payoffs were astronomical. When we go to Mars we must be prepared to

Artist concept of Mars Rover by Jet Propulsion Lab/NASA (JPL/NASA).

pay at least in part with profits gained from space-station production operations and lunar-based mining operations. Private and government funds pushing profitable operations could be the key to a supportable manned Mars exploration program, which could be funded in today's budgetary environment. Of course the United States could always invite a few friends along to cover some of the up-front costs. The key word here is "friends."

There is much to be learned and much to be gained with a concerted effort directed at expanding the human presence in space. We can decide to stay in earth orbit and confine ourselves to the developing International Space Station and learn much in the process, something that would soon become a limited view, or as a nation we could truly begin the further exploration and settlement of our solar system. This generation must make up its mind about how to build on the past to a truly grand future in space and on Earth, or decide to fall back to a secondary position and watch others do the work and gain the high ground. If the United States does not go, someone surely will, and U.S. citizens will be the lesser for it. We are leaders and explorers in the United States and we need to remember that. Do we lead or simply follow? The next great adventure awaits in the twenty-first century, but only if Americans accept the challenge.

> *I don't think we should turn our backs on what this nation has
> accomplished in space and what is still to be accomplished out there.*
> — Walter Cronkite, January 28, 1986

The Next Step

In order to answer the challenge of future space exploration of the moon and on to Mars, NASA has developed the Constellation Program. Primary to this effort is the *Orion* Crew Exploration Vehicle. According to *NASA Facts*, "*Orion* will be capable of carrying crew and cargo to the space station. It will be able to rendezvous with a lunar landing module and an Earth departure stage in low–Earth orbit to carry crews to the moon and, one day, to Mars-bound vehicles assembled in low–Earth orbit. *Orion* will be the Earth entry vehicle for lunar and Mars returns." NASA further reports that "Inside, it will have more than two-and-a-half times the volume of an Apollo capsule. The larger size will allow *Orion* to accommodate four crew members on missions to the moon, and six on missions to the International Space Station or Mars-bound spacecraft. *Orion* is scheduled to fly its first missions to the space station by 2014 and carry out its first sortie to the moon by 2020." So much for future plans, but the question the American people must ask is: Will it be properly funded and will the American people support this next grand exploration of our solar system?

APPENDIX I

Shuttle Flights Pre–Challenger Disaster

A History of Problems

America's Space Shuttles are the most complicated, most sophisticated and most expensive spacecraft ever designed and flown by mankind. Their reusable main engines have the highest thrust-to-weight ratio ever developed. They possess the most sophisticated and lightest thermal protection system ever flown. The computers, although old by today's standards, still serve as the best ever developed for human space flight.

The Shuttles and their crews have deployed, repaired and retrieved orbiting satellites. The crews have conducted many hours of valuable scientific experiments, manufactured pharmaceutical products, produced crystals in orbit, and developed the basis for the future commercialization of space. No country on earth, including the defunct Soviet Union with its abandoned attempt to fly a Shuttle of its own, has ever come close to matching the versatility and promise for the future of the U.S. Shuttle system.

However, with all of these accomplishments, and there are many more to come, the U.S. Space Shuttles are, and always will be, experimental spacecraft. They will never become commercial spaceliners which can be launched fifty to sixty times a year and pay for themselves. The Shuttle is a technological advance which should never have been expected to become fully operational. Any new technology can expect to run into unforeseen problems which need to be worked through, and the Shuttle was no exception. This is something we all simply tried to ignore but by doing so the technology showed — in a very dramatic way — where our thinking had gone wrong.

A Shuttle has yet to crash due to a failure of a system working within its design limits. The Shuttle *Challenger* failed when human beings pushed it to perform in areas that managers and directors knew went well beyond its capabilities. Twin Shuttle losses were management failures, not Shuttle failures. Someday a part will fail and a Shuttle will fail to properly perform as planned. Another Shuttle will probably crash again; it is only a matter of time. We must be prepared to accept this and move on. Space flight is always dangerous and can be very unforgiving.

Listed below are the twenty-four successful flights flown in the pre–Shuttle-disaster period. It is presented as a brief overview of some of the many problems which were encountered during this period of ever-increasing flight frequency. Some of the problems were minor and some were a bit more than that. For NASA and crews who dodged the bullet on several flights, some of the problems could have become deadly. Both NASA and press reports were used in this list's compilation.

Flight	*Launch*	*Orbiter*	Crew	*Flight Time*
STS-1	April 12, 1981	*Columbia*	John W. Young Robert L. Crippen	54 hours, 21 minutes, 57 seconds, 36 orbits

First test flight of the Space Shuttle transportation system.

Computer software problems scrubbed the launch on April 10, 1981. At least fifteen thermal tiles were shaken loose during launch. NASA was concerned that ice build-up on the external tank may have caused damage to the thermal tiles at launch when the ice broke free. Failure of two heaters on one of the APUs (Auxiliary Power Units), as well as a failure of the zero-gravity toilet occurred in orbit. Temperatures fell to 37 degrees inside *Columbia*'s flight deck when the temperature-control system failed. One fourth of one percent drop in thrust of the three main engines after SRB (Solid Rocket Booster) separation was indicated by Mission Control computers. One parachute failed to deploy on each SRB. A discrepancy of between 2,200 and 4,400 pounds of liquid oxygen (LOX) between what NASA believed to have been loaded into the external tank and what sensors within the tank indicated had been used was found. It could be that less LOX was loaded into the tank than had been intended, or it may have been drained from the tank before launch. NASA indicated that loading procedures would be tightened for STS-2. But this would not be the last time this problem would occur. At a postflight inspection it was found that an overpressure wave had occurred when the SRBs were ignited, which resulted in the loss of heat tiles and damage to 148 others. The pressure pulse caused by the ignition of STS-1's solid rockets rebounded off of the launch pad and hit the *Columbia* with a force of several Gs. This pulse bent several of the fuel-tank support struts as well as the orbiter's wing controls. This was the first time solid fuel rockets were used in a manned launch and everyone was on a learning curve.

Flight	*Launch*	*Orbiter*	Crew	*Flight Time*
STS-2	Nov. 12, 1981	*Columbia*	Joe Engle Richard Truly	54 hours, 24 minutes, 4 seconds, 3 orbits

Erosion of O-ring on the right solid booster; joint temperature 70 degrees Fahrenheit.

A forklift accidentally banged into *Columbia*, damaging two of its thermal tiles during the processing. A faulty fuel-loading line dumped 2–4 gallons of corrosive rocket engine propellant over the nose and down the side of *Columbia*, loosening scores of thermal tiles. This occurred when fuel was being loaded into the nose of the Shuttle for its reaction control jets. 67 tiles came loose and 310 others appeared to also be affected. The tiles themselves were not damaged. These had to be recemented back on the Shuttle. On November 4, 1981, the launch was postponed due to oil contamination in one of the APUs. High pressure was also discovered in two of the APUs. *Challenger* was cannibalized of two small telemetry-processing units after *Columbia*'s on-board and ground spares failed. The November 12 launch was delayed when a bad data-transmitting unit had to be replaced. After launch, the SRBs were left floating in the ocean for four days due to high waves. During the flight, a fuel cell failed because it became clogged and stopped producing electrical power and drinking water. This failure caused the five-day mission to be cut to two days. After the flight, six tiles required replacement and twelve others were damaged. It was at this time that external tank construction was pacing the flight schedule. It took fourteen months to manufacture one tank.

Flight	Launch	Orbiter	Crew	Flight Time
STS-3	March 22, 1982	*Columbia*	Jack Lousma Gordon Fullerton	192 hours, 6 minutes, 9 seconds, 130 orbits

The launch was delayed one hour due to a failure of a heater on a nitrogen support line. Halfway toward orbit, one of *Columbia*'s APUs overheated, and it was turned off on instructions from Mission Control. This led to engine number three being shut down 20–30 seconds earlier than planned. On orbit, the toilet motor failed. A transponder also failed, which left only one of four S-band radio channels operational. During the delayed landing at White Sands in New Mexico, the wind picked up and the crew had trouble putting the nose down; the Orbiter hit hard on the front landing gear. Some observers felt that *Columbia* was close to flipping over due to the heavy crosswinds. Thirty-eight of the thermal tiles were lost during launch and nineteen others were damaged. Two television cameras also failed during the mission. The landing site was changed to Northrup Strip, White Sands, New Mexico.

Flight	Launch	Orbiter	Crew	Flight Time
STS-4	June 27, 1982	*Columbia*	Ken Mattingly Henry Hartsfield	169 hours, 10 minutes, 43 seconds, 112 orbits

The solid rocket boosters' performance was less than planned and caused *Columbia* to fly 8000 feet below its planned trajectory. This was later made up by firing the main engines three seconds longer. Both SRBs were destroyed and lost when their main parachutes malfunctioned and the boosters hit the ocean at high speed and then sank. A small leak developed in the primary thrusters in the forward nose of the Orbiter while in orbit. Three hundred thermal tiles were damaged on this flight, although none were lost. Thirty-seven tiles needed to be replaced.

Flight	Launch	Orbiter	Crew	Flight Time
STS-5	Nov. 11, 1982	*Columbia*	Vance D. Brand Robert Overmyer Joseph P. Allen William B. Lenoir	122 hours, 15 minutes, 29 seconds, 81 orbits

Payload: SBS-3, Anik-C3.

A helium tank leaked before launch, but the orbiter was launched without repairs being needed. Both EVA pressure suits failed, which necessitated cancellation of the first Shuttle-program space walk. Hamilton Standard, which had designed and built the space suits, would later be fined for this failure. The inboard wheel on the left main landing gear was damaged on landing and it shredded the tire.

Flight	Launch	Orbiter	Crew	Flight Time
STS-6	April 4, 1983	Challenger	Paul Weitz Karol Bobko Donald Peterson Story Musgrave	120 hours, 24 minutes, 31 seconds, 80 orbits

Payload: TDRS-A. Heat indication on primary O-ring of both solid rockets.

A hydrogen leak was discovered in one of the main engines after its flight-readiness firing. The engine was removed and replaced. A second hydrogen leak was found in another main engine and it turned out to be a ¾-inch crack in the engine. A third engine developed a leak in the LOX heat exchanger. Yet more leaks were found in two of the three main shuttle engines, in two separate hydrogen fuel lines. At 104 percent of thrust, the main engines performed lower than expected. After the Shuttle was in orbit it released the TDRS (Tracking and Data Relay Satellite) and its engine failed to place this payload into its proper orbit. The failure was traced to mechanical and guidance failures in the satellite's booster rocket. This was the first use of the new lighter-weight external tank and new lighter-weight SRBs.

Flight	*Launch*	*Orbiter*	*Crew*	*Flight Time*
STS-7	June 18, 1983	*Challenger*	Robert L. Crippen Frederick Hauck Sally K. Ride John M. Fabian Norman Thagard	146 hours, 25 minutes, 41 seconds, 97 orbits

Payload: Anik-C2, Palapa-B, Spas-1

During the countdown, a leak was discovered in a hydrogen propellant seal at the orbiter's umbilical connection. *Challenger* was scheduled to make the first Shuttle landing at Kennedy Space Center in Florida, but bad weather required a change to Edwards Air Force Base. Upon landing, a brake failure was blamed on washers that failed. As on all previous flights, the SRB's parachutes suffered burn damage.

Flight	*Launch*	*Orbiter*	*Crew*	*Flight Time*
STS-8	August 30, 1983	*Challenger*	Richard H. Truly Dan Brandenstein Guion S. Bluford Dale A. Gardner William Thornton	145 hours, 9 minutes, 32 seconds, 97 orbits

Payload: Insat-1B

One of the SRBs suffered severe damage during launch when the rocket exhaust burned through the booster's protective lining. This problem delayed the flight following this one. This problem could have caused the Shuttle to cartwheel sideways out of control if it had burned all the way through, which could have occurred in 4–14 more seconds of flight. The three-inch coating had burned down to one-fifth of an inch. The problem was traced to the curing process for the carbon cloth insulation. Twenty-seven thermal tiles were damaged during launch.

Flight	*Launch*	*Orbiter*	*Crew*	*Flight Time*
STS-9	Nov. 28, 1983	*Columbia*	John W. Young Brewster Shaw Robert Parker Owen Garriott Byron Lichtenberg Ulf Merbold	247 hours, 47 minutes, 24 seconds, 167 orbits

Payload: Spacelab 1

SRB's aft booster segments were replaced due to the problem on STS-8. This caused a two-month delay when the vehicle was removed from the pad and taken back to the Vehicle

Assembly Building. Two of the on-board computers failed. In orbit, just before reentry, one of three inertial measuring units failed. Problems with two APUs occurred during the flight. The right outboard brake failed after landing when it froze and washers fell out. A fire occurred in two APUs, reaching 400 degrees Fahrenheit, as the orbiter was landing. The fire was caused by a leaking hydrazine pipe which had ruptured inside a rear compartment two minutes before landing. Fifteen minutes after landing the APUs were shut down and an explosion occurred. Jack Riley, a NASA official in Houston, said that the crew was in no danger. After this flight, the *Columbia* was mothballed for two years, for modifications. During this time it was cannibalized for parts needed for other orbiters. This was required to maintain flight operations. Modifications were required to bring the *Columbia* up to the same flight-readiness level of the two new orbiters. It was reported that in some cases there was only one part which had to support all four orbiters and fly all missions. During the flight, a White House script was sent up telling which astronauts could speak on camera during an interview and what they should say.

Flight	*Launch*	*Orbiter*	*Crew*	*Flight Time*
41-B	February 3, 1984	*Challenger*	Vance D. Brand Robert Gibson Bruce McCandless Robert Stewart Ronald McNair	191 hours, 17 minutes, 0 seconds, 127 orbits

Payload: Palapa B, Westar, Spas. Erosion of O-rings on both Solid Rockets, joint temperature 57 degrees Fahrenheit.

One of three parachutes on each of the SRBs failed to open, causing the SRBs to hit the ocean at 75 miles per hour rather than 60 miles per hour, as would be the norm. Both satellites launched failed to reach their required orbits when both Payload Assist Modules (PAM) malfunctioned. A robotic arm malfunctioned at its wrist joint, which caused the SPAS satellite to be kept in the payload bay. A balloon intended to be used as a rendezvous target exploded when it was deployed from the canister which held it in the payload bay. The right outboard brake failed on landing. Thirty-one thermal tiles were damaged. After this flight, NASA reported that all ten missions had brake problems, which would delay landings at Kennedy Space Center in Florida.

Flight	*Launch*	*Orbiter*	*Crew*	*Flight Time*
41-C	April 6, 1984	*Challenger*	Robert Crippen Francis R. Scobee Terry Hart James Van Hoften George Nelson	167 hours, 40 minutes, 54 seconds, 107 orbits

Payload: LDEF. Erosion of the O-ring on the right Solid Rocket; blow-by of O-ring on right SRB; joint temperature 63 degrees Fahrenheit.

The OMS (Orbital Maneuvering System) pod was taken off *Discovery* and placed on *Challenger* so that the orbiter could be launched for this mission. *Challenger*'s OMS pod was damaged on its previous mission when it overheated due to the loss of thermal tiles. The aft skirt of the right SRB was damaged when one of three parachutes failed to deploy. Space-walking astronaut docking failed three times with Solar Max satellite, which had to be grabbed by using the robot arm. One thermal tile was severely damaged on the main landing-gear door. All four brakes were damaged on landing.

258 APPENDIX I

Flight	Launch	Orbiter	Crew	Flight Time
41-D	August 30, 1984	Discovery	Henry Hartsfield Mike Coats Steven A. Hawley Judith A. Resnik Mike Mullane Charles D. Walker	144 hours, 57 minutes, 4 seconds, 96 orbits

Payload: Leasat-2, SBS-4, Telstar-3C. Erosion and blow-by of both SRBs; joint temperature 70 degrees Fahrenheit.

The launch was first scrubbed due to a computer malfunction. A replacement was taken from *Challenger* and put in *Discovery*, as cannibalizing operations continued. At T-4 seconds, the main engines shut down when a valve malfunctioned; the entire engine was replaced. A 25-square-foot section of the orbiter body flap was scorched by a hydrogen fire on the launch pad after the aborted launch. Flames could be clearly seen on the NASA video feed. Launch was postponed again when electrical problems were found in the motor that separates the SRBs from the external tank. While in orbit, ice buildup blocked the wastewater dump line from the space toilet.

Flight	Launch	Orbiter	Crew	Flight Time
41-G	October 5, 1984	Challenger	Robert Crippen Jon A. McBride David Leestma Kathryn Sullivan Sally K. Ride Marc Garneau Paul Scully-Power	197 hours, 24 minutes, 32 seconds, 132 orbits

Payload: ERBS

Problems developed with the Ku-band antenna, which required much of the flight data to be recorded rather than transmitted live to earth as planned. For a time, while in flight, the cabin air-conditioner unit failed, bringing the temperature to 90 degrees. Due to a failure of a 3' × 8' insulation strip, heat caused some serious damage to the OMS pod which was replaced by the right-hand pod from *Atlantis*. Three dozen thermal tiles were replaced and three dozen more tiles needed to be repaired.

Flight	Launch	Orbiter	Crew	Flight Time
51-A	Nov. 8, 1984	Discovery	Frederick Hauck David M. Walker Dale A. Gardner Joseph Allen Anna Fisher	191 hours, 45 minutes, 54 seconds, 126 orbits

Payload: Palapa B2, Westar VI

Launch was postponed at the T-20 minute mark for a day, due to wind shear conditions in the upper atmosphere. One of the factors in the destruction of the *Challenger* on its final flight was a wind shear condition. The left OMS helium isolation valve leaked. Right nose landing gear door bent on landing.

Flight	Launch	Orbiter	Crew	Flight Time
51-C	January 24, 1985	*Discovery*	Ken Mattingly Loren J. Shriver James Buchli Ellison Onizuka Gary Payton	73 hours, 33 minutes, 47 orbits

Payload: Signit (spy satellite). Erosion of left and right SRB, prime and secondary O-rings; joint temperature 53 degrees Fahrenheit.

Thermal-tile problems on *Challenger* forced NASA to change this mission to the orbiter *Discovery*. Launch was postponed one day, due to cold temperatures at the launch pad. The mission may have been shortened due to an unspecified problem with an air force experiment. Thermal tiles received minor damage.

Flight	Launch	Orbiter	Crew	Flight Time
51-D	April 12, 1985	*Discovery*	Karol Bobko Donald Williams M. Rhea Seddon Jeffery Hoffman David Griggs Charles D. Walker Jake Garn	167 hours, 54 minutes, 108 orbits

Payload: Leasat 3, Anik-C. Erosion and blow-by of both SRBs; temperature of joints, 67 degrees Fahrenheit.

A 2,500-pound metal transport bucket slammed into the top of the Shuttle *Discovery* (cargo bay doors), leaving gashes and breaking the leg of a worker. One tire was blown out when the brakes locked and overheated on landing. A second tire was badly damaged. One thermal tile fell off during launch. One hundred twenty-three tiles were damaged, sixty of which needed to be replaced. Leasat 3 booster stage failed to fire, leaving the satellite in an unusable orbit. It would later be saved by another Shuttle mission.

Flight	Launch	Orbiter	Crew	Flight Time
51-B	April 29, 1985	*Challenger*	Robert Overmyer Frederick Gregory Don L. Lind William Thornton Norman Thagard Lodewijk van den Berg Taylor Wang	168 hours, 8 minutes, 47 seconds, 110 orbits

Payload: Spacelab-3. Erosion of right SRB, and erosion and blow-by of left SRBs; joint temperature 75 degrees Fahrenheit.

Monkey feces floated through a quarter-inch crack between an animal experiment cage and the Spacelab, contaminating the lab and the Shuttle. Lab animals were grounded after this mission until the problem could be solved. A fiber panel that held the thermal insulation blanket on the right OMS pod came loose, causing heat damage during reentry. Twenty thermal tiles needed to be replaced. The left main landing gear brakes were damaged on landing. Brake failure almost caused the orbiter to veer off the runway. Brakes and nose gear were redesigned after this mission.

Flight	Launch	Orbiter	Crew	Flight Time
51-G	June 17, 1985	*Discovery*	Dan Brandenstein John Creighton Shannon Lucid John Fabian Steven Nagel Patrick Baudry Salman Al-Saud	169 hours, 39 minutes, 30 seconds, 112 orbits

Payload: Morelos-A, Arabsat-A, Telstar 3-D, Spartan 1. Erosion and blow-by of right SRB, erosion and two blow-bys of left SRB; joint temperature 70 degrees Fahrenheit.

SRB performance was lower than expected. Two small maneuvering rockets on the Shuttle malfunctioned. Due to the many brake problems on past missions, the Shuttle landed on the dry lakebed at Edwards Air Force Base.

Flight	Launch	Orbiter	Crew	Flight Time
51-F	July 29, 1985	*Challenger*	Gordon Fullerton Roy Bridges Karl Henize Anthony England Story Musgrave Loren Acton John-David Bartoe	190 hours, 45 minutes, 26 seconds, 126 orbits

Payload: Spacelab-2. Heat indicated on prime O-ring of right SRB; joint temperature 81 degrees Fahrenheit.

Launch on July 12 was aborted at T-3 seconds when an eight-inch coolant valve failed to close. During launch, engine number one shut down at 5 minutes and 45 seconds, when a sensor indicated that it was overheating. It caused an Abort to Orbit (ATO). A second temperature sensor was turned off to prevent a second engine from shutting down during launch. If a second engine had failed, the crew would have had to try an emergency landing, probably at Edwards Air Force Base. All three sensors were found to be defective. Several thermal tiles were damaged.

Flight	Launch	Orbiter	Crew	Flight Time
51-I	August 27, 1985	*Discovery*	Joe H. Engle Richard Covey James van Hoften John Lounge William Fisher	171 hours, 17 minutes, 42 seconds, 111 orbits

Payload: Aussat-1, ASC-1, Leasat-F4. Erosion in two places on left SRB; joint temperature 76 degrees Fahrenheit.

The August 24 launch attempt was scrubbed due to poor weather. Launch was postponed again on August 25, due to a faulty computer. A problem developed in the RMS elbow joint. A computer from *Challenger* was removed and put on *Discovery*.

Flight	Launch	Orbiter	Crew	Flight Time
51-J	October 3, 1985	*Atlantis*	Karol Bobko Ronald Grabe Robert Stewart David Hilmers William Pailes	97 hours, 44 minutes, 38 seconds, 65 orbits

Payload: Top-Secret Two DSCS-II

Two thermal tiles were reported damaged and needing to be replaced. Brakes were damaged on landing.

Flight	Launch	Orbiter	Crew	Flight Time
61-A	October 30, 1985	*Challenger*	Henry Hartsfield Steven Nagel James Buchli Bonnie Dunbar Guion Bluford Ernst Messerschmid Wubbo Ockels Reinhard Furrer	168 hours, 44 minutes, 51 seconds, 110 orbits

Payload: Spacelab-D1. Erosion on right SRB, two blow-bys of left SRB; joint temperature 75 degrees Fahrenheit.

The helium pressure regulator in the right OMS fuel-feed system failed just after liftoff. Temperature in fuel cell number one did not control correctly and ran 20 degrees too cold. Fire alarm went off six times on board *Challenger*, but these proved to be false alarms. Elliptical mirror furnace experiment failed in Spacelab module. Air leak at around two pounds per hour was detected in one of the experiments. Eighteen thermal tiles damaged, which NASA said was the lowest amount so far in the flight series. Landing gear was damaged when the Shuttle, after landing on the lakebed at Edwards Air Force Base, got stuck in the mud while being towed.

Flight	Launch	Orbiter	Crew	Flight Time
61-B	Nov. 26, 1985	*Atlantis*	Brewster Shaw Bryan O'Conner Mary Cleave Sherwood Spring Jerry Ross Charles Walker Rodolfo Neri Vela	165 hours, 5 minutes, 45 seconds, 108 orbits

Payload: Morelos-B, Aussat-2, Satcom Ku-2

As a crane lifted a left SRB section, a pin was broken that held the ring to the rocket section. A board of inquiry blamed workers for not handling it properly, as well as for using faulty equipment. The inquiry also documented inexperienced and unmotivated workers at the space center. Workers were performing jobs they were not trained to do. An air force study found that NASA's Shuttle ground-support operations were flawed by poor record keeping, disorganized maintenance work and other flaws. Maintenance monitoring was said to be so flawed and ineffective that their "use for long-range planning was not possible." Flights 61-B and STS-2 were cited as flight delay examples. Eleven thermal tiles needed to be replaced due to flight reentry damage.

Flight	Launch	Orbiter	Crew	Flight Time
61-C	January 12, 1986	*Columbia*	Robert Gibson Charles Bolden George Nelson Steven Hawley Franklin Chang-Diaz Robert Cenker William Nelson	146 hours, 4 minutes, 9 seconds, 96 orbits

Payload: Syncom Ku-1, MSL-2. Erosion of right SRB, erosion and blow-by of left SRB; joint temperature 58 degrees Fahrenheit.

The planned December 19 launch was aborted at T-14 seconds due to a false electrical-problem reading in the right SRB. Rescheduled launch on January 6, 1986, was scrubbed at T-31 seconds due to low temperature in the liquid-hydrogen fuel line. A low amount of liquid oxygen was loaded in the external tank due to worker fatigue. This would have caused an abort situation during flight and the orbiter would have had to perform an emergency landing. A liquid-oxygen sensor broke off and was stuck in a pre-valve of engine number two.

Flight	Launch	Orbiter	Crew	Flight Time
51-L	January 28, 1986	*Challenger*	Francis R. Scobee Michael Smith Ellison Onizuka Judith Resnik Ronald McNair Christa McAuliffe Gregory Jarvis	1 minute, 13 seconds, 0 orbits

Payload: TDRS-B, Spartan. Orbiter destroyed with loss of crew and payload; joint temperature 28 degrees Fahrenheit, plus or minus 5 degrees.

It would be thirty-two months before another Shuttle would be launched. The program would not be back on course until the Shuttle *Discovery* lifted off from the Kennedy Space Center on September 29, 1988, carrying a crew of five and a replacement for the TDRS destroyed in the *Challenger* accident.

> "I see this as an extraordinary opportunity and a wonderful year
> [away from] what I would normally have been doing."
> — Payload Specialist Sharon Christa McAuliffe

APPENDIX II

The Flights of Columbia

C = Commander
P = Pilot
MS = Mission Specialist
PS = Payload Specialist

1. April 12, 1981   STS-1   *First Shuttle test flight*
   Crew: C — John W. Young, P — Robert L. Crippen.
   The first launch attempt was scrubbed when a computer failed nine minutes before launch.

2. November 12, 1981   STS-2   *Test flight*
   Crew: C — Joe Engle, P — Richard Truly.
   One Auxiliary Power Unit failed on orbit, causing mission to be cut short.

3. March 22, 1982   STS-3   *Test flight*
   Crew: C — Jack Lousma, P — Gordon Fullerton.
   This was the only flight to land at White Sands, New Mexico. Thirty-eight of the thermal tiles were lost during launch and nineteen others were damaged. Toilet failed in orbit. Two television cameras also failed during the mission, and there was a problem with Shuttle cargo bay doors.

4. June 26, 1982   STS-4   *Test flight*
   Crew: C — Ken Mattingly, P — Henry Hartsfield.
   Hail damage on tiles caused 400 dimples. Both solid rocket boosters sank in the ocean after launch. There was a small leak in the primary thruster and a problem with the Shuttle's cargo bay doors.

5. November 11, 1982   STS-5   *Deploy SBS-3 and ANIK-C3*
   Crew: C — Vance D. Brand, P — Robert Overmyer, MS — Joseph P. Allen, William B. Lenoir.
   EVA was cancelled as both spacesuits malfunctioned.

6. November 28, 1983   STS-9   *Spacelab 1*
   Crew: C — John W. Young, P — Brewster Shaw, MS — Robert Parker, Owen Garriott, Byron Lichtenberg, Ulf Merbold.

One Auxiliary Power Unit leaked and caught fire during reentry. Flight was delayed due to corrosion of a rocket nozzle.

7. January 12, 1986　　　　　STS-61-C　　　　*Deploy Satcom KU-1*
Crew: C — Robert Gibson, P — Charles Bolden, MS — George Nelson, Steven Hawley, Franklin Chang-Diaz, Robert Cenker, William Nelson.
Launch was delayed for twenty-five days. A helium tank leaked and several control thrusters failed during launch. One Auxiliary Power Unit failed during reentry.

8. August 8, 1989　　　　　STS-28　　　　　*Deploy KH-12 Spy satellite*
Crew: C — Brewster H. Shaw, P — Richard N. Richards, MS — David C. Leestma, James C. Adamson, Mark N. Brown.
Smoke appeared in the cockpit when a cable short circuited in orbit. Eight hundred tiles required repair after this mission due to longer exposure to reentry heat related to the roughness of *Columbia*'s left wing.

9. January 9, 1990　　　　　STS-32　　　　　*Deploy Syncom IV F5, LDEF recovery*
Crew: C — Dan Brandenstein, P — James Wetherbee, MS — Bonnie J. Dunbar, David Low, Marsha Ivins.
The Shuttle went into an uncontrolled rotation while the crew was asleep. Reentry was delayed due to a computer failure while on orbit. Computer failure complicated telescope deployment.

10. December 2, 1990　　　　STS-35　　　　　*Spacelab 4, Astro 1*
Crew: C — Vance D. Brand, P — Guy Gardner, MS — John Lounge, Jeff Hoffman, Robert Parker, PS — Ronald Parise, Samuel Durrance.
Mission was postponed for six months due to problems with the cooling system and fuel leaks. Computer failure complicated telescope deployment.

11. June 5, 1991　　　　　　STS-40　　　　　*Spacelab SLS-1*
Crew: C — Bryan D. O'Connor, P — Sidney M. Gutierrez, MS — M. Rhea Seddon, James P. Bagian, Tamara E. Jernigan, PS — F. Andrew Gaffney, Millie Hughes-Fulford.
Computer malfunctions and engine problems delayed launch fourteen days.

12. June 25, 1992　　　　　STS-50　　　　　*U.S. Microgravity Lab—1*
Crew: C — Richard N. Richard, P — Kenneth D. Bowersox, MS — Bonnie J. Dunbar, Ellen S. Baker, Carl J. Meade, PS — Lawrence J. DeLucas, Eugene H. Trinh.
Two defective navigation units were replaced. Temperature transducers on main engine number two became suspect and was replaced.

13. October 22, 1992　　　　STS-52　　　　　*Deploy Lageos-2*
Crew: C — James Wetherbee, P — Michael A Baker, MS — C. Lacy Veach, William M. Shepard, Tamara E. Jernigan, PS — Steven MacLean.
Cracked engine nozzles postponed launch.

14. April 26, 1993　　　　　STS-55　　　　　*Spacelab-D2 (Germany)*
Crew: C — Steven R. Nagel, P — Terence Henricks, MS — Jerry L. Ross, Charles J. Precourt, Bernard A. Harris, Jr., PS — Hans Schlegel, Ulrich Walter.
Contaminated valve on main engine aborted launch as all engines were up and running three seconds before liftoff.

15. October 18, 1993   STS-58   *Spacelab-SLS-2*
Crew: C — John E. Blaha, P — Richard A. Searfoss, MS — M. Rhea Seddon, William S. McArthur Jr., David A. Wolf, Shannon W. Lucid, PS — Martin Fettman.
Computer failure aborted launch thirty-one seconds before scheduled liftoff. Flight was delayed seven days to replace two Auxiliary Power Units found to be defective.

16. March 4, 1994   STS-62   *U.S. Microgravity Payload — 2*
Crew: C — John H. Casper, P — Andrew M. Allen, MS — Pierre J. Thuot, Charles D. Germar, Marsha S. Ivins.
Launch was delayed one day due to high winds at launch site. Unusual high pressure reading appeared in one APU.

17. July 8, 1994   STS-65   *International Microgravity Lab-2*
Crew: C — Robert Cabana, P — James Halsell, Jr., MS — Richard Hieb, Leroy Chiao, Chiaki Mukai, Carl Walz, Donald Thomas.
The water dump system froze on orbit.

18. October 20, 1995   STS-73   *U.S. Microgravity Lab 2*
Crew: C — Kenneth D. Bowersox, P — Kent V. Rominger, MS — Kathryn C. Thornton, Catherine G. Coleman, Michael Lopez-Alegria, PS — Fred W. Leslie, Albert Sacco, Jr.
Launch is scrubbed six times.

19. February 22, 1996   STS-75   *U.S. Microgravity Payload — 3*
Crew: C — Andrew Allen, P — Scott Horowitz, MS — Franklin Chang-Diaz, Claude Nicollier, Jeffrey Hoffman, PS — Maurizio Cheli, Umberto Guidoni.
Contaminates on a satellite tether short circuited, causing the loss of the satellite.

20. June 20, 1996   STS-78   *Life and Microgravity Spacelab*
Crew: C — Terence Henricks, P — Kevin Kregel, MS — Richard Lenneham, Susan Helm, Charles Brady, PS — Jean-Jacques Favier, Robert Thirsk.
Thermal tiles were damaged.

21. November 19, 1996   STS-80   *Wake Shield Facility 3*
Crew: C — Kenneth Cockrell, P — Kent V. Rominger, MS — Story Musgrave, Tamara E. Jernigan, Thomas Jones.
EVA was canceled when the airlock-latching mechanism was jammed by a loose screw. Wake Shield Facility nearly collided with the Shuttle after deployment.

22. April 4, 1997   STS-83   *Microgravity Science Lab — 1*
Crew: C — James Halsell, Jr., P — Susan Still, MS — Mike Gernhardt, Janice Voss, Don Thomas, PS — Roger Crouch, Greg Linteris.
Defective fuel cell cut mission short by twelve days, with very little science conducted. The fuel cell could have exploded.

23. July 1, 1997   STS-94   *Microgravity Science Lab — 1 (Re-flight)*
Crew: C — James Haisell, Jr., P — Susan Still, MS — Mike Gernhardt, Janice Voss, Don Thomas, PS — Roger Crouch, Greg Linteris.
Alarms went off as pressure dropped in one of the APUs. Computer problems developed with a combustion experiment in lab. A micrometeoroid hit a window, causing a small ding.

24. November 19, 1997   STS-87   *Spartan Solar Observation Satellite, USMP-4*

Crew: C — Kevin R. Kregel, P — Steven W. Lindsey, MS — Winston E. Scott, Kalpana Chawla, Takao Doi, PS — Leonid K. Kadenyuk.

Spartan satellite was deployed, failed, and was then pushed by using the Canada arm, putting the satellite into a tumble. Spartan was then grabbed by EVA astronauts and brought back on board.

25. April 17, 1998   STS-90   *Neurolab*

Crew: C — Richard A. Searfoss, P — Scott Altman, MS — Richard M. Linnehan, Dafydd Rhys Williams, Kathryn P. Hire, PS — Jay C. Buckey, James A. Pawelczyk.

Air purification system leaked and was repaired by crew.

26. July 23, 1999   STS-93   *Chandra X-Ray Telescope*

Crew: C — Eileen Collins, P — Jeff Ashby, MS — Steve Hawley, Cady Coleman, Michel Tognini.

One main engine leaked hydrogen fuel during the entire launch into orbit. Controllers on two main engines were shut down due to short circuits.

27. March 1, 2002   STS-109   *Hubble Space Telescope servicing*

Crew: C — Scott Altman, P — Duane Carey, MS — John Grunsfeld, Rick Linneham, Mike Massimino, John Newman.

Launch was delayed by weather, as well as a failure in a landing gear ball bearing.

28. January 16, 2003   STS-107   *SpaceHab RDM*

Crew: C — Rick Husband, P — William McCool, MS — Michael Anderson, David Brown, Kalpana Chawla, Laurel Clark, PS — Ilan Ramon.

Columbia was destroyed during reentry.

APPENDIX III

The Mission and Crew of Challenger *51-L*

"A lot of people don't realize the Shuttle has a lot of valuable things on it. They see people on television floating around in space and that's it. They don't realize the valuable experiments that are done."
— Payload Specialist Sharon Christa McAuliffe

On January 27, 1985, Shuttle *Discovery* with the 51-C crew on board, including Ellison Onizuka, landed at the Kennedy Launch Center to complete this nation's first top-secret military Shuttle mission. Before *Discovery* could land, military aircraft assigned to air security had to chase four private planes out of the space center's dedicated air space. It was the eighteenth anniversary of the Apollo 1 pad fire which had killed astronauts Grissom, White and Chaffee. It was also the day NASA named five of the crew members who would fly on mission 51-L to deploy the third Tracking and Data Relay Satellite (TDRS-C). Selected were Richard Scobee, Mike Smith, Ellison Onizuka, Judith Resnik and Ronald McNair.

MISSION COMMANDER—FRANCIS R. (DICK) SCOBEE

Dick Scobee first flew on the Shuttle during *Challenger*'s fifth orbital flight on mission 41-C launched on April 6, 1984. Scobee flew as pilot on the mission which successfully deployed the Long Duration Exposure Facility (LDEF). This 21,400-pound structure carried fifty-seven experiments, testing what would happen to various structures and surfaces when they were exposed to the rigors of space. Later this crew would retrieve and repair the ailing Solar Maximum Satellite and then redeploys the satellite to continue its survey of the sun. This mission was critical to NASA, as it demonstrated the capability of the space agency to dock with, retrieve and repair satellites in orbit, as it had advertised for years. On board, the crew wore "Ace Satellite Repair Company" T-shirts, and the public cheered.

After the Shuttle landed at Edwards Air Force Base, the news media—using less than honest NASA press reports—hailed the flight as the most trouble free one to date. However, when the solids were inspected at Thiokol's plant, it was found that erosion and blow-by had occurred on the right booster. It was by no means a trouble-free flight. At the same time, no one seemed to notice that *Challenger* was being used for spare parts almost as soon as it had landed. As soon as *Challenger* was ready, it was flown back to Kennedy on top of its 747 carrier aircraft so that one of the OMS rocket pods, which had been borrowed from *Discovery*, could be removed and replaced back on *Discovery* so it could fly its next mission.

Scobee was born on May 19, 1939, in Cle Elum, Washington, and graduated from public high school in Auburn, Washington, in 1957. He later enlisted in the United States Air Force and trained as a mechanic on reciprocating engines. Although he did well in this occupation, he longed to be a pilot, but without a degree that dream would not become a reality. To fulfill his dream he took night classes from the University of Arizona and graduated with a B.S. degree in aerospace engineering. With his new degree it became possible for Scobee to apply for and receive an officer's commission in the air force. He then applied for and was granted pilot training, receiving his wings in 1966. His air force assignments would eventually include a combat tour in Vietnam. In the meantime, Scobee found time to marry June Kent of San Antonio, Texas, and father two children. He later attended the U.S. Air Force Aerospace Research Pilot School at Edwards Air Force Base in 1972, and became involved in several test programs. As a test pilot, Scobee flew over forty-five different types of aircraft and eventually logged more than 6,500 hours of flight time. These aircraft included the Boeing 747, which was chosen as the transport carrier for the Shuttle. He also flew the X-24B, F-111, and the C-5 transport. The X-24B was the last of the lifting-body experimental aircraft tested by the air force, landing on the same runway used by the Shuttle at Edwards Air Force Base. Scobee became a member of the astronaut team in 1978, being selected as a pilot, but he still found time to jog, ride motorcycles and continue his hobbies of woodworking and painting in oils.

Mission Pilot — Captain Michael John Smith

Although he was chosen as an astronaut in 1980, it would be nearly six years before the opportunity to fly in space would be presented to Mike. Born on April 30, 1945, in Beaufort, North Carolina, Smith went on to graduate from the United States Naval Academy in 1967 and receive an MS degree in aeronautical engineering from Naval Postgraduate School in 1968. He completed aviator training at Kingsville, Texas, in May 1969, earning his wings. Smith was then assigned to the navy's Advanced Jet Training Command as an instructor until 1971, when he was assigned to fly A-6 Intruders on board the USS *Kitty Hawk*, stationed off the coast of Vietnam for two years. While on duty in Vietnam, Smith would be awarded the Distinguished Flying Cross, three Air Medals and the Vietnam Cross of Gallantry with Silver Star. After his tours of duty, Smith was assigned to the Strike Aircraft Test Directorate at Patuxent River in Maryland where he worked on cruise-missile guidance systems.

His next job found him instructing navy test pilots, and he later flew jets in the Mediterranean from the aircraft carrier *Saratoga*. It was during this tour of duty when he applied for astronaut training and was accepted. He was married to Jane Jarrel and they had three children. Before the *Challenger* accident, Smith served on several Shuttle missions as capsule communicator, and as NASA media spokesman for several other flights.

Mission Specialist — Judith Arlene Resnik, Ph.D.

Judith Resnik was among the first group of American women selected for astronaut training in January 1978. For the next year Resnik's group trained for positions on the Shuttle as mission specialists. She was later assigned to a number of positions in support of the Shuttle program, working out of the astronaut office at Johnson Space Center, Houston, and becoming one of the first women to serve as capsule communicator for a human space flight. The capsule communicator position is traditionally a working slot for astronauts who soon fly in space.

Judith was born on April 5, 1949, in Akron, Ohio, where she went to public schools before attending Carnegie-Mellon University. She received a B.S. in electrical engineering in 1970, and continued her education at the University of Maryland, where she earned a Ph.D., also in electrical engineering, in 1977. She then went on to work for RCA Corporation during the early 1970s, as well as with the Laboratory of Neurophysiology at the National Institutes of Health in Bethesda, Maryland, from 1974 to 1977.

It would be on Shuttle *Discovery*'s first mission, designated as 41-D, that Resnik would have the opportunity to fly into space. However, this was after she and her fellow crew members experience a heart-stopping launch abort after main-engine ignition, only 4 seconds before lift-off. A computer detected a failure in one of the three main engines, and automatically shut down the launch before Shuttle *Discovery* could leave the launch pad. Once in orbit, this second American woman in space would help deploy three communication satellites and test a new 10-story-tall solar sail which was raised out of the cargo bay twice. This was a vital test of a technology needed to help power the space station *Freedom* then being designed by the United States.

Mission Specialist — Lieutenant Colonel Ellison Shoji Onizuka

When Ellison Onizuka flew on mission 51-C, our first totally dedicated military Shuttle flight, he quietly became the first Asian-American astronaut to orbit the earth. During this much-delayed Defense Department mission, Onizuka performed experiments as a mission specialist and was also involved in the deployment of a top-secret Sigint signals-intelligence-gathering satellite.

Ellison was born in Kealakekua, Kona, Hawaii, on June 24, 1946, of Japanese-American parents. After attending high school in Hawaii, he went on to attend the University of Colorado. It was at Colorado that Ellison was able to focus on his life's ambition to become an astronaut, by studying aerospace engineering under the Air Force ROTC program. He received his B.S. in June 1969 (one month before humans first landed on the moon), and his Masters in December of the same year. Commissioned in the air force in January 1970, Onizuka served on active duty until 1978. His air force assignments first sent him to Sacramento Air Logistics Center as an aerospace flight-test engineer. Later in 1975, he was posted to the Air Force Flight Test Center at Edwards Air Force Base in California, first as a squadron flight-test officer and later as Chief of the Engineering Support Section. He had logged more than 1600 hours of jet flight before applying for the astronaut corps and was accepted as an astronaut trainee in January 1978. At Edwards, Onizuka witnessed the Shuttle *Enterprise* complete its drop tests and landings where he was assigned and from where he applied to become an astronaut.

His first assignments as an astronaut trainee were on Orbiter test and checkout teams as well as working as part of the launch support crew for STS-1 and STS-2. He was able to accomplish his life's dream when he flew into space on mission 51-C.

While attending college, he married Lorna Yoshida, also from Hawaii, and had two children.

Mission Specialist — Ronald Erwin McNair, Ph.D.

Ron McNair was born October 21, 1950, in Lake City, South Carolina, the son of Carl C. McNair and Pearl M. McNair. Before he attended public schools, he had already learned to read. Upon graduating from high school as valedictorian, he attended North Carolina A&T State University, where he earned a B.S. degree in physics in 1971. He then went on to study physics at MIT, specializing in quantum electronics and laser technology. While at MIT he worked on chemical HF/DF and high-pressure CO lasers, publishing several cutting-edge scientific papers on those subjects. He completed his Ph. D. in physics in 1977. During his studies he became a Presidential

Scholar, a Ford Foundation Fellow, and the Omega Psi Phi Scholar of the Year. Becoming an expert in laser physics, he went on to study in France at Ecole d'été de Physique Théorique. It was during this period that he married Cheryl B. Moore of Brooklyn, New York, and fathered two children.

After receiving his Ph.D., McNair began work as a physicist for Hughes Aircraft Research Laboratories in Malibu, California, in the Optical Physics Department. His projects included research on electro-optic laser modulation for satellite-to-satellite space communications. During this period he applied for and was selected for astronaut training in January 1978, the same group as fellow *Challenger* crew members Onizuka, Scobee and Resnik.

On February 3, 1984, McNair began his space career on board *Challenger* for mission 41-B, as he and four other astronauts rocketed into earth orbit. During this mission two communication satellites were deployed from the cargo bay, and McNair used the robotic arm to deploy and recover the Spas payload from a rack in the cargo bay. It was also on this flight that NASA first tested the MMU (Manned Maneuvering Unit) which allow astronauts to fly around in space without being tethered to the spacecraft.

The team had now been selected for Shuttle mission 51-L which would deploy NASA's third relay satellite. Later, two other members would be added to the crew to complete the *Challenger* team, but for now it was up to these five highly trained individuals to prepare for launch and deployment of their payload.

Mission Planning/Payloads

Plans for what would eventually become mission 51-L began in early 1984. The process normally takes from twelve to eighteen months to complete in a complicated system which revolves around progressively more and more details being finalized in repetitive review cycles. For this *Challenger* flight there would be ten major changes or cycles which added or deleted payloads, causing some measure of disruption in the preparation for the flight. As would be expected, the closer to the planned launch date the changes occurred, the more difficult they were to input into the flight.

Mission 51-L had originally been scheduled for a July 1985 launch, but by January 1985, even before the flight crew had been assigned, the date had been postponed to late November of that year. NASA had announced that payload changes would delay 51-L, a problem that would occur again. Finally, NASA announced that *Challenger* would fly with two major satellite payloads and a group of experiments in the crew compartment. The final manifest included:

- Tracking and Data Relay Satellite-B
- Spartan-Halley Satellite
- Comet Halley Active Monitoring Program
- Fluid Dynamics Experiment
- Phase Partitioning Experiment
- Teacher-in-Space Project
- Shuttle Student Involvement Program
- Radiation Monitoring Experiment

TDRS-B

The primary payload to be taken into orbit by the *Challenger* was NASA's second relay satellite. The first had been deployed by *Challenger* on her first mission, designated STS-6. TDRS-B was to have been identical to the first TDRS, with both having the capability to support up to twenty-three user spacecraft simultaneously. Such support included air-to-ground as well as spacecraft-to-spacecraft capability. Both civilian and military communications were to have passed through TDRS-B. As reported by NASA, the satellite would provide "two basic types of service, a multiple access service which can relay data from as many as 19 low-data-rate user spacecraft at the same time and a single access service which will provide two high data rate communications relays from each satellite."

TDRS-B was scheduled to be deployed by the crew using a spring-eject system around ten hours into the flight, approximately 154 nautical miles above the earth. After deployment, Commander Scobee would move the Shuttle away from the satellite and its attached IUS (Inertial Upper Stage) to a safe distance to observe the IUS burn. The IUS would then fire for 2 minutes and 26 seconds, coast for several hours, then fire again for 1 minute 49 seconds, finally placing the TDRS-B into its predetermined geosynchronous orbit.

Spartan-Halley Satellite

The Spartan-Halley was a low-cost, Shuttle-deployed-and-retrieved satellite developed by the Goddard Space Flight Center and the University of Colorado. The system used off-the-shelf instruments placed on a standard frame designed to be reused on many Shuttle missions. Its primary goal on this flight was to record ultraviolet radiation from Comet Halley when it was closest to the sun, in the hope that these observations would show how fast water is broken down by solar radiation. Further, the probe was to search for carbon and sulfur atoms, and to detail the development of Halley's tail. All activities were to be accomplished while the Spartan-Halley was in a free-flight mode away from the Shuttle.

Spartan-Halley used two spectrometers built from instruments made as backups for a Mariner 9 instrument which was launched in 1971 and orbited the planet Mars. Two Nikon 35mm cameras were also part of the satellite, with 105mm and 135mm lenses, loaded with special color film to record the dust and ion tails, as well as debris bursts from the surface of the comet. The program of observation was expected to last twenty orbits, with about 3000 seconds of available observation time per orbit. At the time Comet Halley would have been approximately 139 million miles from the earth and some 59.5 million miles from the sun at perihelion (the closest point in its orbit about the sun), and expected to be most active.

Comet Halley Active Monitoring Program

With a hand-held 35mm camera, one of the crew members would photograph the comet while enclosed in a shroud specially made to shut out all of the lights in the crew cabin. The objective was to target with photography the dynamics and changes of the main body and tail of the comet, as well as imaging the chemical structure of the body. Further, it was planned that spectra of the comet would be photographed using a special grating designed to be used with the 35mm camera and an image intensifier. These images would be taken above most of the earth's atmosphere, thus removing this interfering aspect of cometary study.

Fluid Dynamics Experiment

A Hughes Aircraft Company experiment to understand more fully how liquid fuels interact with the spacecraft was to have flown on *Challenger*. As larger liquid-fuel rockets were being designed to replace the solid rockets used for transfer orbits, it became necessary to understand the dynamics of these fuels. The larger the spacecraft, the greater the problems encountered by the destabilizing effects of sloshing liquid propellant. The experiment involved an enclosed tank, which would be partially filled with liquids to determine the behavior of those liquids in micro gravity, with those behaviors recorded on a video camera. A second set of experiments would study the effects of transmitted motions from the simulated spacecraft tanks to the fluids themselves.

Phase Partitioning Experiment (PPE)

This experiment involved the separation of biomedical materials such as cells and proteins in the micro-gravity environment. It is begun by adding various polymers to a water solution containing the materials to be separated. In theory, the experiment could have been able to separate the materials to a higher quality than can be achieved on earth in one gravity. This was to be the first flight of the PPE and was fully contained in a hand-held device a little larger than a cigarette box.

Teacher-in-Space Project

The Teacher-in-Space Project was initiated by President Ronald Reagan as the start of a program to expand access to space to private citizens. Officially designated the Space Flight Participant Program, it would allow selected private citizens to fly on board a Shuttle. Individuals were to be selected who would be best able to return to earth and tell of their experiences in orbit. It was therefore a logical choice to select a teacher who would not only be able to bring back her experiences to share, but could teach a generation of young people while in space.

On day six of the mission, the teacher/observer was scheduled to conduct two lessons, one in the morning and one in the afternoon. The first was entitled, "The Ultimate Field Trip," which would be a guided tour of the spacecraft and a comparison of daily life on board to that on earth. The second lesson was entitled, "Where We've Been, Where We're Going, Why?" This lesson focused on the technologies being developed, using the micro gravity of earth-orbital space, and what could be expected to be developed in the future. NASA could not have planned a better public-relations operation than these two lessons.

When not engaged in teaching classes, the teacher would be videotaping activities for later classroom use in the areas of magnetism, Newton's laws, effervescence, simple machines/tools, hydroponics in micro gravity, and chromatographic separation of pigments.

Shuttle Student Involvement Program

Continuing a NASA program to involve American students in space research, *Challenger* flew three experiments designed by high school students. The first was an attempt to control the growth of crystals using a semipermeable membrane. The second experiment investigated the effects of weightlessness on grain formation and strength in metals. The final experiment flew twelve white leghorn chicken eggs to study chicken embryo development in space.

Radiation Monitoring Experiment

This passive radiation experiment would continuously monitor the radiation in the crew cabin to help scientists further understand the environment of low-earth-orbit radiation fields in which the Shuttle operates. This type of experiment was a continuation of years of study begun during the Mercury program of the early 1960s.

Summary of Major Activities

The following is taken from Space Shuttle Mission 51-L Press Kit, issued January 1986.

Day One	• Payload bay doors open • Tracking and Data Relay Satellite (TDRS) and Inertial Upper Stage (IUS) checkout • TDRS deployment
Day Two	• Comet Halley Active Monitoring Program (CHAMP) data take • Spartan-Halley satellite check • Fluid dynamics experiment (FDE) • Teacher-in-Space activities
Day Three	• Spartan deploy preparation • Spartan deploy • Student experiments • Fluid dynamics experiment
Day Four	• CHAMP data taken • Student experiments • Fluid dynamics experiment • Teacher-in-Space activities
Day Five	• Spartan rendezvous • Spartan capture • Fluid dynamics experiments • Student experiments
Day Six	• Reaction control system hot fire • FCS checkout • Teacher lesson (field trip) • Teacher lesson (exploration)
Day Seven	• Landing at Kennedy Space Center

The Planning Continues as the Teacher Arrives

The much-delayed cargo integration review for mission 51-L took place on June 18, 1985. This review is one of the major milestones in the planning process. It had been postponed and rescheduled six times due to payload changes. At the review, all payload requirements were checked to insure that they could be accomplished within the capabilities of the vehicle and to insure that the crew would be able to perform all of the flight requirements. This included a review of crew members' backgrounds and specific mission training. After the cargo integration review, mission planners are able to develop the final flight design products. For mission 51-L, most of the payload changes occurred before the review, so there were no serious disruptions in the mission-planning process.

The next step in planning involves a procedure called the "flight design process," which becomes the driving force for the rest of the mission development until a flight-readiness review is conducted very near the launch date. This process reviews all of the objectives of the mission and organizes them into a detailed sequence of events from launch through landing. It includes a point-by-point schedule-of-events check, listing all of the trajectory data expected to occur, any consumable requirements and their periods of use. Also, any communications requirements and computer programming for the orbiter, Mission Control and the Shuttle training simulator are considered as part of the "flight design process." Before this process could be completed, a new addition to the crew was announced—Sharon Christa McAuliffe.

A Ticket to Ride

On August 27, 1984, President Reagan announced that the first private citizen to fly in space on the Shuttle would be a teacher. The president made the decision with a mind toward encouraging interest in high-technology education throughout the United States, in students from grade school through high school. NASA selected the Council of Chief State School Offices, through the Department of Education, to organize the selection process and to establish the criteria for selection. The target flight date was set for late 1985 or early 1986. It was decided that since college instructors were already able to apply as payload specialists through the university system or individually, that they would not qualify under this new program.

Exactly 11,146 teachers from across the country, Puerto Rico, Guam, the U.S. Virgin Islands, Department of Defense overseas schools, Department of State overseas schools, and Bureau of Indian Affairs schools applied for the opportunity to fly in space. From each state, territory and agency two individuals were selected, for a total of 114 nominees who met in Washington, D.C., June 22–27, 1985, for a conference focusing on space education. As part of the conference, all of the nominees met with the National Review Panel which selected ten finalists. The panel consisted of three former astronauts, three university presidents, former basketball player Wes Unseld, Dr. Robert Jarvik (who invented the artificial heart) and actress Pam Dawber. (A basketball player and an actress?) On July 1, 1985, NASA and the Council of Chief State School Officers announced the ten finalists:

- Kathleen Anne Beres—Kenwood High School, Baltimore, Maryland
- Robert S. Foerster—Cumberland Elementary School, West Lafayette, Indiana
- Judith Marie Garcia—Thomas Jefferson School for Science & Technology, Alexandria, Virginia
- Peggy J. Lathlaen—Westwood Elementary School, Friendwood, Texas
- Sharon Christa McAuliffe—Concord High School, Concord, New Hampshire
- David M. Marquart—Boise High School, Boise, Idaho
- Michael W. Metcalf—Hazen Union School, Hardwick, Vermont
- Richard A. Methia—New Bedford High School, New Bedford, Massachusetts
- Barbara R. Morgan—McCall-Donnelly Elementary School, McCall, Idaho
- Niki Mason Wenger—Vandevender Jr. High School, Parkersburg, West Virginia

For three days in July, the finalists met with a NASA evaluation committee, which consisted of senior NASA officials. They made recommendations to the administration of NASA who would make the final selection for the Teacher-in-Space Project. On July 19, 1985, Vice President George Bush announced that high school social studies teacher Sharon Christa McAuliffe had been selected to fly the Space Shuttle and that elementary school teacher Barbara R. Morgan would act as her backup for the flight. McAuliffe said, "When that Shuttle goes up, there might be one body, but [there will] be ten souls that I'm taking with me."

PAYLOAD SPECIALIST—SHARON CHRISTA MCAULIFFE

Christa McAuliffe was born on September 2, 1948, the oldest child of Edward and Grace Corrigan. During the Apollo program Christa was very excited over the moon landing and space flight development, which never faded as the years went on. She wrote on her astronaut application form, "I watched the space age being born and I would like to participate." She graduated from Framingham State College in 1970. A few weeks after graduation she married Steven McAuliffe and moved to Washington, D.C., so that her new husband could attend Georgetown Law School. In the Washington area she taught in secondary schools and specialized in American history and social studies. Later, Christa completed her Masters degree at Bowie State University in Maryland, while she continued to teach. In 1978, she moved with her husband to Concord, New Hampshire, and took a teaching post at Concord High School. She had two children, Scott and Caroline.

Christa, when interviewed after her selection, said, "Just as the pioneer travelers of the Conestoga wagon days kept personal journals, I, as a pioneer space traveler, would do the same.... Future historians would use my eyewitness accounts to help in their studies of the impact of the space age on the general population." McAuliffe's hand-held voice recorder used during the flight was recovered and is now located at one of the NASA centers.

After the "Teacher-in-Space" flight was completed, the publicity-minded NASA had decided that the next individual to fly on board a Shuttle as part of its Space Flight Participant Program would be a full-time journalist. Forty candidates were to have been selected by March 1986, with the primary and backup journalist selected by April 17. Reports were flying, mostly among the press, that venerable space reporter Walter Cronkite had the inside track, but NASA insisted that the selection process was wide open.

After the addition of the teacher to the crew of 51-L the launch-minus-five-months Flight Planning and Stowage Review was conducted to review and address any unresolved issues in the mission planning. The review was conducted on August 20, 1985. Normally mission events are firmly established by the time this review is conducted, but there would be one more addition to the 51-L crew after this review was completed. During this review the crew activity plan and the formal flight requirements were gone over once again. Also addressed were the photo and television requirements and crew-compartment stowage issues. It was found that there was a need for further launch window reviews due to the addition of Christa McAuliffe to the crew. The mission was starting to get a bit complicated and would continue to do so with the further addition of the Hughes Aircraft Company payload specialist.

PAYLOAD SPECIALIST—GREGORY BRUCE JARVIS

On July 6, 1984, Hughes Aircraft Company newsletter *Hughes News* announced, "Lucky Two Picked for Ride in Space." Picked from among 600 company applications, Greg Jarvis and John Konrad were chosen to become payload specialists on upcoming Shuttle flights. Originally these two Hughes engineers were selected to fly on missions to deploy a new generation of Hughes Communication satellite, but schedule changes would place Jarvis on a flight with no Hughes payload other than the fluid dynamics experiments which could be flown on any Shuttle flight. Along with Jarvis and Konrad, Hughes Aircraft selected two backups, Bill Butterworth and Steve Cunningham.

Greg Jarvis was an 11-year employee of Hughes and was a project manager in the Systems Applications Laboratory, involved with the Leasat program since it was begun. Born on August 24, 1944, in Detroit, Michigan, he had received his B.S. in electrical engineering in 1967, from the State University of New York, Buffalo, and later received an M.S. in the same field in 1969, from Northeastern University in Boston. Later, Jarvis received an M.S. in management science from West Coast University, Los Angeles, in 1973. He was assigned to mission 51-L on October 25, 1985.

The December 3, 1984 issue of *Aviation Week and Space Technology* reported, "Corporate astronauts will require only minimal Shuttle flight training, similar to the payload specialists training required for the Shuttle flight observer participants." In that issue a list of NASA training areas was published.

- Medical examination   8 hours
- Environmental familiarization/altitude-chamber run   8 hours

- Shuttle vehicle familiarization workbook — 2 hours
- Space flight physiology — 3 hours
- Flight familiarization with the T-38 and Boeing KC-135 — 4 hours
- Kennedy Space Center launch and landing oriented training — 8 hours
- First Aid training — 8 hours
- Orbiter internal location codes with workbook — 2 hours
- Orbiter light switches with workbook — 2 hours
- Flight data file with workbook — 5 hours
- Toilet workbook — 2 hours
- Food and dining with workbook — 2 hours
- Crew equipment with workbook — 2 hours
- Orbiter ingress/egress with workbook — 2 hours
- Cabin habitability issues — 5 hours
- Orbiter audio system — 2 hours
- Zero-G toilet training — 1 hour
- Ascent/reentry cabin procedures & ingress/egress mockup train — 19 hours
- Medical procedures — 3 hours
- Shuttle mission simulator ascent runs — 4 hours
- Shuttle mission simulator in orbit runs — 4 hours
- Shuttle mission simulator reentry runs — 4 hours
- Simulator/Mission Control integrated simulations — 8 hours
- Mission timeline review — 4 hours

For many within the halls of NASA, the prospect of flying private citizens on Shuttle flights did not sit well. Many of the astronauts shared these views, but Administrator James Beggs felt NASA needed this type of publicity. He wanted to continue the propagandized impression that the Shuttle was fully operational and therefore prove that nearly anyone could fly on board.

Members of Congress saw this as a golden opportunity to find their own way to the stars. Both Congressman Bill Nelson and Senator Jake Garn put as much pressure as they could muster to force NASA's hand. When NASA Administrator Beggs invited Senator Garn to fly in November 1984, he justified the decision by saying that it was "appropriate for members of Congress to participate in flight demonstrations of the hardware for which they have oversight responsibility." Many felt that it was a very weak explanation. Garn, who was then Chairman of the Senate Appropriations Subcommittee on Housing and Urban Development, also oversaw NASA's budget. In politics, pressure does not get any greater than when it comes to funding. In one Senate hearing held on May 12, 1981, Garn stated, "My first question is: when do I get to ride the Space Shuttle?" In 1982 he said, "My continuing desire some time in the future [is] to personally check out the Space Shuttle, so we can provide continuing appropriations." In other words, no flight for Garn could mean no money for NASA. Other legislators saw the opportunity to do a little high-tech sightseeing on the Shuttle, including Representative Eldon Rudd (R–AZ) and Representative Beverly Byron (D–MD). Unlike many of these "passengers," including one who was said to have been restrained on an earlier flight, McAuliffe and Jarvis were considered professionals by the other crew members; they trained hard for their upcoming flight.

With Hughes Payload Specialist Greg Jarvis now on board, the crew was finally complete. The crew then began its final period of training thirty-seven weeks before the launch. NASA soon found that training time for this mission would become more and more compressed due to the limited availability of the Shuttle mission simulator and the time pressures by other crews also training for future missions. The 51-L crew required Shuttle simulator training on the robot arm, rendezvous in space, Inertial Upper Stage deployment, and ascent and entry procedures before they could be certified for flight. On average, the crew put in 48.7 hours per week during the final nine weeks before launch, reaching as high as 70 hours a week. Launch and landing delays for mission 61-C also become a training issue when that crew went back to the simulator for additional training during launch delays for their flight. Due to problems encountered in the launch of 61-C, the *Challenger* launch was delayed several times. Training for 51-L would become increasingly compressed and difficult to schedule, but NASA was reporting that throughout the trying period no training time was lost.

Appendix IV

The Crew of Columbia STS-107

This information was obtained from the *Columbia* Accident Report.

Mission Commander—Rick Husband

Rick Husband, 45, was a colonel in the U.S. Air Force, a test pilot, and a veteran of STS-96. He received a B.S. in mechanical engineering from Texas Tech University and a M.S. in Mechanical Engineering from California State University, Fresno. He was a member of the Red Team, working on experiments including the European Research in Space and Terrestrial Osteoporosis and the Shuttle Ozone Limb Sounding Experiment.

Pilot—William C. McCool

William C. McCool, 41, was a commander in the U.S. Navy and a test pilot. He received a B.S. in applied science from the U.S. Naval Academy, an M.S. in computer science from the University of Maryland, and an M.S. in aeronautical engineering from the U.S. Naval Postgraduate School. A member of the Blue Team, McCool worked on experiments including the Advanced Respiratory Monitoring System, Biopack, and Mediterranean Israeli Dust Experiment.

Payload Commander/Mission Specialist Michael P. Anderson

Michael P. Anderson, 43, was a lieutenant colonel in the U.S. Air Force, a former instructor pilot and tactical officer, and a veteran of STS-89. He received a B.S. in physics/astronomy from the University of Washington, and a M.S. in physics from Creighton University. A member of the Blue Team, Anderson worked with experiments including the Advanced Respiratory Monitoring System, Water Mist Fire Suppression, and Structures of Flame Balls at Low Lewis-number.

Mission Specialist—David M. Brown

David M. Brown, 46, was a captain in the U.S. Navy, a naval aviator, and a naval flight surgeon. He received a B.S. in biology from the College of William and Mary and an M.D. from Eastern Virginia Medical School. A member of the Blue Team, Brown worked on the Laminar Soot Processes, Structures of Flame Balls at Low Lewis-number, and Water Mist Fire Suppression experiments.

Mission Specialist — Kalpana Chawla

Kalpana Chawla, 41, was an aerospace engineer, a FAA certified flight instructor, and a veteran of STS-87. She received a B.S. in Aeronautical Engineering from Punjab Engineering College, India, an M.S. in aerospace engineering from the University of Texas, Arlington, and a Ph.D. in Aerospace Engineering from the University of Colorado, Boulder. A member of the Red Team, Chawla worked with experiments on Astroculture, Advanced Protein Crystal Facility, Mechanics of Granular Materials, and the Zeolite Crystal Growth Furnace.

Mission Specialist — Laurel Clark

Laurel Clark, 41, was a commander in the U.S. Navy and a naval flight surgeon. She received both a B.S. in zoology and an M.D. from the University of Wisconsin, Madison. A member of the Red Team, Clark worked on experiments including the Closed Equilibrated Biological Aquatic System, Sleep-Wake Actigraphy and Light Exposure During Spaceflight, and the Vapor Compression Distillation Flight Experiment.

Payload Specialist — Ilan Ramon

Ilan Ramon, 48, was a colonel in the Israeli Air Force, a fighter pilot, and Israel's first astronaut. Ramon received a B.S. in Electronics and Computer Engineering from the University of Tel Aviv, Israel. As a member of the Red Team, Ramon was the primary crew member responsible for the Mediterranean Israeli Dust Experiment (MEIDEX). He also worked on the Water Mist Fire Suppression and the Microbial Physiology Flight Experiments Team experiments, among others.

Appendix V

Personal Observations on the Reliability of the Shuttle

*Internal report by Richard P. Feynman,
Member of the Presidential Commission on
the Space Shuttle* Challenger *Accident*

Introduction

It appears that there are enormous differences of opinion as to the probability of a failure with loss of vehicle and of human life. The estimates range from roughly 1 in 100 to 1 in 100,000. The higher figures come from the working engineers, and the very low figures from management. What are the causes and consequences of this lack of agreement? Since 1 part in 100,000 would imply that one could put a Shuttle up each day for 300 years expecting to lose only one, we could properly ask "What is the cause of management's fantastic faith in the machinery?"

We have also found that certification criteria used in flight readiness reviews often develop a gradually decreasing strictness. The argument that the same risk was flown before without failure is often accepted as an argument for the safety of accepting it again. Because of this, obvious weaknesses are accepted again and again, sometimes without a sufficiently serious attempt to remedy them, or to delay a flight because of their continued presence.

There are several sources of information. There are published criteria for certification, including a history of modifications in the form of waivers and deviations. In addition, the records of the flight readiness reviews for each flight document the arguments used to accept the risks of the flight. Information was obtained from the direct testimony and the reports of the range safety officer, Louis J. Ullian, with respect to the history of success of solid fuel rockets. There was a further study by him (as chairman of the launch abort safety panel (LASP)) in an attempt to determine the risks involved in possible accidents leading to radioactive contamination from attempting to fly a plutonium power supply (RTG) for future planetary missions. The NASA study of the same question is also available. For the history of the Space Shuttle main engines, interviews with management and engineers at Marshall, and informal interviews with engineers at Rocketdyne, were made. An independent (Cal Tech) mechanical engineer who consulted for NASA about engines was also interviewed informally. A visit to Johnson was made to gather information on the reliability of the avionics (computers, sensors, and effectors). Finally there is a report—"A Review of Certification Practices, Potentially Applicable to Man-Rated Reusable Rocket Engines"—

prepared at the Jet Propulsion Laboratory by N. Moore, et al., in February 1986, for NASA Headquarters, Office of Space Flight. It deals with the methods used by the FAA and the military to certify their gas-turbine and rocket engines. These authors were also interviewed informally.

Solid Rockets (SRB)

As estimate of the reliability of solid rockets was made by the range safety officer, by studying the experience of all previous rocket flights. Out of a total of nearly 2,900 flights, 121 failed (1 in 25). This includes, however, what may be called early errors, rockets flown for the first few times in which design errors are discovered and fixed. A more reasonable figure for the mature rockets might be 1 in 50. With special care in the selection of parts and in inspection, a figure of below 1 in 100 might be achieved but 1 in 1,000 is probably not attainable with today's technology. (Since there are two rockets on the Shuttle, these rocket failure rates must be doubled to get Shuttle failure rates from solid rocket booster failure).

NASA officials argue that the figure is much lower. They point out that these figures are for unmanned rockets but since the Shuttle is a manned vehicle, "the probability of mission success is necessarily very close to 1.0." It is not very clear what this phrase means. Does it mean it is close to 1 or that it ought to be close to 1? They go on to, explain "Historically this extremely high degree of mission success has given rise to a difference in philosophy between human space flight programs and unmanned programs; i.e., numerical probability usage versus engineering judgment." (These quotations are from "Space Shuttle Data for Planetary Mission RTG Safety Analysis," Page 3-1, February 15, 1985, NASA, JSC). It is true that if the probability of failure was as low as 1 in 100,000 it would take an inordinate number of tests to determine it (you would get nothing but a string of perfect flights from which no precise figure [could be obtained], other than that the probability is likely less than the number of such flights in the string so far). But, if the real probability is not so small, flights would show troubles, near failures, and possible actual failures with a reasonable number of trials, and standard statistical methods could give a reasonable estimate. In fact, previous NASA experience had shown, on occasion, just such difficulties, near accidents, and accidents, all giving warning that the probability of flight failure was not so very small. The inconsistency of the argument not to determine reliability through historical experience, as the range safety officer did, is that NASA also appeals to history, beginning "Historically this high degree of mission success...."

Finally, if we are to replace standard numerical probability usage with engineering judgment, why do we find such an enormous disparity between the management estimate and the judgment of the engineers? It would appear that, for whatever purpose, be it for internal or external consumption, the management of NASA exaggerates the reliability of its product, to the point of fantasy.

The history of the certification and flight readiness reviews will not be repeated here. (See other part of Commission reports.) The phenomenon of accepting for flight, seals that had shown erosion and blow-by in previous flights, is very clear. The *Challenger* flight is an excellent example. There are several references to flights that had gone before. The acceptance and success of these flights is taken as evidence of safety. But erosion and blow-by are not what the design expected. They are warnings that something is wrong. The equipment is not operating as expected, and therefore there is a danger that it can operate with even wider deviations in this unexpected and not thoroughly understood way. The fact that this danger did not lead to a catastrophe before is no guarantee that it will not the next time, unless it is completely understood. When playing Russian roulette, the fact that the first shot got off safety is little comfort for the next. The origin and consequences of the erosion and blow-by were not understood. They did not occur equally on all flights and all joints; sometimes more, and sometimes less. Why not sometime, when whatever conditions determined it were right, still more leading to catastrophe?

In spite of these variations from case to case, officials behaved as if they understood it, giving apparently logical arguments to each other often depending on the "success" of previous flights. For example, in determining if flight 51-L was safe to fly in the face of ring erosion in flight 51-C, it was noted that the erosion depth was only one-third of the radius. It had been noted in an experiment cutting the ring, that cutting it as deep as one radius was necessary before the ring failed. Instead of being very concerned that variations of poorly understood conditions might reasonably create deeper erosion this time, it was asserted [that] there was "a safety factor of three." This is a strange use of the engineer's term "safety factor." If a bridge is built to withstand a certain load without the beams permanently deforming, cracking, or breaking, it may be designed for the materials used to actually stand up under three times the load. This "safety factor" is to allow for uncertain excesses of load, or unknown extra loads, or weaknesses in the material that might have unexpected flaws, etc. If now the expected load comes on to the new bridge and a crack appears in a beam, this is a failure of the design. There was no safety factor at all; even though the bridge did not actually collapse because the crack went one-third of the way through the beam. The O-ring of the solid rocket boosters was not designed to erode. Erosion was a clue that something was wrong. *Erosion was not something from which safety can be inferred.*

There was no way, without full understanding, that one could have confidence that conditions the next time might not produce erosion three times more severe than the time before. Nevertheless, officials fooled themselves into thinking they had such understanding and confidence, in spite of the peculiar variations from case to case. A mathematical model was made to calculate erosion. This was a model based not on physical understanding but on empirical curve fitting. To be more detailed, it was supposed a stream of hot gas impinged on the O-ring material, and the heat was determined at the point of stagnation (so far, with reasonable physical, thermodynamic laws). But to determine how much rubber eroded it was assumed this depended only on this heat by a formula suggested by data on a similar material. A logarithmic plot suggested a straight line, so it was supposed that the erosion varied as the .58 power of the heat, the .58 being determined by a nearest fit. At any rate, adjusting some other numbers, it was determined that the model agreed with the erosion (to depth of one-third the radius of the ring). There is nothing much so wrong with this as believing the answer! Uncertainties appear everywhere. How strong the gas stream might be was unpredictable, it depended on holes formed in putty. Blow-by showed that the ring might fail even though not, or only partially eroded through. The empirical formula was known to be uncertain, for it did not go directly through the very data points by which it was determined. There were a cloud of points some twice above, and some twice below the fitted curve, so erosions twice predicted were reasonable from that cause alone. Similar uncertainties surrounded the other constants in the formula, etc., etc. When using a mathematical model, careful attention must be given to uncertainties in the model.

Liquid Fuel Engine (SSME)

During the flight of 51-L, the three Space Shuttle main engines all worked perfectly, even, at the last moment, beginning to shut down the engines as the fuel supply began to fail. The question arises, however, as to whether, had it failed, and we were to investigate it in as much detail as we did the solid rocket booster, we would find a similar lack of attention to faults and a deteriorating reliability. In other words, were the organization weaknesses that contributed to the accident confined to the solid rocket booster sector or were they a more general characteristic of NASA? To that end the Space Shuttle main engines and the avionics were both investigated. No similar study of the orbiter or the external tank was made.

The engine is a much more complicated structure than the solid rocket booster, and a great deal more detailed engineering goes into it. Generally, the engineering seems to be of high quality and apparently considerable attention is paid to deficiencies and faults found in operation.

The usual way that such engines are designed (for military or civilian aircraft) may be called the component system, or bottom-up design. First it is necessary to thoroughly understand the properties and limitations of the materials to be used (for turbine blades, for example), and tests are begun in experimental rigs to determine those. With this knowledge larger component parts (such as bearings) are designed and tested individually. As deficiencies and design errors are noted they are corrected and verified with further testing. Since one tests only parts at a time these tests and modifications are not overly expensive. Finally, one works up to the final design of the entire engine, to the necessary specifications. There is a good chance, by this time that the engine will generally succeed, or that any failures are easily isolated and analyzed because the failure modes, limitations of materials, etc., are so well understood. There is a very good chance that the modifications to the engine to get around the final difficulties are not very hard to make, for most of the serious problems have already been discovered and dealt with in the earlier, less expensive, stages of the process.

The Space Shuttle main engine was handled in a different manner, top down, we might say. The engine was designed and put together all at once with relatively little detailed preliminary study of the material and components. Then when troubles [were] found in the bearings, turbine blades, coolant pipes, etc., it [was] more expensive and difficult to discover the causes and make changes. For example, cracks have been found in the turbine blades of the pressure oxygen turbo pump. Are they caused by flaws in the material, the effect of the oxygen atmosphere on the properties of the material, the thermal stresses of startup or shutdown, the vibration and stresses of steady running, or mainly at some resonance at certain speeds? How long can we run from crack initiation to crack failure, and how does this depend on power level? Using the completed engine as a test bed to resolve such questions is extremely expensive. One does not wish to lose an entire engine in order to find out where and how failure occurs. Yet, an accurate knowledge of this information is essential to acquire a confidence in the engine reliability in use. Without detailed understanding, confidence cannot be attained.

A further disadvantage of the top-down method is that, if an understanding of a fault is obtained, a simple fix, such as a new shape for the turbine housing, may be impossible to implement without a redesign of the entire engine.

The Space Shuttle main engine is a very remarkable machine. It has a greater ratio of thrust to weight than any previous engine. It is built at the edge of, or outside of, previous engineering experience. Therefore, as expected, many different kinds of flaws and difficulties have turned up. Because, unfortunately, it was built in the top-down manner, they are difficult to find and fix. The design aim of a lifetime of 55 missions, equivalent firings (27,000 seconds of operation, either in a mission of 500 seconds, or on a test stand) has not been obtained. The engine now requires very frequent maintenance and replacement of important parts, such as turbo pumps, bearings, sheet metal housings, etc. The high-pressure fuel turbo pump had to be replaced every three or four mission equivalents (although that may have been fixed, now) and the high-pressure oxygen turbo pump every five or six. This is at most ten percent of the original specification. But our main concern here is the determination of reliability.

In a total of about 250,000 seconds of operation, the engines have failed seriously perhaps sixteen times. Engineering pays close attention to these failings and tries to remedy them as quickly as possible. This it does by test studies on special rigs experimentally designed for the flaws in question, by careful inspection of the engine for suggestive clues (like cracks), and by considerable study and analysis. In this way, in spite of the difficulties of top-down design, through hard work, many of the problems have apparently been solved.

A list of some of the problems follows. Those followed by an asterisk (*) are probably solved:

1. Turbine blade cracks in high-pressure fuel turbo pumps (HPFTP). (May have been solved)
2. Turbine blade cracks in high-pressure oxygen turbo pumps (HPOTP).
3. Augmented Spark Igniter (ASI) line rupture.*
4. Purge check valve failure.*
5. ASI chamber erosion.*
6. HPFTP turbine sheet metal cracking.
7. HPFTP coolant liner failure.*
8. Main combustion chamber outlet elbow failure.*
9. Main combustion chamber inlet elbow weld offset.*
10. HPOTP sub-synchronous whirl.*
11. Flight acceleration safety cutoff system (partial failure in a redundant system).*
12. Bearing spalling (partially solved).
13. A vibration at 4,000 Hertz making some engines inoperable, etc.

Many of these solved problems are the early difficulties of a new design, for thirteen of them occurred in the first 125,000 seconds and only three in the second 125,000 seconds. Naturally, one can never be sure that all the bugs are out, and, for some, the fix may not have addressed the true cause. Thus, it is not unreasonable to guess there may be at least one surprise in the next 250,000 seconds, a probability of 1/500 per engine per mission. On a mission there are three engines, but some accidents would possibly be contained, and only affect one engine. The system can abort with only two engines. Therefore let us say that the unknown surprises do not, even of themselves, permit us to guess that the probability of mission failure do to the Space Shuttle Main Engine is less than 1/500. To this we must add the chance of failure from known, but as yet unsolved, problems (those without the asterisk in the list above). These we discuss below. (Engineers at Rocketdyne, the manufacturer, estimate the total probability as 1/10,000. Engineers at Marshall estimate it as 1/300, while NASA management, to whom these engineers report, claims it is 1/100,000. An independent engineer consulting for NASA thought 1 or 2 per 100 a reasonable estimate.)

The history of the certification principles for these engines is confusing and difficult to explain. Initially the rule seems to have been that two sample engines must each have had twice the time operating without failure as the operating time of the engine to be certified (rule of 2x). At least that is the FAA practice, and NASA seems to have adopted it, originally expecting the certified time to be ten missions (hence twenty missions for each sample). Obviously the best engines to use for comparison would be those of greatest total (flight plus test) operating time — the so-called "fleet leaders." But what if a third sample and several others fail in a short time? Surely we will not be safe because two were unusual in lasting longer. The short time might be more representative of the real possibilities, and in the spirit of the safety factor of 2, we should only operate at half the time of the short-lived samples.

The slow shift toward decreasing safety factor can be seen in many examples. We take that of the HPFTP turbine blades. First of all the idea of testing an entire engine was abandoned. Each engine number has had many important parts (like the turbo pumps themselves) replaced at frequent intervals, so that the rule must be shifted from engines to components. We accept an HPFTP for a certification time if two samples have each run successfully for twice that time (and of course, as a practical matter, no longer insisting that time be as large as ten missions). But what is '"successfully"? The FAA calls a turbine blade crack a failure, in order, in practice, to really provide a safety factor greater than 2. There is some time that an engine can run between the time a crack originally starts until the time it has grown large enough to fracture. (The FAA is contemplating new rules that take this extra safety time into account, but only if it is very carefully analyzed through known models within a known range of experience and with materials thoroughly tested. None of these conditions apply to the Space Shuttle main engine.)

Cracks were found in many second-stage HPFTP turbine blades. In one case three were found after 1,900 seconds, while in another they were not found after 4,200 seconds, although usually these longer runs showed cracks. To follow this story further we shall have to realize that the stress depends a great deal on the power level. The *Challenger* flight was to be at, and previous flights had been at, a power level called 104 percent of rated power level during most of the time the engines were operating. Judging from some material data it is supposed that at the level 104 percent of rated power level, the time to crack is about twice that at 109 percent or full power lever (FPL). Future flights were to be at this level because of heavier payloads, and many tests were made at this level. Therefore dividing time at 104 percent by 2, we obtain units called equivalent full power level (EFPL). (Obviously, some uncertainty is introduced by that, but it has not been studied.) The earliest cracks mentioned above occurred at 1,375 EFPL.

Now the certification rule becomes "limit all second stage blades to a maximum of 1,375 seconds EFPL." If one objects that the safety factor of 2 is lost, it is pointed out that the one turbine ran for 3,800 seconds EFPL without cracks, and half of this is 1,900, so we are being more conservative. We have fooled ourselves in three ways. First we have only one sample, and it is not the fleet leader, for the other two samples of 3,800 or more seconds had seventeen cracked blades between them. (There are fifty-nine blades in the engine.) Next we have abandoned the 2x rule and substituted equal time. And finally, 1,375 is where we did see a crack. We can say that no crack had been found below 1,375, but the last time we looked and saw no cracks was 1,100 seconds EFPL. We do not know when the crack formed between these times, for example cracks may have formed at 1,150 seconds EFPL. (Approximately two-thirds of the blade sets tested in excess of 1,375 seconds EFPL had cracks. Some recent experiments have, indeed, shown cracks as early as 1,150 seconds.) It was important to keep the number high for the *Challenger* was to fly an engine very close to the limit by the time the flight was over.

Finally, it is claimed that the criteria are not abandoned, and the system is safe, by giving up the FAA convention that there should be no cracks, and considering only a completely fractured blade a failure. With this definition no engine has yet failed. The idea is that since there is sufficient time for a crack to a fracture we can insure that all is safe by inspecting all blades for cracks. If they are found, replace them, and if none are found we have enough time for a safe mission. This makes the crack problem not a flight-safety problem, but merely a maintenance problem.

This may in fact be true. But how well do we know that cracks always grow slowly enough that no fracture can occur in a mission? Three engines have run for long times with a few cracked blades (about 3,000 seconds EFPL) with no blades broken off.

But a fix for this cracking may have been found. By changing the blade shape, shot peening the surface, and covering with insulation to exclude thermal shock, the blades have not cracked so far.

A very similar story appears in the history of certification of the HPOTP, but we shall not give the details here.

It is evident, in summary, that the flight readiness reviews and certification rules show a deterioration for some of the problems of the Space Shuttle main engine that is closely analogous to the deterioration seen in the rules for the solid rocket booster.

Avionics

By "avionics" is meant the computer system on the orbiter as well as its input sensors and output actuators. At first we will restrict ourselves to the computers proper and not be concerned with the reliability of the input information from the sensors of temperature, pressure, etc., nor with whether the computer output is faithfully followed by the actuators of rocket firings, mechanical controls, displays to astronauts, etc.

The computer system is very elaborate, having over 250,000 lines of code. It is responsible, among many other things, for the automatic control of the entire ascent to orbit, and for the descent until well into the atmosphere (below Mach 1) once one button is pushed deciding the landing site desired. It would be possible to make the entire landing automatically (except that the landing gear lowering signal is expressly left out of computer control, and must be provided by the pilot, ostensibly for safety reasons) but such an entirely automatic landing is probably not as safe as a pilot-controlled landing. During orbital flight it is used in the control of payloads, in displaying information to the astronauts, and the exchange of information to the ground. It is evident that the safety of flight requires guaranteed accuracy of this elaborate system of computer hardware and software.

In brief, the hardware reliability is ensured by having four essentially independent computer systems. Where possible each sensor also has multiple copies, usually four, and each copy feeds all four of the computer lines. If the inputs from the sensors disagree, depending on circumstances, certain averages, or a majority selection is used as the effective input. The algorithm used by each of the four computers is exactly the same, so their inputs (since each sees all copies of the sensors) are the same. Therefore, at each step the results in each computer should be identical. From time to time they are compared, but because they might operate at slightly different speeds a system of stopping and waiting at specific times is instituted before each comparison is made. If one of the computers disagrees, or is too late in having its answer ready, the three which do agree are assumed to be correct and the errant computer is taken completely out of the system. If, now, another computer fails, as judged by the agreement of the other two, it is taken out of the system, and the rest of the flight canceled, and descent to the landing site is instituted, controlled by the two remaining computers. It is seen that this is a redundant system since the failure of only one computer does not affect the mission. Finally, as an extra feature of safety, there is a fifth independent computer, whose memory is loaded with only the programs of ascent and descent, and which is capable of controlling the descent if there is a failure of more than two of the computers of the mainline four.

There is not enough room in the memory of the mainline computers for all the programs of ascent, descent, and payload programs in flight, so the memory is loaded about four times from tapes, by the astronauts.

Because of the enormous effort required to replace the software for such an elaborate system, and for checking a new system out, no change has been made to the hardware since the system began about fifteen years ago (1971). The actual hardware is obsolete; for example, the memories are of the old ferrite core type. It is becoming more difficult to find manufacturers to supply such old-fashioned computers reliably and of high quality. Modern computers are very much more reliable, can run much faster, simplifying circuits, and allowing more to be done, and would not require so much loading of memory, for the memories are much larger.

The software is checked very carefully in a bottom-up fashion. First, each new line of code is checked, then, sections of code or modules with special functions are verified. The scope is increased step by step until the new changes are incorporated into a complete system and checked. This complete output is considered the final product, newly released. But completely independently there is an independent verification group that takes an adversary attitude to the software development group, and tests and verifies the software as if it were a customer of the delivered product. There is additional verification in using the new programs in simulators, etc. A discovery of an error during verification testing is considered very serious, and its origin studied very carefully to avoid such mistakes in the future. Such unexpected errors have been found only about six times in all the programming and program changing (for new or altered payloads) that has been done. The principle that is followed is that all the verification is not an aspect of program safety, it is merely a test of that safety in a non-catastrophic verification. Flight safety is to be judged solely on how well the programs do in the verification tests. A failure here generates considerable concern.

To summarize then, the computer software checking system and attitude is of the highest quality. There appears to be no process of gradually fooling oneself while degrading standards so characteristic of the solid rocket booster or Space Shuttle main engine safety systems. To be sure, there have been recent suggestions by management to curtail such elaborate and expensive tests as being unnecessary at this late date in Shuttle history. This must be resisted for it does not appreciate the mutual subtle influences, and sources of error generated by even small changes of one part of a program on another. There are perpetual requests for changes as new payloads and new demands and modifications are suggested by the users. Changes are expensive because they require extensive testing. The proper way to save money is to curtail the number of requested changes, not the quality of testing for each.

One might add that the elaborate system could be very much improved by more modern hardware and programming techniques. Any outside competition would have all the advantages of starting over, and whether that is a good idea for NASA now should be carefully considered.

Finally, returning to the sensors and actuators of the avionics system, we find that the attitude to system failure and reliability is not nearly as good as for the computer system. For example, a difficulty was found with certain temperature sensors sometimes failing. Yet eighteen months later the same sensors were still being used, still sometimes failing, until a launch had to be scrubbed because two of them failed at the same time. Even on a succeeding flight this unreliable sensor was used again. Again reaction control systems, the rocket jets used for reorienting and control in flight still are somewhat unreliable. There is considerable redundancy, but a long history of failures, none of which has yet been extensive enough to seriously affect flight. The action of the jets is checked by sensors, and, if they fail to fire the computers choose another jet to fire. But they are not designed to fail, and the problem should be solved.

Conclusions

If a reasonable launch schedule is to be maintained, engineering often cannot be done fast enough to keep up with the expectations of originally conservative certification criteria designed to guarantee a very safe vehicle. In these situations, subtly, and often with apparently logical arguments, the criteria are altered so that flights may still be certified in time. They therefore fly in a relatively unsafe condition, with a chance of failure of the order of a percent (it is difficult to be more accurate).

Official management, on the other hand, claims to believe the probability of failure is a thousand times less. One reason for this may be an attempt to assure the government of NASA perfection and success in order to ensure the supply of funds. The other may be that they sincerely believed it to be true, demonstrating an almost incredible lack of communication between themselves and their working engineers.

In any event this has had very unfortunate consequences, the most serious of which is to encourage ordinary citizens to fly in such a dangerous machine, as if it had attained the safety of an ordinary airliner. The astronauts, like test pilots, should know their risks, and we honor them for their courage. Who can doubt that McAuliffe was equally a person of great courage, who was closer to an awareness of the true risk than NASA management would have us believe?

Let us make recommendations to ensure that NASA officials deal in a world of reality in understanding technological weaknesses and imperfections well enough to be actively trying to eliminate them. They must live in reality in comparing the costs and utility of the Shuttle to other methods of entering space. And they must be realistic in making contracts, in estimating costs, and the difficulty of the projects. Only realistic flight schedules should be proposed, schedules that have a reasonable chance of being met. If in this way the government would not support them, then so be it. NASA owes it to the citizens from whom it asks support to be frank, honest, and informative, so that these citizens can make the wisest decisions for the use of their limited resources.

For a successful technology, reality must take precedence over public relations, for nature cannot be fooled.

APPENDIX VI

The Columbia *Emails:*
Opportunities Missed

E-mail Series Discussing Debris Hit and Reentry of Columbia
(Documents from Columbia Report).

-----Original Message-----
From: Stoner-1, Michael D
Sent: Friday, January 17, 2003 4:03 PM
To: Woodworth, Warren H; Reeves, William D
Cc: Wilder, James; White, Doug; Bitner, Barbara K; Blank, Donald E; Cooper, Curt W; Gordon, Michael P.
Subject: RE: STS 107 Debris

Just spoke with Calvin and Mike Gordon (RCC SSM) about the impact.

Basically the RCC is extremely resilient to impact type damage. The piece of debris (most likely foam/ice) looked like it most likely impacted the WLE RCC and broke apart. It didn't look like a big enough piece to pose any serious threat to the system and Mike Gordon the RCC SSM concurs. At T +81seconds the piece wouldn't have had enough energy to create a large damage to the RCC WLE system. Plus they have analysis that says they have a single mission safe re-entry in case of impact that penetrates the system.

As far as the tile go in the wing leading edge area they are thicker than required (taper in the outer mold line) and can handle a large area of shallow damage which is what this event most likely would have caused. They have impact data that says the structure would get slightly hotter but still be OK.

Mike Stoner
USA TPS SAM

[RCC=Reinforced Carbon-Carbon, SSM=Sub-system Manager, WLE=Wing Leading Edge, TPS=Thermal Protection System, SAM= Sub-system Area Manager]

January 17, 2003, Impact recognized.

```
-----Original Message-----
From:      ROCHA, ALAN R. (RODNEY) (JSC-ES2) (NASA)
Sent:      Tuesday, January 21, 2003 4:41 PM
To:        SHACK, PAUL E. (JSC-EA42) (NASA); HAMILTON, DAVID A. (DAVE) (JSC-EA) (NASA); MILLER, GLENN J. (JSC-EA) (NASA)
Cc:        SERIALE-GRUSH, JOYCE M. (JSC-EA) (NASA); ROGERS, JOSEPH E. (JOE) (JSC-ES2) (NASA); GALBREATH, GREGORY F. (GREG) (JSC-ES2) (NASA)
Subject:   STS-107 Wing Debris Impact, Request for Outside Photo-Imaging Help
```

Paul and Dave,

The meeting participants (Boeing, USA, NASA ES2 and ES3, KSC) all agreed we will always have big uncertainties in any transport/trajectory analyses and applicability/extrapolation of the old Arc-Jet test data until we get definitive, better, clearer photos of the wing and body underside. Without better images it will be very difficult to even bound the problem and initialize thermal, trajectory, and structural analyses. Their answers may have a wide spread ranging from acceptable to not-acceptable to horrible, and no way to reduce uncertainty. Thus, giving MOD options for entry will be very difficult.

Can we petition (beg) for outside agency assistance? We are asking for Frank Benz with Ralph Roe or Ron Dittemore to ask for such. Some of the old timers here remember we got such help in the early 1980's when we had missing tile concerns.

Despite some nay-sayers, there are some options for the team to talk about: On-orbit thermal conditioning for the major structure (but is in contradiction with tire pressure temp. cold limits), limiting high cross-range de-orbit entries, constraining right or left had turns during the Heading Alignment Circle (only if there is struc. damage to the RCC panels to the extent it affects flight control.

Rodney Rocha
Structural Engineering Division (ES-SED)
- ES Div. Chief Engineer (Space Shuttle DCE)
- Chair, Space Shuttle Loads & Dynamics Panel

Mail Code ES2

[USA=United Space Alliance, NASA ES2, ES3=separate divisions of the Johnson Space Center Engineering Directorate, KSC=Kennedy Space Center, MOD=Missions Operations Directorate, or Mission Control]

January 21, 2003, 4:41 P.M., Photo support request.

```
-----Original Message-----
From:      SCHOMBURG, CALVIN (JSC-EA) (NASA)
Sent:      Tuesday, January 21, 2003 9:26 AM
To:        SHACK, PAUL E. (JSC-EA42) (NASA); SERIALE-GRUSH, JOYCE M. (JSC-EA) (NASA); HAMILTON, DAVID A. (DAVE) (JSC-EA) (NASA)
Subject:   FW: STS-107 Post-Launch Film Review - Day 1
```

FYI-TPS took a hit-should not be a problem-status by end of week.

[FYI=For Your Information, TPS=Thermal Protection System]

January 21, 2003, 9:26 A.M., Not a problem for reentry.

> -----Original Message-----
> **From:** SHACK, PAUL E. (JSC-EA42) (NASA)
> **Sent:** Tuesday, January 21, 2003 9:33 AM
> **To:** ROCHA, ALAN R. (RODNEY) (JSC-ES2) (NASA); SERIALE-GRUSH, JOYCE M. (JSC-EA) (NASA)
> **Cc:** KRAMER, JULIE A. (JSC-EA4) (NASA); MILLER, GLENN J. (JSC-EA) (NASA); RICKMAN, STEVEN L. (JSC-ES3) (NASA); MADDEN, CHRISTOPHER B. (CHRIS) (JSC-ES3) (NASA)
> **Subject:** RE: STS-107 Debris Analysis Team Plans
>
> This reminded me that at the STS-113 FRR the ET Project reported on foam loss from the Bipod Ramp during STS-112. The foam (estimated 4X5X12 inches) impacted the ET Attach Ring and dented an SRB electronics box cover.
>
> Their charts stated "ET TPS foam loss over the life of the Shuttle program has never been a 'Safety of Flight' issue". They were severely wire brushed over this and Brian O'Conner (Associate Administrator for Safety) asked for a hazard assessment for loss of foam.
>
> The suspected cause for foam loss is trapped air pockets which expand due to altitude and aerothermal heating.
>
> *[FRR=Flight Readiness Review, ET=External Tank, SRB=Solid Rocket Booster, TPS=Thermal Protection System]*

January 21, 2003, 9:33 A.M., Bipod ramp loss.

> -----Original Message-----
> **From:** DITTEMORE, RONALD D. (JSC-MA) (NASA)
> **Sent:** Wednesday, January 22, 2003 9:14 AM
> **To:** HAM, LINDA J. (JSC-MA2) (NASA)
> **Subject:** RE: ET Briefing - STS-112 Foam Loss
>
> You remember the briefing! Jerry did it and had to go out and say that the hazard report had not changed and that the risk had not changed...But it is worth looking at again.
>
> -----Original Message-----
> **From:** HAM, LINDA J. (JSC-MA2) (NASA)
> **Sent:** Tuesday, January 21, 2003 11:14 AM
> **To:** DITTEMORE, RONALD D. (JSC-MA) (NASA)
> **Subject:** FW: ET Briefing - STS-112 Foam Loss
>
> You probably can't open the attachment. But, the ET rationale for flight for the STS-112 loss of foam was lousy. Rationale states we haven't changed anything, we haven't experienced any 'safety of flight' damage in 112 flights, risk of loss of bi-pod ramp TPS is same as previous flights...So ET is safe to fly with no added risk
>
> Rationale was lousy then and still is....
>
> -----Original Message-----
> **From:** MCCORMACK, DONALD L. (DON) (JSC-MV6) (NASA)
> **Sent:** Tuesday, January 21, 2003 9:45 AM
> **To:** HAM, LINDA J. (JSC-MA2) (NASA)
> **Subject:** FW: ET Briefing - STS-112 Foam Loss
> **Importance:** High
>
> FYI - it kinda says that it will probably be all right
>
> *[ORR=Operational Readiness Review, VAB=Vehicle Assembly Building, IFA=In-Flight Anomaly, TPS=Thermal Protection System, ET=External Tank]*

January 21, 22, 2003, Bipod ramp loss.

> -----Original Message-----
> **From:** LEE, TIMOTHY F., LTCOL. (JSC-MT) (USAF)
> **Sent:** Wednesday, January 22, 2003 9:01 AM
> **To:** MCCORMACK, DONALD L. (DON) (JSC-MV6) (NASA)
> **Subject:** NASA request for DOD
>
> Don,
>
> FYI: Lambert Austin called me yesterday requesting DOD photo support for STS-107. Specifically, he is asking us if we have a ground or satellite asset that can take a high resolution photo of the shuttle while on-orbit--to see if there is any FOD damage on the wing. We are working his request.
>
> Tim
>
> [DOD=Department of Defense, FOD=Foreign Object Debris]

January 22, 2003, 9:01 A.M., Photo support.

> -----Original Message---
> **From:** HAM, LINDA J. (JSC-MA2) (NASA)
> **Sent:** Wednesday, January 22, 2003 9:33 AM
> **To:** AUSTIN, LAMBERT D. (JSC-MS) (NASA); ROE, RALPH R. (JSC-MV) (NASA)
> **Subject:** ET Foam Loss
>
> Can we say that for any ET foam lost, no 'safety of flight' damage can occur to the Orbiter because of the density?

Responses included the following:

> -----Original Message-----
> **From:** ROE, RALPH R. (JSC-MV) (NASA)
> **Sent:** Wednesday, January 22, 2003 9:38 AM
> **To:** SCHOMBURG, CALVIN (JSC-EA) (NASA)
> **Subject:** FW: ET Foam Loss
>
> Calvin,
>
> I wouldn't think we could make such a generic statement but can we bound it some how by size or acreage?
>
> [Acreage=larger areas of foam coverage]

January 22, 2003, 9:33 A.M., 9:38 A.M., Flight safety issue.

> -----Original Message-----
> **From:** DITTEMORE, RONALD D. (JSC-MA) (NASA)
> **Sent:** Wednesday, January 22, 2003 10:15 AM
> **To:** HAM, LINDA J. (JSC-MA2) (NASA)
> **Subject:** RE: ET Briefing - STS-112 Foam Loss
>
> Another thought, we need to make sure that the density of the ET foam cannot damage the tile to where it is an impact to the orbiter...Lambert and Ralph need to get some folks working with ET.

January 22, 2003, 10:15 A.M., Possible tile damage.

-----Original Message-----
From: SCHOMBURG, CALVIN (JSC-EA) (NASA)
Sent: Wednesday, January 22, 2003 10:53 AM
To: ROE, RALPH R. (JSC-MV) (NASA)
Subject: RE: ET Foam Loss

No-the amount of damage ET foam can cause to the TPS material-tiles is based on the amount of impact energy-the size of the piece and its velocity(from just after pad clear until about 120 seconds-after that it will not hit or it will not enough energy to cause any damage)-it is a pure kinetic problem-there is a size that can cause enough damage to a tile that enough of the material is lost that we could burn a hole through the skin and have a bad day-(loss of vehicle and crew -about 200-400 tile locations(out of the 23,000 on the lower surface)-the foam usually fails in small popcorn pieces-that is why it is vented-to make small hits-the two or three times we have been hit with a piece as large as the one this flight-we got a gouge about 8-10 inches long about 2 inches wide and 3/4 to an 1 inch deep across two or three tiles. That is what I expect this time-nothing worst. If that is all we get we have have no problem-will have to replace a couple of tiles but nothing else.

[ET=External Tank, TPS=Thermal Protection System]

January 22, 2003, 10:53 A.M., Best guess is "no problem."

-----Original Message-----
From: AUSTIN, LAMBERT D. (JSC-MS) (NASA)
Sent: Wednesday, January 22, 2003 3:22 PM
To: HAM, LINDA J. (JSC-MA2) (NASA)
Cc: WALLACE, RODNEY O. (ROD) (JSC-MS2) (NASA); NOAH, DONALD S. (DON) (JSC-MS) (NASA)
Subject: RE: ET Foam Loss

NO. I will cover some of the pertinent rationale....there could be more if I spent more time thinking about it. Recall this issue has been discussed from time to time since the inception of the basic "no debris" requirement in Vol. X and at each review the SSP has concluded that it is not possible to PRECLUDE a potential catastrophic event as a result of debris impact damage to the flight elements. As regards the Orbiter, both windows and tiles are areas of concern.

You can talk to Cal Schomberg and he will verify the many times we have covered this in SSP reviews. While there is much tolerance to window and tile damage, ET foam loss can result in impact damage that under subsequent entry environments can lead to loss of structural integrity of the Orbiter area impacted or a penetration in a critical function area that results in loss of that function. My recollection of the most critical Orbiter bottom acreage areas are the wing spar, main landing gear door seal and RCC panels...of course Cal can give you a much better rundown.

We can and have generated parametric impact zone characterizations for many areas of the Orbiter for a few of our more typical ET foam loss areas. Of course, the impact/damage significance is always a function of debris size and density, impact velocity, and impact angle--these latter 2 being a function of the flight time at which the ET foam becomes debris. For STS-107 specifically, we have generated

this info and provided it to Orbiter. Of course, even this is based on the ASSUMPTION that the location and size of the debris is the same as occurred on STS-112------this cannot be verified until we receive the on-board ET separation photo evidence post Orbiter landing. We are requesting that this be expedited. I have the STS-107 Orbiter impact map based on the assumptions noted herein being sent down to you. Rod is in a review with Orbiter on this info right now.

[SSP=Space Shuttle Program, ET=External Tank]

January 22, 2003, 3:22 A.M., Potential catastrophic event.

Opposite, bottom: January 26, 2003, 7:45 P.M., Safety of flight issue.

-----Original Message-----
From: CAIN, LEROY E. (JSC-DA8) (NASA)
Sent: Thursday, January 23, 2003 12:07 PM
To: JONES, RICHARD S. (JSC-DM) (NASA); OLIVER, GREGORY T. (GREG) (JSC-DM4) (NASA); CONTE, BARBARA A. (JSC-DM) (NASA)
Cc: ENGELAUF, PHILIP L. (JSC-DA8) (NASA); AUSTIN, BRYAN P. (JSC-DA8) (NASA); BECK, KELLY B. (JSC-DA8) (NASA); HANLEY, JEFFREY M. (JEFF) (JSC-DA8) (NASA); STICH, J. S. (STEVE) (JSC-DA8) (NASA)
Subject: Help with debris hit

The SSP was asked directly if they had any interest/desire in requesting resources outside of NASA to view the Orbiter (ref. the wing leading edge debris concern).

They said, No.

After talking to Phil, I consider it to be a dead issue.

[SSP=Space Shuttle Program]

January 23, 2003, 12:07 P.M., "Dead issue."

-----Original Message-----
From: ROCHA, ALAN R. (RODNEY) (JSC-ES2) (NASA)
Sent: Sunday, January 26, 2003 7:45 PM
To: SHACK, PAUL E. (JSC-EA42) (NASA); MCCORMACK, DONALD L. (DON) (JSC-MV6) (NASA); OUELLETTE, FRED A. (JSC-MV6) (NASA)
Cc: ROGERS, JOSEPH E. (JOE) (JSC-ES2) (NASA); GALBREATH, GREGORY F. (GREG) (JSC-ES2) (NASA); JACOBS, JEREMY B. (JSC-ES4) (NASA); SERIALE-GRUSH, JOYCE M. (JSC-EA) (NASA); KRAMER, JULIE A. (JSC-EA4) (NASA); CURRY, DONALD M. (JSC-ES3) (NASA); KOWAL, T. J. (JOHN) (JSC-ES3) (NASA); RICKMAN, STEVEN L. (JSC-ES3) (NASA); SCHOMBURG, CALVIN (JSC-EA) (NASA); CAMPBELL, CARLISLE C., JR (JSC-ES2) (NASA)
Subject: STS-107 Wing Debris Impact on Ascent: Final analysis case completed

As you recall from Friday's briefing to the MER, there remained open work to assess analytically predicted impact damage to the wing underside in the region of the main landing gear door. This area was considered a low probability hit area by the image analysis teams, but they admitted a debris strike here could not be ruled out.

As with the other analyses performed and reported on Friday, this assessment by the Boeing multi-technical discipline engineering teams also employed the system integration's dispersed trajectories followed by serial results from the *Crater* damage prediction tool, thermal analysis, and stress analysis. It was reviewed and accepted by the ES-DCE (R. Rocha) by Sunday morning, Jan. 26. The case is defined by a large area gouge about 7 inch wide and about 30 inch long with sloped sides like a crater, and reaching down to the densified layer of the TPS.

SUMMARY: Though this case predicted some higher temperatures at the outer layer of the honeycomb aluminum face sheet and subsequent debonding of the sheet, there is no predicted burn-through of the door, no breeching of the thermal and gas seals, nor is there door structural deformation or thermal warpage to open the seal to hot plasma intrusion. Though degradation of the TPS and door structure is likely (if the impact occurred here), there is no safety of flight (entry, descent, landing) issue.

Note to Don M. and Fred O.: On Friday I believe the MER was thoroughly briefed and it was clear that open work remained (viz., the case summarized above), the message of open work was not clearly given, in my opinion, to Linda Ham at the MMT. I believe we left her the impression that engineering assessments and cases were all finished and we could state with finality no safety of flight issues or questions remaining. This very serious case could not be ruled out and it was a very good thing we carried it through to a finish.

Rodney Rocha (ES2)
- Division Shuttle Chief Engineer (DCE), ES-Structural Engineering Division
- Chair, Space Shuttle Loads & Dynamics Panel

[MER=Mission Evaluation Room, ES-DCE=Structural Engineering-Division Shuttle Chief Engineer, TPS=Thermal Protection System]

> -----Original Message-----
> **From:** Robert H. Daugherty
> **Sent:** Monday, January 27, 2003 3:35 PM
> **To:** CAMPBELL, CARLISLE C., JR (JSC-ES2) (NASA)
> **Subject:** Video you sent
>
> WOW!!!
> I bet there are a few pucker strings pulled tight around there!
> Thinking about a belly landing versus bailout...... (I would say that if there is a question about main gear well burn thru that its crazy to even hit the deploy gear button...the reason being that you might have failed the wheels since they are aluminum..they will fail before the tire heating/pressure makes them fail..and you will send debris all over the wheel well making it a possibility that the gear would not even deploy due to ancillary damage...300 feet is the wrong altitude to find out you have one gear down and the other not down...you're dead in that case)
> Think about the pitch-down moment for a belly landing when hitting not the main gear but the trailing edge of the wing or body flap when landing gear up...even if you come in fast and at slightly less pitch attitude...the nose slapdown with that pitching moment arm seems to me to be pretty scary...so much so that I would bail out before I would let a loved one land like that.
> My two cents.
> See ya.
> Bob

January 27, 2003, 3:35 P.M., Bail out and belly landing options.

The following reply from Campbell to Daugherty was sent at 4:49 P.M.:

> -----Original Message-----
> **From:** "CAMPBELL, CARLISLE C., JR (JSC-ES2) (NASA)"
> **To:** "'Bob Daugherty'"
> **Subject:** FW: Video you sent
> **Date:** Mon, 27 Jan 2003 15:59:53 -0600
> **X-Mailer:** ßInternet Mail Service (5.5.2653.19)
>
> Thanks. That's why they need to get all the facts in early on--such as look at impact damage from the spy telescope. Even then, we may not know the real effect of the damage.
>
> The LaRC ditching model tests 20 some years ago showed that the Orbiter was the best ditching shape that they had ever tested, of many. But, our structures people have said that if we ditch we would blow such big holes in the lower panels that the orbiter might break up. Anyway, they refuse to even consider water ditching any more--I still have the test results[Bailout seems best.

[LaRC=Langley Research Center]

On the next day, Tuesday, Daugherty sent the following to Campbell:

> -----Original Message-----
> **From:** Robert H. Daugherty
> **Sent:** Tuesday, January 28, 2003 12:39 PM
> **To:** CAMPBELL, CARLISLE C., JR (JSC-ES2) (NASA)
> **Subject:** Tile Damage
>
> Any more activity today on the tile damage or are people just relegated to crossing their fingers and hoping for the best?
> See ya,
> Bob

Campbell's reply:

```
-----Original Message-----
From:     "CAMPBELL, CARLISLE C., JR (JSC-ES2) (NASA)"
To:       "'Robert H. Daugherty'"
Subject:  RE: Tile Damage
Date:     Tue, 28 Jan 2003 13:29:58 -0600
X-Mailer: Internet Mail Service (5.5.2653.19)

I have not heard anything new. I'll let you know if I do.

CCC
```

Opposite and above: January 27 and 28, 2003, "Crossing their fingers."

Appendix VII

Mission 51-L Press Kit

*This appendix is designed to resemble the original press kit.
Original page breaks are indicated by rules.*

NASA
National Aeronautics and
Space Administration

Space Shuttle Mission 51-L

Press Kit   January 1986

RELEASE NO: 86-1   January 1986

CONTACTS

Charles Redmond / Sarah Keegan
Headquarters, Washington, D.C.
(Phone: 202/453-8536)

Leon Perry
Headquarters, Washington, D.C.
(Phone: 202/453-1547

Ed Medal / Terry Eddleman
Marshall Space Flight Center, Huntsville, Alabama
(Phone: 205/453-0034)

Barbara Schwartz
Johnson Space Center, Houston, Texas
(Phone: 713/483-5111)

Jim Ball
Kennedy Space Center, Florida
(Phone: 305/867-2468)

Jim Elliott
Goddard Space Flight Center, Greenbelt, Maryland
(Phone: 301/344-6256)

CONTENTS

51-L Mission — Quick Look	298
General Release	298
General Information	300
51-L Briefing Schedule	301
Trajectory Sequence of Events	301
Summary of Major Activities	302
51-L Payload and Vehicle Weights Summary	302
Tracking and Data Relay Satellite System and TDRS-B	303
Inertial Upper Stage	304
Spartan-Halley Mission	307
Teacher in Space Project	310
Shuttle Student Involvement Program	311
Comet Halley Active Monitoring Program	313
Phase Partitioning Experiment	313
Fluid Dynamics in Space	314
U.S. Liberty Coins	315
51-L Flight Crew Data	315

Shuttle Mission 51-L — Quick Look

Crew: Francis R. Scobee, Commander
Michael J. Smith, Pilot
Judith A. Resnik, Mission Specialist
Ellison S. Onizuka, Mission Specialist
Ronald E. McNair, Mission Specialist
Gregory Jarvis, Payload Specialist
S. Christa McAuliffe, Teacher Observer

Orbiter: *Challenger* (099)
Launch Site: Pad 39-B, Kennedy Space Center, Florida
Launch Date/Time: January 25, 1986 — 3:21 P.M. CST
Orbital Inclination: 28:45 degrees
Insertion Orbit: 153.5 n.mi circular
Mission Duration: 6 days, 34 minutes
Orbits: 96 full orbits; landing on 97
Landing Date/Time: January 31, 1986, 4:55 P.M. CST

Primary Landing Site: Kennedy Space Center, Florida, Runway 33
  Weather Alternate: Edwards Air Force Base, California, Runway 17
  Return to Launch Site: Kennedy Space Center
  Abort-Once-Around: Edwards Air Force Base
  Trans-Atlantic Abort: Dakar, Senegal

Cargo and Payloads:
  Tracking and Data Relay Satellite (TDRS-B)
  Spartan-Halley Mission
  Teacher in Space Project
  Comet Halley Active Monitoring Program
  Fluid Dynamics Experiment
  Phase Partitioning Experiment
  Radiation Monitoring Experiment
  3 Student Experiments

National Aeronautics and
Space Administration

George C. Marshall Space Flight Center
Huntsville, Alabama 35812

Release No: 86-1

For Release
January 1986

Teacher in Space and Comet Halley Study Highlight 51-L Flight

The launch of a high school teacher as America's first private citizen to fly aboard the Shuttle in NASA's Space Flight Participant Program will open a new chapter in space travel when *Challenger* lifts off on the 25th Space Shuttle mission.

A science payload programmed for 40 hours of comet Halley observations and the second of NASA's Tracking and Data Relay Satellites (TDRS-B) will be aboard for *Challenger*'s 10th flight, targeted for launch at 3:21 P.M. CST on January 25.

Challenger's liftoff will mark the first use of Pad 39-B for a Shuttle launch. Pad B was last used for the Apollo Soyuz Test Project in July 1975 and has since been modified to support the Shuttle program.

Four Shuttle veterans will be joined by rookie astronaut Michael Smith, teacher observer Christa McAuliffe and Hughes payload specialist Gregory Jarvis for a mission that will extend just beyond 6 days.

Commanding the seven-member crew will be Francis R. Scobee, who served as pilot aboard *Challenger* on mission 41-C. Michael Smith will be 51-L pilot.

Mission specialists Judith Resnik, Ellison Onizuka and Ronald McNair each will be making their second trip into space.

Challenger will be launched into a 177-statute-mile circular orbit inclined 28.45 degrees to the equator for the 6-day, 34-minute mission. The orbiter is scheduled to make its end-of-mission landing on the 3-mile-long Shuttle Landing Facility at the Kennedy Space Center.

Deployed on the first day of the flight, TDRS-B will join TDRS-1 in geosynchronous orbit to provide high-capacity communications and data links between Earth and the Shuttle, as well as other spacecraft and launch vehicles.

After deployment from the Shuttle cargo bay, TDRS-B will be boosted to geosynchronous transfer orbit by the Inertial Upper Stage (IUS). Its orbit will be circularized and it will be positioned over the Pacific Ocean at 171 degrees west longitude.

TDRS-1, launched from *Challenger* in April 1983 on the sixth Space Shuttle flight, is located over the Atlantic Ocean at 41 degrees west longitude.

With the addition of the second satellite, real-time coverage through the single ground station at White Sands, New Mexico, is expected to be available for about 85 percent of each orbit of a user spacecraft.

The TDRS satellites, built by TRW Space Systems, are owned by Space Communications Company (SPACECOM) and leased by NASA for a period of 10 years. A third TDRS satellite will be launched on a later mission to serve as an in-orbit spare.

Spartan-Halley is the second payload in the NASA-sponsored Spartan program for flying low-cost experiment packages aboard the Shuttle.

The scientific objective of Spartan-Halley is to measure the ultraviolet spectrum of comet Halley as the comet approaches the point of its orbit that will be closest to the sun.

The Spartan mission peculiar support structure will be deployed from the Shuttle cargo bay and retrieved later in the mission for return to Earth.

Ultraviolet measurements and photographs of comet Halley will be made by instruments on the Spartan support structure during 40 hours of free flying in formation with the Shuttle.

Several mid-deck experiments, including those associated with the Teacher in Space Project, and three student experiments complete *Challenger*'s payload manifest.

Teacher observer Christa McAuliffe will perform experiments that will demonstrate the effects of microgravity on hydroponics, magnetism, Newton's laws, effervescence, chromatography and the operation of simple machines.

The Teacher in Space experiments will be filmed for use after the flight in educating students.

McAuliffe also will assist in operating three student experiments being carried aboard *Challenger*. These experiments include a study of chicken embryo development in space, research on how microgravity affects a titanium alloy and an experiment in crystal growth.

The Fluid Dynamics Experiment, a package of six experiments, will be flown on the mid-deck. They involve simulating the behavior of liquid propellants in low gravity. The fluid dynamics experiments will be conducted by Hughes payload specialist Gregory Jarvis.

Among the fluid investigations will be simulations to understand the motion of propellants during Shuttle Frisbee deployments, which have been employed for the Hughes Leasat satellites.

Another mid-deck experiment will be the Radiation Monitoring Experiment consisting of handheld and pocket monitors to measure radiation levels at various times in orbit. This is the seventh flight for the RME.

Challenger will perform its deorbit maneuver and burn over the Indian Ocean on orbit 96 with landing at Kennedy occurring on orbit 97 at a mission elapsed time of 6 days, 34 minutes.

Touchdown on the Florida runway should come at 4:55 P.M. CST on January 31.

(END OF GENERAL RELEASE; BACKGROUND INFORMATION FOLLOWS.)

GENERAL INFORMATION

NASA Select Television Transmission

NASA-Select television coverage of Shuttle mission 51-L will be carried on a full satellite transponder:

Satcom F-2R, Transponder 13, C-Band
Orbital Position: 72 degrees west longitude
Frequency: 3954.5 MHz vertical polarization
Audio Monaural: 6.8 MHz

NASA-Select video also is available at the AT&T Switching Center, Television Operation Control in Washington, D.C., and at the following NASA locations:

NASA Headquarters, Washington, D.C.
Langley Research Center, Hampton, Virginia
John F. Kennedy Space Center, Florida
Marshall Space Flight Center, Huntsville, Alabama
Johnson Space Center, Houston, Texas
Dryden Flight Research Facility, Edwards, California
Ames Research Center, Mountain View, California
Jet Propulsion Laboratory, Pasadena, California

The schedule for television transmission from the orbiter and for the change-of-shift briefings from Johnson Space Center will be available during the mission at Kennedy Space Center, Marshall Space Flight Center, Johnson Space Center and NASA Headquarters.

The television schedule will be updated daily to reflect changes dictated by mission operations. Television schedules also may be obtained by calling COMSTOR (713/280-8711). COMSTOR is a computer data-base service requiring the use of a telephone modem.

Special Note to Broadcasters

Beginning January 22 and continuing throughout the mission, approximately 7 minutes of audio interview material with the crew of 51-L will be available to broadcasters by calling 202/269-6572.

Briefings

Flight control personnel will be on 8-hour shifts. Change-of-shift briefings by the off-going flight director will occur at approximately 8-hour intervals.

51-L BRIEFING SCHEDULE

Time (EST)	Briefing	Origin
T-1 Day		
9:00 A.M.	TDRS-B	KSC
9:45 A.M.	Spartan-Halley Mission	KSC
10:30 A.M.	Teacher in Space	KSC
11:15 A.M.	Shuttle Student Involvement Program	KSC
12:00 noon	Fluid Dynamics Experiment	KSC
3:00 P.M.	Pre-Launch Press Conference	KSC
T-Day		
Launch + 1 hour	Post-Launch Briefing	KSC
Launch Through End-of-Mission		
Times announced on NASA Select	Flight Director Change-of-Shift Briefings	JSC
Landing Day		
Landing + 1 hour	Post-Landing Briefing	KSC

51-L TRAJECTORY SEQUENCE OF EVENTS

Event	Orbit	TIG MET (D:H:M)	Duration (Min:Sec)	Burn Delta V (fps)	Post Burn (Apogee/Perigee (S.Mi.)
Launch		0:00:00			
OMS-1	1	0:00:10	2:26	223	
OMS-2	1	0:00:44	2:03	185	153 × 154
Deploy TDRS	7	0:10:02			153 × 154
RCS-1 Separation	7	0:10:03	08	2.2	153 × 154
OMS-3 Separation	8	0:10:21	26	40.0	153 × 177
OMS-4	21	1:06:00	27	43.3	152 × 153
Deploy Spartan	37	05:51			151 × 153
RCS Separation	37	2:06:01	08	2.0	152 × 154
Aft RCS Separation	53	3:06:06	16	4.3	150 × 153
Aft RCS	63	3:21:08	16	4.2	148 × 152
Aft RCS	64	3:23:12	00	0.2	148 × 152
Aft RCS	65	4:00:08	12	3.2	150 × 153
TPF	66	4:01:28	19	5.0	150 × 154
Deorbit				285.0	
Entry Interface		6:04:19			
Landing	97	6:00:34			

Summary of Major Activities

	MET
Day 1	
Payload bay doors open	
Tracking and Data Relay Satellite (TDRS) and Inertial Upper Stage (IUS) checkout	
TDRS deploy	0/10:01
Day 2	
Comet Halley Active Monitoring Program (CHAMP) data take	
Spartan health check	
Fluid Dynamics Experiment (FDE)	
Teacher in Space activities	
Day 3	
Spartan deploy preparations	2/04:40
Spartan deploy	2/06:51
Student Experiments	
FDE	
Day 4	
CHAMP data take	
Student Experiments	
FDE	
Teacher in Space activities	
Day 5	
Spartan rendezvous	4/01:32
Spartan capture	4/02:15
FDE	
Student Experiments	
Day 6	
RCS hot fire	
FCS checkout	
Teacher lesson (Field Trip)	4/21:00
Teacher lesson (Exploration)	4/23:00
Day 7	
Landing at KSC	6/00:34

STS 51-L Payload and Vehicle Weights Summary

Orbiter Without Consumables	176,403
TDRS-B/IUS	37,636
Total Payload Including Other Experiments	48,361
Orbiter Including Cargo at SRB Ignition	268,471
Total Vehicle at SRB Ignition	4,529,122
Orbiter Landing Weight	199,700

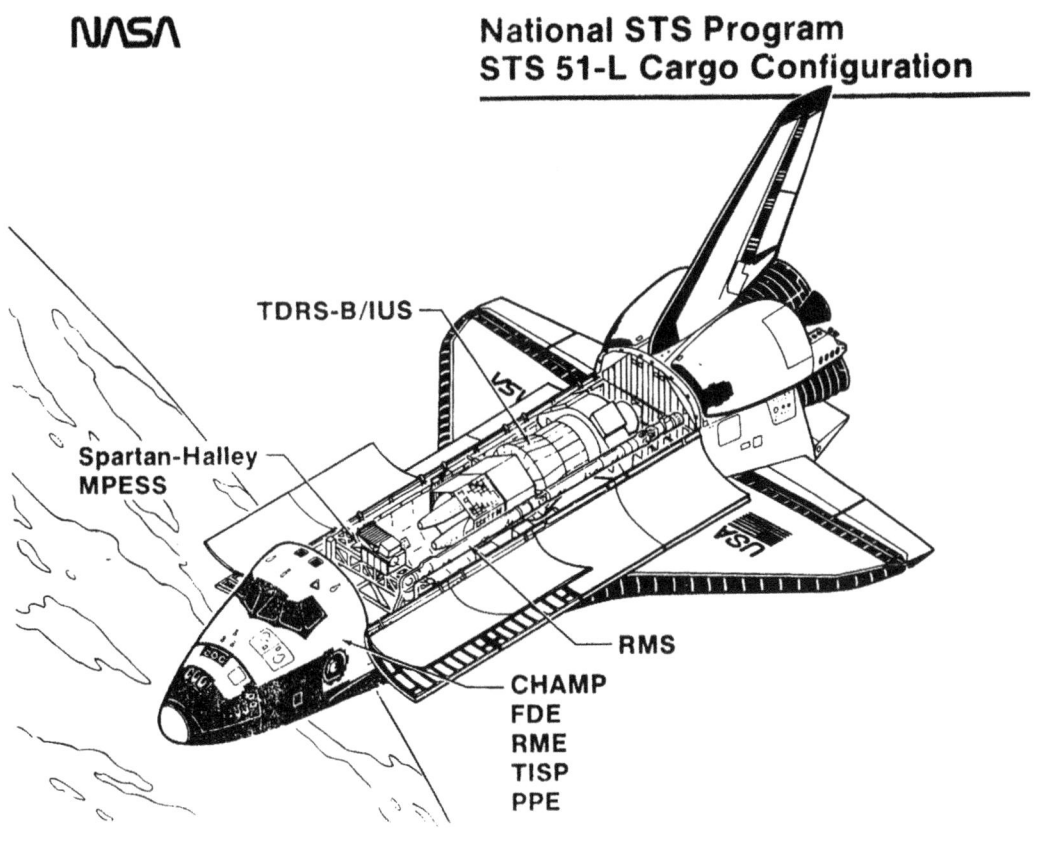

Tracking and Data Relay Satellite System (TDRSS) and TDRS-B

The Tracking and Data Relay Satellite (TDRS-B) is the second TDRSS advanced communications spacecraft to be launched from the orbiter *Challenger*. The first was launched during *Challenger*'s maiden flight in April 1983.

TDRS-1 is now in geosynchronous orbit over the Atlantic Ocean just east of Brazil (41 degrees west longitude). It initially failed to reach its desired orbit following successful Shuttle deployment because of booster rocket failure. A NASA-industry team conducted a series of delicate spacecraft maneuvers over a 2-month period to place TDRS-1 into the desired 22,300-mile altitude.

Following its deployment from the obiter, TDRS-B will undergo a series of tests prior to being moved to its operational geosynchronous position over the Pacific Ocean south of Hawaii (171 degrees W. longitude).

A third TDRSS satellite is scheduled for launch in July 1986, providing the Tracking and Data Relay Satellite System with an on-orbit spare located between the two operational satellites.

TDRS-B will be identical to its sister satellite and the two-satellite configuration will support up to 23 user spacecraft simultaneously, providing two basic types of service: a multiple access service which can relay data from as many as 19 low-data-rate user spacecraft at the same time and a single access service which will provide two high data rate communications relays from each satellite.

TDRS-B will be deployed from the orbiter approximately 10 hours after launch. Transfer to geosynchronous orbit will be provided by the solid propellant Boeing/U.S. Air Force Inertial Upper Stage (IUS). Separation from the IUS occurs approximately 17 hours after launch.

The concept of using advanced communication satellites was developed following studies in the early 1970s which showed that a system of communications satellites operated from a single ground terminal could support Space Shuttle and other low Earth-orbit space missions more effectively than a worldwide network of ground stations.

NASA's Space Tracking and Date Network ground stations eventually will be phased out. Three of the network's present 12 ground stations—Madrid, Spain; Canberra, Australia; and Goldstone, California—have been transferred to the Deep Space Network managed by the Jet Propulsion Laboratory in Pasadena, California, and the remainder—except for two stations considered necessary for Shuttle launch operations—will be closed or transferred to other agencies after the successful launch and checkout of the next two TDRS satellites.

The ground station network, managed by the Goddard Space Flight Center, Greenbelt, Maryland, provides communications support for only a small fraction (typically 15–20 percent) of a spacecraft's orbital period. The TDRSS network of satellites, when established, will provide coverage for almost the entire orbital period of user spacecraft (about 85 percent).

A TDRSS ground terminal has been built at White Sands, New Mexico, a location that provides a clear view to the TDRSS satellites and weather conditions generally good for communications.

The NASA Ground Terminal at White Sands provides the interface between the TDRSS and its network elements, which have their primary tracking and communication facilities at Goddard. Also located at Goddard is the Network Control Center, which provides system scheduling and is the focal point for NASA communications with the TDRSS satellites and network elements.

The TDRSS satellites are the larges privately-owned telecommunications spacecraft ever built, each weighing about 5,000 pounds. Each satellite spans more than 57 feet, measured across its solar panels. The single-access antennas, fabricated of molybdenum and plated with 14K gold, each measure 16 feet in diameter, and, when deployed, span more than 42 feet from tip to tip.

The satellite consists of two modules. The equipment module houses the subsystems that operate the satellite. The telecommunications payload module has electronic equipment for linking the user spacecraft with the ground terminal. The spacecraft has seven antennas.

The TDRS spacecraft are the first designed to handle communications through S, Ku and C frequency bands.

Under contract, NASA has leased the TDRSS service from the Space Communications Company (Spacecom), Gaithersburg, Maryland, the owner, operator and prime contractor for the system.

TRW Space and Technology Group, Redondo Beach, California, and the Harris Government Communications System Division, Melbourne, Florida, are the two primary subcontractors to Spacecom for spacecraft and ground terminal equipment, respectively. TRW also provided the total software for the ground segment operation and did the integration and testing for the ground terminal and the TDRSS, as well as the systems engineering.

Primary users of the TDRSS satellite have been the Space Shuttle, Landsat Earth resources satellites, the Solar Mesosphere Explorer, the Earth Radiation Budget Satellite, the Solar Maximum Mission satellite and Spacelab.

Future users include the Hubble Space Telescope, scheduled for launch October 27, 1986; the Gamma Ray Observatory, due to be launched in 1988; and the Upper Atmosphere Research Satellite in 1989.

Inertial Upper Stage

The Inertial Upper Stage (IUS) will be used to place NASA's second Tracking and Date Relate Satellite (TDRS-B) into geosynchronous orbit. The first TDRS was launched by an IUS aboard *Challenger* in April 1983 during mission STS-6.

SPACECRAFT CONFIGURATION

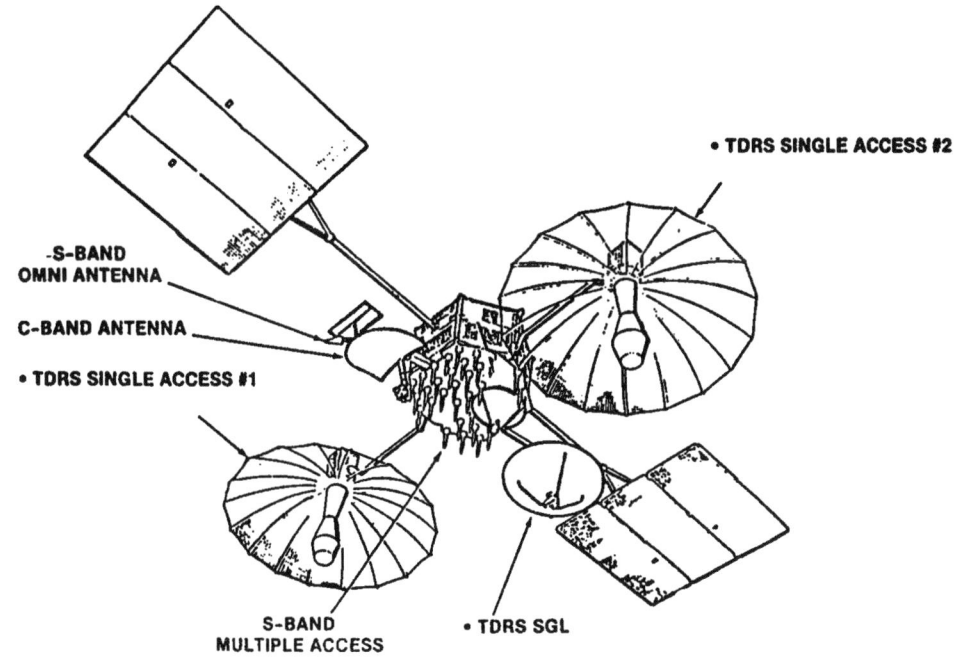

TDRSS CONCEPT

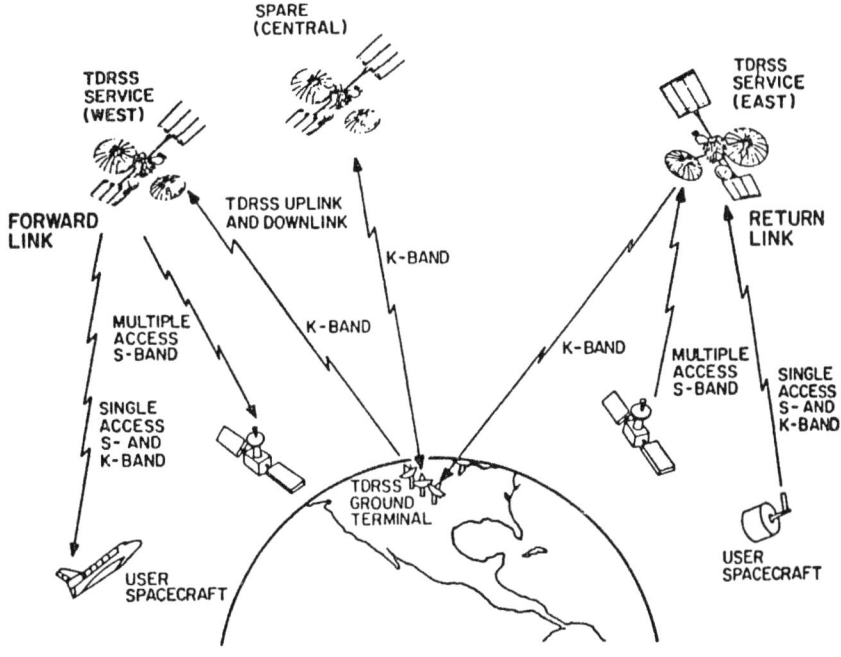

TDRSS SYSTEM ELEMENTS

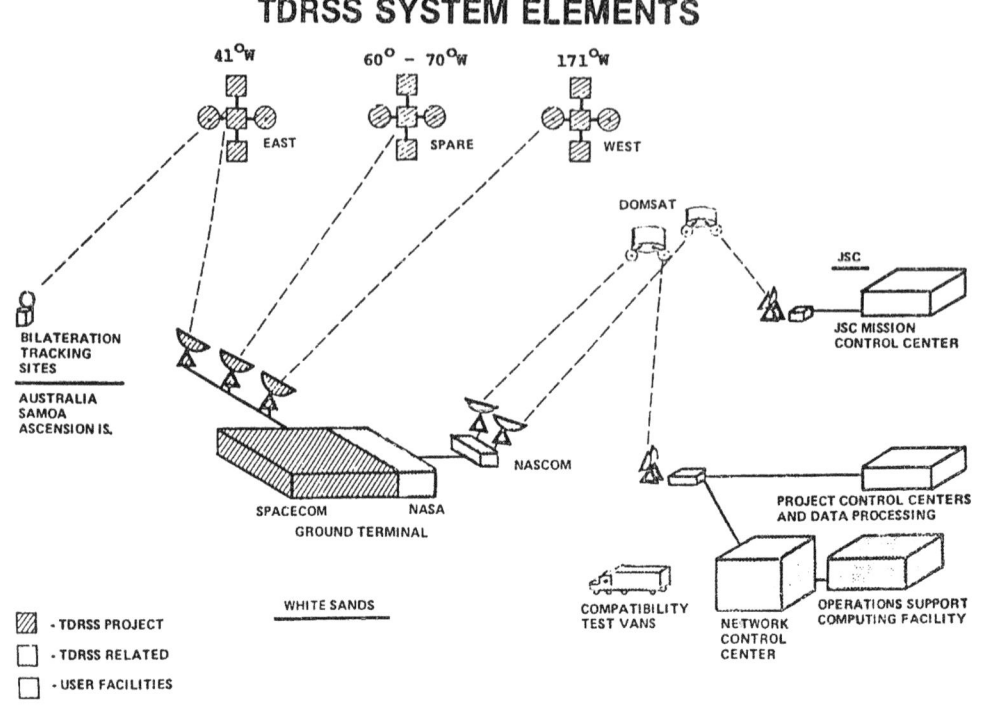

The 51-L crew will deploy IUS/TDRS-B approximately 10 hours after liftoff from a low–Earth orbit of 153.5 nautical miles. Upper stage airborne support equipment, located in the orbiter payload bay, positions the combined IUS/TDRS-B into the proper deployment attitude — an angle of 59 degrees — and ejects it into low–Earth orbit. Deployment from the orbiter will be by a spring eject system.

Following deployment from the payload bay, the orbiter will move away from the IUS/TDRS-B to a safe distance. The first stage will fire about 55 minutes after deployment.

Following the aft (first) stage burn of 2 minutes, 26 seconds, the solid fuel motor will shut down and the two stages will separate. After coasting for several hours, the forward (second) stage motor will ignite at 6 hours, 14 minutes after deployment to place the spacecraft in its desired orbit. Following a 1-minute, 49-second burn, the forward stage will shut down as the IUS/TDRS-B reaches the predetermined geosynchronous orbit position.

Six hours, 54 minutes after deployment from *Challenger*, the forward stage will separate from TDRS-B and perform an anti-collision maneuver with its onboard reaction control system.

After the IUS reaches a safe distance from TDRS-B, the upper state will relay performance data back to a NASA tracking station and then shut itself down 7 hours, 5 minutes after deployment from the payload bay.

As with the first NASA IUS launched in 1983, the second has a number of features which distinguish it from other previous upper stages. It has the first completely redundant avionics system ever developed for an unmanned space vehicle. The system has the capability to correct in-flight features within milliseconds.

Other advanced features include a carbon composite nozzle throat that makes possible the high-temperature, long-duration firing of the IUS motors and a redundant computer system in which the second computer is capable of taking over functions from the primary computer if necessary.

The IUS is 17 feet long, 9 feet in diameter and weighs more than 32,000 pounds, including 27,000 pounds of solid fuel propellant. The IUS consists of an aft skirt; an aft stage containing 21,000 pounds of solid propellant fuel, generating 45,000 pounds of thrust; an interstage; a forward stage containing 6,000 pounds of propellant, generating 18,500 pounds of thrust; and an equipment support section. The equipment support section contains the avionics which provide guidance, navigation, telemetry, command and date management, reaction control and electrical power.

Solid propellant rocket motors were selected in the design of the IUS because of their compactness, simplicity, inherent safety, reliability and lower cost.

The IUS is built by Boeing Aerospace Corp., Seattle, under contract to the U.S. Air Force Systems Command. Marshall Space Flight Center, Huntsville, Alabama, is NASA's lead center for IUS development and program management of NASA-configured IUSs procured from the Air Force.

SPARTAN-HALLEY MISSION

For the Spartan-Halley mission, NASA's Goddard Space Flight Center and the University of Colorado's Laboratory for Atmospheric and Space Physics (LASP) have recycles several instruments and designs to produce a low-cost, high-yield spacecraft to watch Halley's Comet when it is too close to the sun for other observatories to do so.

It will record ultraviolet light emitted by the comet's chemistry when it is closest to the sun and most active so that scientists may determine how fast water is broken down by sun light, search for carbon and sulfur atoms and related compounds, and understand how the tail evolves.

Principal investigator is Dr. Charles Barth of the University of Colorado LASP. Mission manager is Morgan Windsor of Goddard Space Flight Center.

The Instruments

Two spectrometers, derived from backups for a Mariner 9 instrument which studied the Martian atmosphere in 1971, have been rebuilt to survey Halley's Comet in ultraviolet light from 128 to 340 nanometers (nm) wavelength, stopping just above the human eye's limit of about 400 nm.

Each spectrometer uses the Ebert-Fastie design: an off-axis reflector telescope, with magnesium fluoride coatings to enhance transmission which focuses light from Halley, via a spherical mirror and a spectral grating, on a coded anode converter with 1,024 detectors in a straight line. The grating is ruled at 2,400 lines per millimeter.

The detectors are made of cesium iodide (CsI) for the G-spectrometer (128–168 nm) and cesium telluride (CsTe) for the F-spectrometer (180–340 nm). The system has a focal length of 250 mm and an aperture of 50 mm.

The F-spectrometer grating can be rotated to cover its wider range in six 40 nm sections. A slit limits its field of view to a strip of sky 1 by 80 arc-minutes (the apparent diameter of the moon is about 30 arc-minutes). The G-spectrometer has a 3 × 80 arc-minute slit because emissions are fainter at shorter wavelengths.

With Halley as little a 10 degrees away from the sun, two sets of battles must be used to reduce stray light. An internal set is part of the Mariner design. A new external set serves both instruments. It has two knife-edge baffles 38.5 inches away from the spectrometer entrances, and 20 secondary baffles to stop earthlight. Together the two baffle sets reduce stray light by a factor of a trillion. It is this system that will make it possible for Spartan-Halley to observe the comet while so close to the sun. In addition, internal filters reduce solar lyman-alpha light (121.6 nm), scattered by the Earth's hydrogen corona, which would saturate the instruments.

Two film cameras, boresighted with the spectrometers, will photograph Halley to assure pointing accuracy in post-flight analysis and to match changes in the tail with spectral changes. The 35mm Nikon F3 cameras have 105 mm and 135 mm lenses and are loaded with 65-frame rolls of QX-851 thin-base color film. The cameras will capture large-scale activity such as the separation angle between the dust and ion tails, bursts from the nucleus, and asymmetries in the shape of the coma.

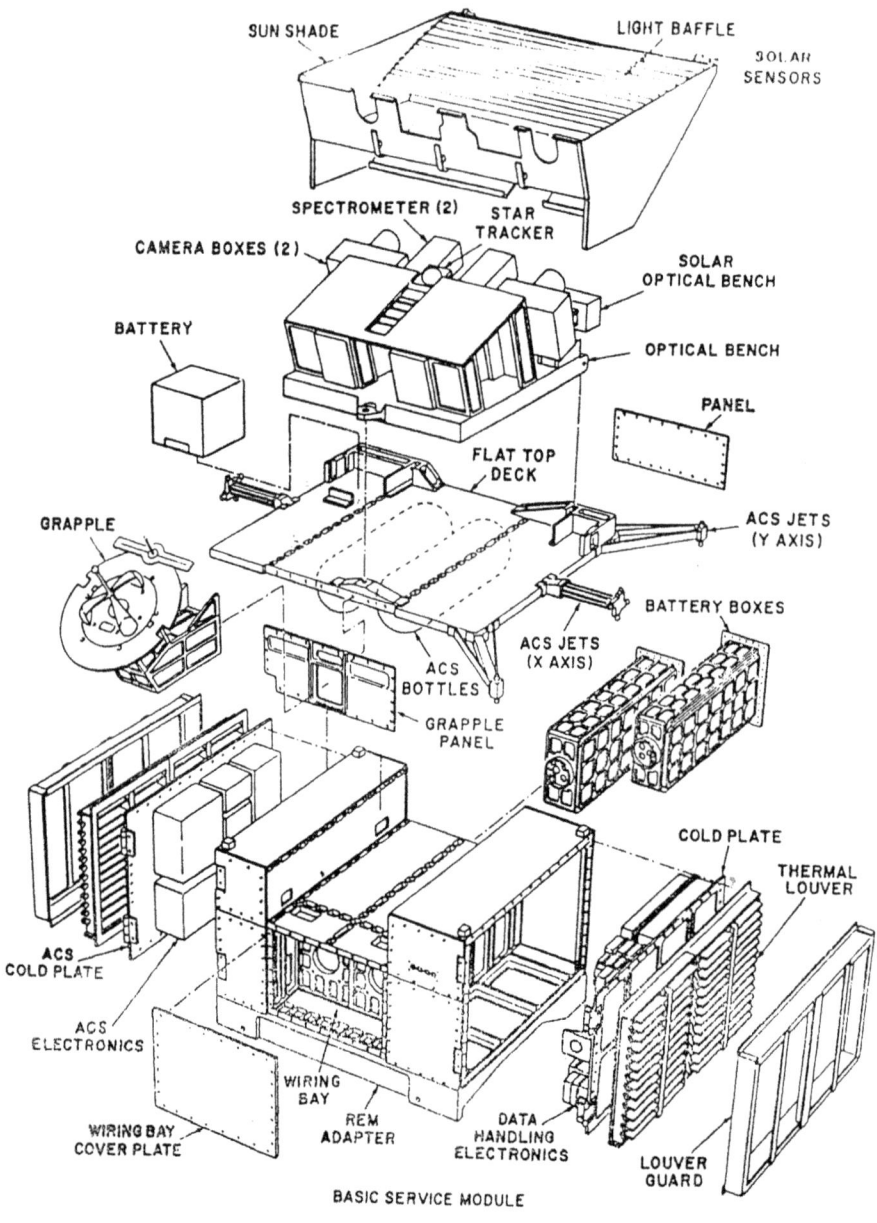

SPARTAN-HALLEY

The whole instrument package is mounted on an aluminum optical bench — 35 × 37 inches and weighing 175 pounds — attached to the Spartan carrier. This provides a clean interface with the carrier and aligns the spectrometers with the Spartan attitude control sensors. A 15-inch high housing covers the spectrometers and the cameras.

The instrument package is controlled by a LASP-developed microprocessor which stores the comet Halley ephemeris and directs the Spartan carrier attitude control system.

Mission Operations

Halley's Comet will be of greatest scientific interest from January 20 to February 22; perihelion is on February 9. At that time Halley will be 139.5 million miles from Earth and 59.5 million miles from the sun. The Shuttle will go into an orbit 176 miles high and inclined 28.5 degrees to the equator. This will have Halley visible for more than 3,000 seconds per orbit (about 56 percent of the orbit), including more than 90 seconds with the sun occulted by the Earth.

After a predeployment health check of Spartan voltages and currents, the Shuttle robot arm will pick up the spacecraft and hold it over the side. Upon release, Spartan will perform a 90-second "pirouette" to confirm that it is working and the Shuttle will back away to at least 5 miles so light reflected from the Shuttle does not confuse Spartan's sensors. After two orbits of preparation, the 40-hour science mission will begin. A backup timer will ensure that the spectrometer doors open 70 minutes after release.

Spartan-Halley will conduct 20 orbits of science observations interspersed with five orbits of attitude control updates. A typical science orbit will start with four 100-second calibration scans of Earth's atmosphere, followed by a 9900-second tail scan. Observing will be interrupted for 15 minutes of pointing updates and housekeeping. It then resumes with four 200-second scans of the coma, followed by sunset and four coma scans while the sun is occulted. At the end of the mission, Spartan-Halley will be retrieved by the Shuttle robot arm and placed in the payload bay.

After the mission, the processed film and data tapes will be returned to the University of Colorado team for scientific analysis.

The Science

Current theories hold that comets are "dirty snowballs" made up largely of water ice and lightweight elements and compounds left over form the creating of the solar system. Remote sensing of the chemistry of Halley's Comet, by measuring how sunlight is reflected, will help in assaying the comet. The "dirt" in the snowball is detectable in visible light, and the "snow" (water ice) and other gases are detectable, indirectly, in ultraviolet.

The most important objective of the Spartan-Halley mission is to obtain ultraviolet spectra of comet Halley when it is less than 67 million miles from the sun. As Halley nears the sun, temperatures rise, releasing ices and calthrates, compounds trapped in ice crystals.

The highest science priority for Spartan is to determine the rate at which water is broken up (dissociated) by sunlight. This must be measured indirectly from the spectra of hydroxyl radicals (OH) and atomic oxygen which are the primary and secondary products. The hydroxyl coma of the comet will be more compact than the atomic oxygen coma because of its short life when exposed to sunlight. Hydrogen, the other product, will not be detectable because of the lyman-alpha filters in the spectrometers.

Heavier compounds will be sought by measuring spectral lines unique to carbon, carbon monoxide (CO), carbon dioxide (CO_2), sulfur, carbon sulfide (CS), molecular sulfur (S_2), nitric oxide (NO) and cyanogen (CN), among others.

Spartan-Halley's spectrometers will not produce images, but will reveal the comet's chemistry through the ultraviolet spectral lines they record. With these data, scientists will gain a better understanding of how:

- Chemical structure of the comet evolves from the coma and proceeds down the tail;
- Species change with relation to sunlight and dynamic processes within the comet; and
- Dominant atmospheric activities at perihelion relate to the comet's long-term evolution.

Other observations will be studying Halley's comet, but only Spartan can observe near perihelion.

The Spartan Program

The Spartan-Halley 200 carrier measures 52 by 43 by 51 inches and weighs 2,250 pounds without payload. The attitude control system and other components use elements from the sounding rocket program. All data are stored on a Bell & Howell MARS 1400 recorder; 500 megabytes of storage are available to the experimenter. The spacecraft sits on the Spartan Flight Support Structure and is help in place by a release-and-engage mechanism.

TEACHER IN SPACE PROJECT

Project History

President Reagan announced on August 27, 1984, that a teacher would be chosen as the first private citizen to fly no the Space Shuttle. The Teacher in Space Project is part of NASA's Space Flight Participant Program designed to expand Shuttle opportunities to a wider segment of private citizens with the purpose of communicating the experience and flight activities to the public through educational and public information programs.

The Council of Chief State School Officers (CCSSO) was selected by NASA to coordinate the selection process. The Council is a nonprofit organization comprised of the public officials responsible for education in each state.

In November 1984, the Announcement of Opportunity was distributed, listing the eligibility requirements and a description of the selection process and criteria, medical requirements and responsibilities of the teacher selected to fly on the Shuttle mission.

Applications were accepted from December 1, 1984, to February 1, 1985. More than 11,000 teachers from the 50 states, the District of Columbia, Puerto Rico, Guam, the Virgin Islands, Department of Defense overseas schools, Department of State overseas schools and Bureau of Indian Affairs schools applied for the flight. State, territorial and agency review panels each selected two nominees for a total of 114 nominees.

The 114 nominees met in Washington, D.C., from June 22 to June 27, 1985, for a National Awards Conference which focused on various aspects of space education.

During their stay in Washington, all nominees met formally and informally with the National Review Panel, which selected the 10 finalists.

On July 1, 1985, the 10 finalists were announced and on July 7 they traveled to Johnson Space Center for a week of thorough medical examinations and briefings about space flight.

From July 15 to 18, each of the finalists was interviewed by a NASA Evaluation Committee made up of senior NASA officials. The committee made recommendations to the NASA Administrator who selected Christa McAuliffe and Barbara Morgan as the primary and backup candidates for the NASA Teacher in Space Project.

Live Lessons

Teacher observer Christa McAuliffe will conduct two lessons on Flight Day 6 of the mission. The first lesson will begin at approximately 11:40 A.M. EST; the second is scheduled for approximately 1:40 P.M. EST.

The first lesson entitled "The Ultimate Field Trip" will allow students to compare daily life on the Shuttle with that on Earth. Conducting a tour of the Shuttle for viewers, McAuliffe will explain crewmembers' roles, show the location of computers and controls and describe experiments being conducted on the mission. She also will demonstrate how daily life in space is different from that on Earth in the preparation of food, movement, exercise, personal hygiene, sleep and the use of leisure time.

The second lesson, "Where We've Been, Where We're Going, Why?" will help the audience understand why people use and explore space by demonstrating the advantages of manufacturing in the microgravity environment, highlighting technological advances that evolve from the space program and projecting the future of humans in space.

Mission Watch

Classrooms with access to a satellite dish or cable network that carries NASA Select television will have, in addition to the live broadcast, the opportunity to participate in a "Mission Watch" which covers aspects of the entire Shuttle flight. The Mission Watch during 51-L will start the day before the launch and continue through the conclusion of the mission and will be carried only on NASA Select. Barbara Morgan, backup candidate for the NASA Teacher in Space Project, will act s moderator for the Mission Watch broadcast to schools. Morgan will give daily briefings on that day's planned events and update viewers on McAuliffe's activities.

Classroom Earth, a Spring Valley, Illinois, organization dedicated to direct satellite transmission to elementary and secondary schools, will serve as the center for information and materials for schools that wish to use satellite dish antennas to receive both the live broadcast and the Mission Watch.

Specific information about the satellite transmission is available by writing to Classroom Earth, Spring Valley, Illinois 61362 or by calling 815/664-4500.

In-Flight Activities

During the 51-L mission, McAuliffe will be involved in several activities which will be filmed and later used in educational products.

- Magnetism — Photograph and observe the lines of magnetic force in three dimensions in a microgravity environment.
- Newton's Law — Demonstrate Newton's first, second and third laws in microgravity.
- Effervescence — Understand why products may or may not effervesce in a microgravity environment.
- Simple Machines/Tools — Understand the use of simple machines/tools and the similarities and differences between their uses in space and on Earth.
- Hyponics in Microgravity — Show the effect of microgravity on plant growth, growth of plants without soil (hydroponics) and capillary action.
- Chromatographic Separation of Pigments — Demonstrate chromatography in a microgravity environment and show capillary action (the mechanism by which plants transport water and nutrients).

SHUTTLE STUDENT INVOLVEMENT PROGRAM

- Utilizing a Semi-Permeable Membrane to Direct Crystal Growth

This is an experiment proposed by Richard S. Cavoli, formerly of Marlboro Central High School, Marlboro, New York. Cavoli is now enrolled at Union College, Schenectady, New York.

The experiment will attempt to control crystal growth through the use of semi-permeable membrane. Lead iodide crystals will be formed as a result of a double replacement reaction. Lead acetate and potassium iodide will react to form insoluble lead iodide crystals, potassium ions and acetate ions. As the ions travel across a semi-permeable membrane, the lead and iodide ions will collide, forming the lead iodide crystal.

Cavoli's hypothesis states that the shape of the semi-permeable membrane and the concentrations of the two precursor compounds will determine the growth rate and shape of the resultant crystal without regard to other factors experienced in earthbound crystal growing experiments.

Following return of the experiment apparatus to Cavoli, an analysis will be performed on the crystal color, density, hardness, morphology, refractive index, and electrical and thermal characteristics. Crystals of this type are useful in imaging systems for detecting gamma and X-rays and could be used in spacecraft sensors for astrophysical research purposes.

Cavoli's high school advisor is Annette M. Saturnelli of Marlboro Central High School and his college advisor and experiment sponsor is Dr. Charles Seaise of Union College.

• Effects of Weightlessness on Grain Formation and Strength in Metals

This is an experiment proposed by Lloyd C. Bruce formerly of Sumner High School, St. Louis. Bruce is now a sophomore at the University of Missouri.

The experiment proposes to heat a titanium alloy metal filament to near the melting point to observe the effect that weightlessness has on crystal reorganization within the metal. It is expected that heating in microgravity will produce larger crystal grains and thereby increase the inherent strength of the metal filament. The experiment uses a battery supply, a timer and thermostat to heat a titanium alloy filament to 1,000 degrees C. At a temperature of 882 degrees C, the titanium-aluminum alloy crystal lattice network undergoes a metamorphosis from closely packed hexagonal crystals to centered cubic crystals.

Following return of the experiment gear to Sumner, he will perform an analysis comparing the space-tested alloy sample with one heated on Earth to analyze any changes in strength, size and shape of the crystal grains, and any change in the homogeneity of the alloy. If necessary microscopic examination, stress testing and X-ray diffraction analysis also will be used. Any changes between the two samples could lead to variations on this experiment to be proposed for future Shuttle flights. A positive test might lead to a new and stronger titanium-aluminum alloy or a new type of industrial process.

Bruce's student advisor is Vaughan Morrill of Sumner High School. His sponsor is McDonnell Douglas Corp., St. Louis, and his experiment advisor is Julius Bonini of McDonnell Douglas.

• Chicken Embryo Development in Space

This is an experiment of John C. Vellinger, formerly of Jefferson High School, Lafayette, Indiana, to determine any effects of spaceflight on the development of a fertilized chicken embryo. Vellinger is now a sophomore at Purdue University.

The experiment will fly 12 White Leghorn chicken eggs which have been fertilized immediately prior to launch to see if any changes in the developing embryo can be attributed to weightlessness or space radiation effects. The development of a chicken embryo is greatest during the first several weeks following fertilization.

Eight eggs will be subjected to both weightlessness and radiation from space. The four remaining eggs will have lead shields placed around them to assist in determining any peculiar effects from space radiation. All 12 eggs will be placed in an incubator box and send aboard *Challenger* while a similar group of 12 eggs will remain on Earth as a control group. Vellinger's hypothesis is that chickens could form a basis for food developed and grown in space but only if their development from fertilized eggs proceeds normally.

Upon return to Earth, the space incubators will remain at KSC for a period of about 10 days to allow the chicks to hatch. Vellinger will attend to the earthbound eggs much as a mother hen would, turning them five times a day to counter the effects of Earth's gravity on the yolk sack. Following hatching of both groups, Vellinger will attempt to determine if there are any statistically significant differences between the bone structure, nervous systems and internal organs of the two groups.

Vellinger's student advisor is Stanley W. Poelstra of Jefferson High School. Kentucky Fried Chicken, Louisville, is sponsoring the experiment and Dr. Lisbeth Kraft, NASA Ames Research Center, Mountain View, California, is serving as technical advisor.

Comet Halley Active Monitoring Program (CHAMP)

Objectives of the CHAMP payload include investigating the dynamical/morphological behavior as well as the chemical structure of Halley's Comet. Photographic images and spectra will be obtained through Columbia's windows using a handheld 35mm camera and associated equipment. A crew member will enclose himself in a camera shroud to eliminate all cabin light interference. Using International Halley Watch stand comet filters, several image-intensified monochromatic exposures will be made. In addition, spectra of the comet will be photographed using a grating and image intensifier.

Similar observations were made on the last flight and will be made on the March flight to study the variations of the comet with time. CHAMP requires no spacecraft power or other systems support and is stored in two-thirds of one mid-deck locker.

The principal investigators for CHAMP are S. Alan Sterm, Laboratory for Atmospheric and Space Physics (LASP), University of Colorado–Boulder, and Dr. Stephen Mende, Lockheed Palo Alto Research Laboratory. Mission management support is provided by the Engineering Directorate, Johnson Space Center for the Office of Space Science and Applications, NASA Headquarters.

Phase Partitioning Experiment

Phase partitioning is a selective, gentle and inexpensive technique, ideal for the separation of biomedical materials such as cells and proteins. It involves establishing a two-phase system by adding various polymers to a water solution containing the materials to be separated. Two phase systems most familiar are oil and water or cream and milk. When two phase polymer systems are established, the biomedical materials they contain tend to separate or "partition" into the different phases.

Theoretically, phase partitioning should separate cells with significantly higher resolution than is presently obtained in the laboratory. It is believed that when the phases are emulsified on Earth, the rapid, gravity-driven fluid movements, occurring as the phases coalesce, tend to randomize the separation process. It is expected that the theoretical capabilities of phase partitioning systems can be closely approached in the weightlessness of orbital spaceflight where gravitational effect of buoyancy and sedimentation are minimized.

The first exploratory flight of Phase Partitioning Experiment (PPE) equipment involves the use of a small, handheld device, a little larger than a cigarette box and weighing about one pound. This unit will fit within a small part of a standard mid-deck locker. The unit has 15 chambers to allow the test of different volume rations and compositions of the phases and differences in wall coatings within the chambers.

The Microgravity Science and Applications Division of the Office of Space Science and Applications, NASA Headquarters, sponsors the experiment. Marshall Space Flight Center is responsible for mission implementation.

Fluid Dynamics in Space

Hughes payload specialist Greg Jarvis will perform experiments to investigate fluid dynamics in microgravity. These experiments will improve Hughes' understanding of how fluids act in orbiting spacecraft and may lead to the design of more efficient and less costly spacecraft.

Hughes is now designing larger, more massive spacecraft to take advantage of the Shuttle capabilities. This evolution in design has led to the replacement of solid rocket motors with highly efficient liquid propulsion systems for transfer orbit maneuvers. Spacecraft design has always taken into account the possible destabilizing effects of liquid propellants. However, as the quantity of liquid increases it becomes more important to understand the interaction between the fluid and the spacecraft.

There are two categories of experiments to be performed: fluid behavior in enclosed tanks and fluid motion interactions with spacecraft motions.

For the enclosed tank experiments, Jarvis will try to determine the behavior of a fluid in a partially-filled tank for different levels of fill. A metal base plate will be used to support the experiment and will be attached with Velcro tape to the cabin walls of *Challenger*. A metal shaft attaches to the base plate and supports a hub assembly connecting to two 6-inch diameter tanks.

The tanks and hub do not move in one of these experiments and are set spinning at about 10 revolutions/minute in a second experiment. Throughout these experiments, Jarvis will record the fluid motions using orbiter video cameras and recorders. A third experiment in this series will use a spring to spin up the hub assembly (much as a helicopter toy uses a spring to impart a one-time spin-up). Once spun up, the hub assembly will be videotaped to record the transfer of motion from the simulated spacecraft to the fluid inside. The spheres are transparent so the motions of the fluid inside can be seen readily.

Another set of experiments will observe the effects of energy dissipation between spacecraft motions and fluid motions within the spacecraft. In these experiments, a four-tank spherical tank model will first be attached to the hub and used for observations. This model consists of four 3¼ inch-diameter clear plastic tanks. Following those, an ellipsoidal centerline tank model will be used. The ellipsoidal tank is 3¼ inches long by 4¼ inches in diameter.

This experiment will measure transmitted motions from the simulated spacecraft tanks to the fluids in the tanks using very sensitive accelerometers which transmit their information via infrared light emitting diodes (similar to the way most television and video recorder remote control units work).

This technique removes any potential drag which might result from cabling between the moving tank models and the instrument recorders. For this series of experiments, the models will be spun up to about 100 revolutions a minute by Jarvis.

A third series of experiments will measure the fluid dynamics of "Frisbee" style deployed satellites using a cradle model in lieu of the base plate used in earlier experiments.

The videotapes and recorded accelerometer data will be analyzed by Hughes engineers once the mission is over.

U.S. LIBERTY COINS

Two complete sets of the newly-minted U.S. Liberty coins will become the first legal tender American coinage to make a trip into orbit on mission 51-L.

The Liberty coins—a silver dollar, half dollar and $5 gold coin—are being minted by authorization from Congress to honor the Statue of Liberty's centennial anniversary in 1986 and to help raise funds for the restoration and future maintenance of the statue and Ellis Island. They are the first and only government-issue coins to feature the Statue of Liberty.

School children in grades 4 through 8 will be given the chance to learn about coins and American history in a special education package—Commemorating Liberty Through Coins—which will be delivered to every public, private and parochial school in the nation in early March 1986.

The U.S. Liberty coins will be available at financial institutions and department stores across the country in April 1986.

51-L FLIGHT CREW

FRANCIS R. (DICK) SCOBEE is a spacecraft commander. Born May 19, 1939, in Cle Elum, Washington, he became a NASA astronaut in 1978. He received a B.S. degree in aerospace engineering from the University of Arizona in 1965.

Scobee was a reciprocating engine mechanic in the Air Force. He was commissioned in 1965, and after receiving his wings in 1966, completed a number of assignments including a combat tour in Vietnam. He attended the Aerospace Research Pilot School at Edwards Air Force Base, flying such varied aircraft as the Boeing 747, the X-24B, the transonic aircraft technology (TACT) F-111 and the C-5. He has logged more than 6,500 hours in 45 types of aircraft.

Scobee was pilot of STS 41-C in 1984. During this mission the crew deployed the Long Duration Exposure Facility (LDEF); and retrieved, repaired aboard the orbiting *Challenger*, and returned to orbit, the ailing Solar Maximum Mission satellite.

MICHAEL J. SMITH, Commander, USN, is pilot. Born April 30, 1945, in Beaufort, N.C., he became a NASA astronaut in 1980.

Smith received a B.S. degree in naval science from the U.S. Naval Academy and an M.S. degree in aeronautical engineering from the U.S. Naval Postgraduate School.

Smith flew A-6 Intruders and completed a Vietnam cruise while assigned to Attack Squadron 52 aboard the USS Kitty Hawk. He was awarded the Navy Distinguished Flying Cross, 3 Air Medals, 13 Strike Flight Air Medals, the Navy Commendation Medal with "V," the Navy Unit Citation and the Vietnamese Cross of Gallantry with Silver Star.

He has flown 28 different types of civilian and military aircraft, logging over 4,300 hours—4,000 in jet aircraft.

JUDITH A. RESNIK, Ph.D., is one of three mission specialists aboard 51-L. She was born April 5, in Akron, Ohio. She received a B.S. degree in electrical engineering from Carnegie-Mellon University and a Ph.D. in electrical engineering from the University of Maryland. She became an astronaut in 1978.

Resnik worked for RCA, Moorestown, New Jersey, designing circuits and developing custom integrated circuitry for phased-array radar control systems. She was a biomedical engineer and staff fellow in the laboratory of neurophysiology at the National Institutes of Health, Bethesda, Maryland. She also served as senior systems engineer in product development with Xerox Corporation, El Segundo, California.

Resnik was mission specialist on STS 41-D. During this mission, the crew deployed three satellites. Resnik has logged 144 hours, 57 minutes in space.

RONALD E. MCNAIR, Ph.D., mission specialist, became an astronaut in 1978. Born October 21, 1950, in Lake City, South Carolina, he received a B.S. degree in physics from North Carolina A&T State University and a Ph.D. in physics from Massachusetts Institute of Technology.

While at MIT, McNair performed some of the earliest development of chemical HF/DF and high pressure CO lasers. He became a staff physicist with Hughes Research Laboratories in Malibu, California, and conducted research on electro-optic laser modulation for satellite-to-satellite space communications.

McNair was a mission specialist on Shuttle mission 41-B. During the flight, two Hughes 376 communications satellites were deployed. It was the first flight of the Manned Maneuvering Unit and first use of the Canadian arm (operated by McNair) to position EVA crewman around *Challenger*'s payload bay. McNair has 191 hours in space.

ELLISON S. ONIZUKA, Lt. Col., USAF, is a mission specialist. He became an astronaut in 1978. Born June 24, 1946, in Kealakekua, Kona, Hawaii, he received B.S. and M.S. degrees in aerospace engineering from the University of Colorado. He became a NASA astronaut in 1978.

Onizuka was an aerospace flight test engineer with the Sacramento Air Logistics Center at McClellan Air Force Base. He participated in flight test programs and systems safety engineering for the F-84, F-100, F-105, F-111, EC-121T, T-33, T-39, T-28 and A-1 aircraft. He has logged more than 1,700 hours flying time.

Onizuka was a mission specialist on STS 51-C, the first dedicated Department of Defense mission. He has logged 74 hours in space.

GREGORY B. JARVIS, payload specialist, was born August 24, 1944, in Detroit. He received a B.S. degree in electrical engineering from the State University of New York at Buffalo; an M.S. degree in electrical engineering from Northeastern University, Boston; and has completed the course work for an M.S. degree in management science at West Coast University, Los Angeles. Jarvis was selected as a payload specialist candidate in 1984.

Jarvis worked at Raytheon, Bedford, Massachusetts, designing circuits on the SAM-D missile. Later, as a communications payload engineer in the Satellite Communications Program Office, he worked on advanced tactical communications satellites. He later joined Hughes Aircraft Company's Space and Communications Group where he worked as subsystem engineer on the MARISTAT Program.

He was test and integration manager for the F-1, F-2 and FD-3 spacecraft and cradle in 1983. The F-1 and F-2 Leasat spacecraft were successfully deployed.

S. CHRISTA CORRIGAN MCAULIFFE is the Teacher in Space participant. Born September 2, 1948, in Boston, she received a B.A. degree from Framingham State College and a masters degree in education from Bowie State College, Bowie, Maryland.

McAuliffe has taught English and American history since 1970. Until her selection as NASA Teacher in Space, she taught economics, law, American history, and a course she developed, "The American Woman," to 10th through 12th grade students.

McAuliffe was selected as primary candidate for the NASA Teacher in Space Project in July 1985.

Abbreviations and Definitions

Abbreviations

AFB	Air Force Base
APU	Auxiliary Power Units
AOA	Abort once around
ATO	Abort to Orbit
ASTP	Apollo Soyuz Test Project
BSM	Booster separation motor
CCAFS	Cape Canaveral Air Force Station
DOD	Department of Defense
DM	Demonstration motor
EMU	Extravehicular mobility unit
ETM	Engineering test motor
ET	External tank
EVA	Extravehicular activity
ELV	Expendable launch vehicle
EAFB	Edwards Air Force Base
ESA	European Space Agency
ETR	Eastern Test Range, Kennedy
ESA	European Space Agency
FAA	Federal Aviation Administration
GAO	General Accounting Office
GMT	Greenwich Mean Time
GSFC	Goddard Space Flight Center, Maryland
HAC	Heading Alignment Circle
IVA	Intra-vehicular activity
IUS	Inertial Upper Stage
JSC	Lyndon B. Johnson Space Center, Houston
KSC	John F. Kennedy Space Center, Cape Canaveral, Florida
LC-39	Launch Complex 39, Kennedy
MECO	Main engine cutoff
MCC-H	Mission Control Center at Houston
MLP	Mobile launch platform
MOL	Manned Orbital Laboratory
MMU	Manned maneuvering unit

MSFC	Marshall Space Flight Center, Alabama
MPTA	Main propulsion test article
NASA	National Aeronautics and Space Administration
NTSB	National Transportation Safety Board
OPF	Orbiter Processing Facility
OMS	Orbital Maneuvering System
OTV	Orbital transfer vehicle
OV	Orbiter vehicle
POCC	Payload Operations Control Center
PEAP	Personal egress air packs
QM	Qualification motor
RPV	Remotely piloted vehicle
RTLS	Return to Launch Site
RDSM	Redesigned solid rocket motor
SCA	Shuttle Carrier Aircraft, 747
SLF	Shuttle Landing Facility at Kennedy
SRB	Solid rocket booster
SRM	Solid rocket motor
SSME	Space Shuttle main engine
SMS	Shuttle Mission Simulator
STA	Structural test article
STS	Space Transportation System
SSTS	Space Shuttle Transportation System
TDRS	Tracking and Data Relay Satellite
TDRSS	Tracking and Data Relay Satellite System
TPS	Thermal protective system
TVC	Thrust vector control
TAL	Transoceanic abort landing
USCG	United States Coast Guard
VAB	Vehicle Assembly Building at Kennedy
VAFB	Vandenberg Air Force Base
VPF	Vertical Processing Facility at Kennedy
WSTF	White Sands Test Facility
WTR	Western Test Range, Vandenberg

Definitions

Ablative material Thermal protective material which burns off mass during reentry, taking the frictional heat with it.

Abort Ending countdown of a launch vehicle after main engine ignition or intentionally ending a launch after liftoff but before entry into orbit.

Apogee motor Final rocket stage used to place payloads into geosynchronous orbit.

Apogee High point in an orbit around the earth.

Blackout The point during a reentry of a spacecraft where communication is blocked by the ionized layer of the atmosphere and shock wave created by the spacecraft.

Cryogenic fuels Super-cooled fuels used to power spacecraft engines.

Deorbit burn Engine burn opposite to the direction of orbital spacecraft sufficient to allow the spacecraft to reenter the atmosphere.

Drogue parachute Small parachute used to pull out the main parachute of a landing spacecraft.

Geosynchronous orbit An orbit of the earth by a satellite whose orbital speed matches the earth's rotation, thus keeping the satellite positioned over one point on the earth.

Hypergolic fuel Fuel which ignites spontaneously when it comes into contact with an oxidizer.
Launch azimuth Compass direction of a spacecraft at launch.
Lifting body Wingless aircraft which acquires its aerodynamic lift by the shape of the craft.
Mock-up Full-scale model of a spacecraft normally used as a test bed for future construction of flight vehicles.
OV-099 Space Shuttle *Challenger*, rolled out on June 30, 1982; first flown on April 4, 1983; destroyed January 28, 1986.
OV-101 Space Shuttle *Enterprise*, rolled out on January 31, 1977; never flew in space.
OV-102 Space Shuttle *Columbia*, rolled out on March 8, 1979; first flown on April 12, 1981; destroyed February 1, 2003.
OV-103 Space Shuttle *Discovery*, rolled out on October 16, 1983; first flown on August 30, 1984.
OV-104 Space Shuttle *Atlantis*, rolled out on April 6, 1984; first flown on October 3, 1985.
OV-105 Space Shuttle *Endeavour*, rolled out on April 25, 1991; first flown on May 7, 1992.
Perigee Low point in an orbit around the earth.
Pitch over Change of direction of a spacecraft from a vertical to a more horizontal direction.
Spacelab European-built science lab specifically constructed for use on the Shuttles.
Thrust vector Manipulation of the thrust of a vehicle in any direction required.

Chapter Notes

Chapter One

1. *Air & Space*, April/May 1991, "The Space Shuttle Family Tree."
2. *NASA Facts*, February 1994, "HL-10 Lifting Body."
3. *Air & Space*, April/May 1991, "The Space Shuttle Family Tree."
4. *NASA Facts*, February 1994, "HL-10 Lifting Body."
5. *Air & Space*, April/May 1991, "The Space Shuttle Family Tree."
6. *NASA Facts*, "The Lifting Bodies."
7. *Spaceflight News*, May 1988, "X-Plane Veteran Still Going Strong."
8. *Air & Space*, October/November 1993, "At the Threshold of Space."
9. *Air & Space*, April/May 1991, "The Space Shuttle Family Tree."
10. *Daily Breeze*, October 21, 1969, "Shuttle Bus Is Vital to Space Plans."
11. *Air & Space*, April/May 1991, "The Space Shuttle Family Tree."
12. *NASA Facts*, February 1994, "HL-10 Lifting Body."
13. *NASA Facts*, "The Lifting Bodies."
14. *NASA Facts*, February 1994, "HL-10 Lifting Body."
15. *NASA Facts*, "The Lifting Bodies."
16. *Ibid.*
17. Northrop Corporation History, pages 228–236, M2-F2/F3, HL-10 Lifting Bodies.
18. *NASA Facts*, February 1994, "HL-10 Lifting Body."
19. *Air & Space*, April/May 1991, "The Space Shuttle Family Tree."
20. *Los Angeles Times*, May 11, 1967, "Wingless Craft Crashes During NASA Glide Test."
21. *Air & Space*, April/May 1991, "The Space Shuttle Family Tree."
22. *Ibid.*
23. *Los Angeles Times*, October 24, 1968, "Rocket Failure Spoils Wingless Craft's Test."
24. *Daily Breeze*, October 21, 1969, "Shuttle Bus Is Vital to Space Plans."
25. *Daily Breeze*, February 20, 1970, "Edwards Air Force Base."
26. *NASA Facts*, February 1994, "HL-10 Lifting Body."
27. *NASA Facts*, "The Lifting Bodies."
28. *Ibid.*
29. *Herald Examiner*, August 26, 1971, "Supersonic Launch a Success."
30. *NASA Facts*, "The Lifting Bodies."
31. *Daily Breeze*, October 21, 1969, "Shuttle Bus Is Vital to Space Plans."
32. *NASA Facts*, "The Lifting Bodies."
33. *Daily Breeze*, October 21, 1969, "Shuttle Bus Is Vital to Space Plans."
34. *NASA Facts*, "The Lifting Bodies."
35. *Daily Breeze*, October 21, 1969, "Shuttle Bus Is Vital to Space Plans."
36. *Aviation Week & Space Technology*, September 19, 1966, "Prime Lifting Body Undergoes Systems Test."
37. *Daily Breeze*, October 21, 1969, "Shuttle Bus Is Vital to Space Plans."
38. *NASA Facts*, "The Lifting Bodies."
39. Associated Press, August 2, 1973, "X-24B Makes Debut."
40. *Spaceflight News*, May 1988, "X-Plane Veteran Still Going Strong."
41. *Daily Breeze*, August 21, 1975, "Space Shuttle Test Works."
42. Associated Press, August 2, 1973, "X-24B Makes Debut."
43. *NASA Activities*, October 1975, "Rocket-Powered Aircraft Makes Final Flight."
44. *NASA Facts*, "The Lifting Bodies."
45. *Aviation Week & Space Technology*, January 21, 1974, "Shuttle Orbiter Nears Test Flight Decision."
46. *Air & Space*, April/May 1991, "The Space Shuttle Family Tree."
47. *Report of the Presidential Commission on the Space Shuttle* Challenger *Accident*, Washington, D.C., Volume 1.
48. *Ibid.*
49. *Ibid.*
50. Richard Lewis. *The Voyages of* Columbia. Columbia University Press, 1984. Chapter 9.
51. *Report of the Presidential Commission on the Space Shuttle* Challenger *Accident*, Volume 1, Chapter 1.
52. Lewis, Chapter 4.
53. *Daily Breeze*, April 15, 1972, "Space Payload Seen as Big Area Boost."
54. *Report of the Presidential Commission on the Space Shuttle* Challenger *Accident*, Volume 1.
55. Lewis, Chapter 2.
56. *Air & Space*, April/May 1991, "The Space Shuttle Family Tree."
57. Lewis, Chapter 2.
58. *Daily Breeze*, February 5, 1974, "Plans to Build the Virtus."
59. *Report of the Presidential Commission on the Space Shuttle* Challenger *Accident*, Volume 1.
60. *Time*, January 17, 1972, "A Boost for NASA."
61. United Press International, January 6, 1972, "Shuttle Job Manna from Washington."
62. *Daily Breeze*, January 8, 1972, "Space Shuttle Jobs May Number 25,000."
63. Lewis, Chapter 2.
64. *Ibid.*
65. *Report of the Presidential Commission on the Space Shuttle* Challenger *Accident*, Volume 1, Chapter 1.
66. *National Space Transportation System Reference*, June 1988.

67. *Daily Breeze*, April 2, 1972, "NAR Gets Okay on Shuttle Pact."
68. Joseph J. Trento. *Prescription for Disaster*. Crown Publishers, 1987. Chapter 6.
69. United Press International, April 20, 1972, "House Votes on Shuttle."
70. *Daily Breeze*, April 21, 1972, "Shuttle Vote Elates Bayans."
71. NASA live coverage of Apollo 16 moonwalk, April 20, 1972.
72. Trento.
73. *Air & Space*, December 1993/January 1994, "Securing the High Ground."
74. *National Space Transportation System Reference*, June 1988.
75. *Daily Breeze*, July 27, 1972, "Shuttle Contract News Helps Industry Morale."
76. Lewis, Chapter 4.
77. *Ibid.*
78. *Daily Breeze*, June 27, 1973, "NASA Defends Shuttle program."
79. Lewis, Chapter 3.
80. *Daily Breeze*, July 30, 1973, "Space Plan Under Fire."
81. Trento.
82. *Air & Space*, April/May 1991, "The Space Shuttle Family Tree."
83. *National Space Transportation System Reference*, June 1988.
84. *Los Angeles Times*, January 3, 1979, "Space Shuttle Engine Explodes in Test."
85. Lewis, Chapter 5.
86. *Ibid.*
87. *Ibid.*
88. *National Space Transportation System Reference*, June 1988.
89. *Report of the Presidential Commission on the Space Shuttle* Challenger *Accident*, Volume 1.
90. *Science News*, Volume 119, March 21, 1981, "*Columbia* Countdown."
91. Lewis, Chapter 7.
92. *Ibid.*
93. *Ibid.*
94. Lewis, Chapter 6.
95. *Ibid.*
96. *Ibid.*
97. *Ibid.*
98. *National Space Transportation System Reference*, June 1988.
99. *Los Angeles Times*, January 21, 1979, "Space Shuttle Launching Set for November 9."
100. *National Space Transportation System Reference*, June 1988.
101. *Los Angeles Times*, September 18, 1976, "Space Shuttle Put on Display."
102. Associated Press, September 9, 1976, "Star Trek Fans Torpedo Name for Space Shuttle."
103. NASA Press Kit: Space Shuttle Orbiter Test Flight Series.
104. *Ibid.*
105. *NASA Activities*, May 1976, "Test Pilots Named for Space Shuttle."
106. NASA Press Kit: Space Shuttle Orbiter Test Flight Series.
107. *NASA Activities*, May 1976, "Test Pilots Named for Space Shuttle."
108. NASA Press Kit, August 1977, "First Shuttle Orbiter Free Flight Test."
109. *Daily Breeze*, February 27, 1977, "Shuttle's Third Piggyback Flight a Success."
110. NASA Press Kit, August 1977, "First Shuttle Orbiter Free Flight Test."
111. *Los Angeles Times*, August 13, 1977, "Space Shuttle Sails Through Solo Flight."
112. *National Space Transportation System Reference*, June 1988.
113. Associated Press, March 11, 1978, "Arrives at Houston."
114. Associated Press, May 2, 1979, "Inching Forward."
115. *National Space Transportation System Reference*, June 1988.
116. Lewis, Chapter 9.
117. Lewis, Chapter 7.
118. Trento, Chapter 7.
119. *Ibid.*
120. *National Geographic* 159, no. 3, March 1981, "When the Space Shuttle."
121. Lewis, Chapter 7.
122. *Ibid.*
123. *Los Angeles Times*, March 21, 1979, "Heading East."
124. *Los Angeles Times*, March 21, 1979, "Storms Force Shuttle to Land in El Paso."
125. *National Space Transportation System Reference*, June 1988.
126. Lewis, Chapter 7.
127. *Aviation Week & Space Technology*, July 21, 1980, "NASA Presses to Hold Shuttle Schedule."
128. Lewis, Chapter 9.
129. NASA Press Kit, April 1981, "STS-1 First Space Shuttle Mission."
130. Associated Press, December 30, 1980, "A Giant Step."
131. Lewis, Chapter 9.
132. *Los Angeles Times*, February 21, 1981, "Officials Jubilant as Space Shuttle Engine Passes Test."
133. NASA Press Kit, April 1981, "STS-1 First Space Shuttle Mission."
134. *Ibid.*
135. *Los Angeles Times*, March 22, 1981, "The Investigation."
136. *Los Angeles Times*, April 13, 1981, "Shuttle Loses Some of Heat Shield."
137. Trento, Chapter 8.
138. *Los Angeles Times*, April 13, 1981, "Shuttle Loses Some of Heat Shield."
139. *National Space Transportation System Reference*, June 1988.
140. *Ibid.*
141. *Ibid.*
142. Lewis, Epilogue.
143. Trento.

Chapter Two

1. Joseph J. Trento. *Prescription for Disaster*. Crown Publishers, 1987, Chapter 5.
2. *Ibid.*
3. *Los Angeles Times*, March 15, 1972, "Space Shuttle."
4. *Report of the Presidential Commission on the Space Shuttle* Challenger *Accident*, Volume 1.
5. *Ibid.*
6. *Countdown* 4, no. 7, July 1986.
7. *Report of the Presidential Commission on the Space Shuttle* Challenger *Accident*, Volume 1.
8. *Ibid.*
9. *Ibid.*
10. *Ibid.*
11. *Countdown* 4, no. 7, July 1986.
12. *Report of the Presidential Commission on the Space Shuttle* Challenger *Accident*, Volume 1.
13. *Countdown* 4, no. 7, July 1986.
14. *Ibid.*
15. *Report of the Presidential Commission on the Space Shuttle* Challenger *Accident*, Volume 1.
16. *Countdown* 4, no. 7, July 1986.
17. *Ibid.*
18. *Report of the Presidential Commission on the Space Shuttle* Challenger *Accident*, Volume 1.
19. *Countdown* 4, no. 7, July 1986.
20. *Ibid.*
21. *Ibid.*
22. *Report of the Presidential Commission on the Space Shuttle* Challenger *Accident*, Volume 1.
23. *Ibid.*
24. *Countdown* 4, no. 7, July 1986.
25. *Report of the Presidential Commission on the Space Shuttle* Challenger *Accident*, Volume 1.
26. *Countdown* 4, no. 7, July 1986.
27. *Report of the Presidential Commission on the Space Shuttle* Challenger *Accident*, Volume 1.
28. *Ibid.*
29. *Ibid.*
30. *Ibid.*
31. *Countdown* 4, no. 7, July 1986.
32. *Ibid.*
33. *Ibid.*
34. *Ibid.*
35. *Ibid.*
36. *Report of the Presidential Commission on the Space Shuttle* Challenger *Accident*, Volume 1.
37. *National Space Transportation System Reference*, Volume 2, Operations, September 1988.
38. *Countdown* 4, no. 7, July 1986.
39. *Report of the Presidential Com-

mission on the Space Shuttle *Challenger Accident*, Volume 1.
40. *Ibid.*
41. *Countdown* 4, no. 7, July 1986.
42. *Ibid.*
43. *Ibid.*
44. *Ibid.*
45. *Report of the Presidential Commission on the Space Shuttle Challenger Accident*, Volume 1.
46. *Countdown* 4, no. 7, July 1986.
47. *Ibid.*
48. *Ibid.*

Chapter Three

1. *Aviation Week & Space Technology*, Oct. 27, 1969, "Reusable Space Shuttle Effort Gains Momentum."
2. *Air & Space*, December 1993/January 1994, "Securing the High Ground."
3. Joseph J. Trento. *Prescription for Disaster*. Crown Publishers, 1987, Chapter 4.
4. Richard Lewis. *The Voyages of* Columbia. Columbia University Press, 1984, Chapter 4.
5. *Air & Space*, December 1993/January 1994, "Securing the High Ground."
6. *Ibid.*
7. *Spaceflight News*, May 1988, "X-Plane Veteran Still Going Strong."
8. *Aviation Week & Space Technology*, Sept. 12, 1966, "Five Orbital Shots Planned in MOL Tests."
9. Lewis, Chapter 4.
10. *Aviation Week & Space Technology*, Nov. 14, 1966, "Titan 3C Passes Sixth Test, Furnishes MOL Support."
11. Lewis, Chapter 4.
12. *Air & Space*, December 1993/January 1994, "Securing the High Ground."
13. Trento, Chapter 4.
14. *Air & Space*, December 1993/January 1994, "Securing the High Ground."
15. *Daily Breeze*, July 15, 1974, "Space Shuttle Will Meet Future Needs."
16. *Daily Breeze*, April 14, 1972, "Two Bases Win Space Shuttles."
17. *Report of the Presidential Commission on the Space Shuttle Challenger Accident*, Volume 1.
18. *Ibid.*
19. *Aviation Week & Space Technology*, Feb. 17, 1986, "Presidential Accident Commission to Broaden Scope of Investigation."
20. *Countdown* 4, no. 7, July 1986.
21. *Aviation Week & Space Technology*, Feb. 17, 1986, "Presidential Accident Commission to Broaden Scope of Investigation."
22. *Countdown* 4, no. 7, July 1986.
23. *Report of the Presidential Commission on the Space Shuttle Challenger Accident*, Volume 1.
24. *Countdown* 4, no. 7, July 1986.
25. *Ibid.*
26. *Ibid.*
27. *Report of the Presidential Commission on the Space Shuttle Challenger Accident*, Volume 1.

Chapter Four

1. Associated Press, June 2, 1986, "NASA Accepted Reports on 'Fail-Safe' Boosters."
2. *Countdown* 4, no. 7, July 1986.
3. *New York Times*, May 13, 1986, "NASA Knew of Rocket Seal Danger, Document Shows."
4. *Countdown* 4, no. 7, July 1986.
5. *Ibid.*
6. *Aviation Week & Space Technology*, February 17, 1986, "Presidential Accident Commission to Broaden Scope of Investigation."
7. *Ibid.*
8. *Countdown* 4, no. 7, July 1986.
9. *Aviation Week & Space Technology*, February 17, 1986, "Presidential Accident Commission to Broaden Scope of Investigation."
10. *Time*, June 9, 1986, "Fixing NASA."
11. Joseph J. Trento. *Prescription for Disaster*. Crown Publishers, 1987.
12. *Time*, June 9, 1986, "Fixing NASA."
13. *National Space Transportation System Reference*, Volume 2, Operations, September 1988.
14. *Ibid.*
15. *Countdown* 4, no. 7, July 1986.
16. *Aviation Week & Space Technology*, February 17, 1986, "Presidential Accident Commission to Broaden Scope of Investigation."
17. *National Space Transportation System Reference*, Volume 2, Operations, September 1988.
18. *Ibid.*
19. *Countdown* 4, no. 7, July 1986.
20. *Aviation Week & Space Technology*, May 19, 1986, "Commission Evidence Shows Marshall Problems Contributed to Shuttle Failure."
21. *Countdown* 4, no. 7, July 1986.
22. *Ibid.*
23. *Ibid.*
24. *Countdown* 4, no. 2, February 1986.
25. *Countdown* 4, no. 7, July 1986.
26. *Ibid.*
27. *Time*, June 9, 1986, "Fixing NASA."
28. *Aviation Week & Space Technology*, May 19, 1986, "Commission Evidence Shows Marshall Problems Contributed to Shuttle Failure."
29. *Countdown* 4, no. 7, July 1986.
30. *Ibid.*
31. *Countdown* 4, no. 2, February 1986.

Chapter Five

1. *Countdown* 4, no. 7, July 1986.
2. *Countdown* 4, no. 3, March 1986.
3. *Ibid.*
4. *Countdown* 6, no. 5, May 1988.
5. *Countdown* 4, no. 3, March 1986.
6. *Aviation Week & Space Technology*, February 10, 1986, "Booster Investigation Forces Examination of Procedures by NASA and Contractors."
7. *Countdown* 4, no. 3, March 1986.
8. *Countdown* 4, no. 7, July 1986.
9. *Countdown* 4, no. 3, March 1986.
10. Joseph J. Trento. *Prescription for Disaster*. Crown Publishers, Inc., 1987.
11. *National Space Transportation System Reference*, Volume 2, Operations, September 1988.
12. *Countdown* 4, no. 3, March 1986.
13. *Ibid.*
14. *Report of the Presidential Commission on the Space Shuttle Challenger Accident*, Volume 1.
15. *Ibid.*
16. *Countdown* 4, no. 3, March 1986.
17. *National Space Transportation System Reference*, Volume 2, Operations, September 1988.
18. *Countdown* 4, no. 7, July 1986.
19. *Countdown* 4, no. 3, March 1986.
20. *Aviation Week & Space Technology*, March 17, 1986, "Ground Controllers Canceled Mission 61-C Thruster Tests Because of Explosion Danger."
21. *National Space Transportation System Reference*, Volume 2, Operations, September 1988.
22. *Countdown* 4, no. 3, March 1986.
23. *Ibid.*
24. *Countdown* 4, no. 7, July 1986.
25. *Aviation Week & Space Technology*, February 3, 1986, "Search Continues for Shuttle Debris off Coast of Florida."
26. *Countdown* 4, no. 3, March 1986.
27. *Ibid.*

Chapter Six

1. *Report of the Presidential Commission on the Space Shuttle Challenger Accident*, Volume 1.
2. *Countdown* 4, no. 7, July 1986.

3. *Ibid.*
4. *Ibid.*
5. *Ibid.*
6. *Ibid.*
7. *Ibid.*
8. *Ibid.*
9. Associated Press, February 21, 1986, "Thiokol Opposed Launch."
10. *Countdown* 4, no. 7, July 1986.
11. *Ibid.*
12. *Ibid.*
13. *Ibid.*
14. *Aviation Week & Space Technology*, February 24, 1986, "Wind Tunnel Data Indicated Potential for Frozen Seals."
15. *Ibid.*
16. *Ibid.*
17. David J. Shayler. *Disasters and Accidents in Manned Space Flight.* Praris Publishing, 2000.
18. *Countdown* 4, no. 3, March 1986.
19. Shayler.
20. *Countdown* 4, no. 3, March 1986.
21. *Ibid.*
22. *Countdown* 4, no. 7, July 1986.
23. *Countdown* 4, no. 3, March 1986.
24. *Ibid.*
25. *Countdown* 4, no. 7, July 1986.
26. *Ibid.*
27. Shayler.
28. *Ibid.*
29. *Countdown* 4, no. 7, July 1986.
30. *Countdown* 4, no. 3, March 1986.
31. *Ibid.*
32. *Countdown* 4, no. 9, September 1986.

Chapter Seven

1. *Countdown* 4, no. 7, July 1986.
2. *Ibid.*
3. *Ibid.*
4. *Ibid.*
5. *Ibid.*
6. *Ibid.*
7. *Ibid.*
8. *Ibid.*
9. *Ibid.*
10. *Ibid.*
11. *Ibid.*
12. *Ibid.*
13. *Ibid.*
14. *Ibid.*
15. David J. Shayler. *Disasters and Accidents in Manned Space Flight.* Praris Publishing, 2000.
16. *Countdown* 4, no. 7, July 1986.
17. *Ibid.*
18. *Ibid.*
19. *Ibid.*
20. *Ibid.*
21. *Ibid.*
22. *Countdown* 4, no. 3, March 1986.
23. Shayler.
24. *Countdown* 4, no. 7, July 1986.
25. *Countdown* 4, no. 9, September 1986.
26. *Aviation Week & Space Technology*, April 14, 1986, "Booster Certification Process Raises Further Shuttle Design Flaw Concerns."
27. Shayler.
28. *Countdown* 4, no. 7, July 1986.
29. *Aviation Week & Space Technology*, February 3, 1986, "Search Continues for Shuttle Debris off Coast of Florida."
30. *Ibid.*
31. *Ibid.*
32. Shayler.
33. *Countdown* 4, no. 9, September 1986.
34. Shayler.
35. *Ibid.*
36. *Ibid.*
37. *Countdown* 4, no. 3, March 1986.
38. *Ibid.*
39. *Aviation Week & Space Technology*, February 3, 1986, "Search Continues for Shuttle Debris off Coast of Florida."
40. *Ibid.*
41. *Ibid.*
42. *Ibid.*
43. *Ibid.*
44. *Ibid.*
45. *Ibid.*
46. *Countdown* 4, no. 3, March 1986.
47. *Aviation Week & Space Technology*, February 3, 1986, "Search Continues for Shuttle Debris off Coast of Florida."
48. *Ibid.*
49. *Ibid.*

Chapter Eight

1. *Countdown* 4, no. 12, December 1986.
2. *Aviation Week & Space Technology*, February 3, 1986, "Search Continues for Shuttle Debris off Coast of Florida."
3. *Countdown* 4, no. 12, December 1986.
4. *Ibid.*
5. *Ibid.*
6. *Countdown* 4, no. 4, April 1986.
7. *Time*, February 10, 1986, "'They Slipped the Surly Bonds of Earth to Touch the Face of God.'"
8. *Countdown* 4, no. 12, December 1986.
9. *Countdown* 4, no. 4, April 1986.
10. *Ibid.*
11. *Countdown* 4, no. 12, December 1986.
12. *Countdown* 4, no. 4, April 1986.
13. *Countdown* 4, no. 12, December 1986.
14. *Ibid.*
15. *Ibid.*
16. *Ibid.*
17. *Ibid.*
18. *Ibid.*
19. *Countdown* 4, no. 4, April 1986.
20. *Ibid.*
21. *Ibid.*
22. *Countdown* 4, no. 12, December 1986.
23. *Ibid.*
24. *Ibid.*
25. *Ibid.*
26. *Countdown* 4, no. 4, April 1986.
27. *Countdown* 4, no. 12, December 1986.
28. *Aviation Week & Space Technology*, February 10, 1986, "Search Team Focuses Efforts on Retrieving Orbiter Wreckage."
29. *Los Angeles Times*, February 3, 1986, "Shuttle Carried No Booster Sensors."
30. *Countdown* 4, no. 4, April 1986.
31. *Aviation Week & Space Technology*, February 10, 1986, "Search Team Focuses Efforts on Retrieving Orbiter Wreckage."
32. Associated Press, February 5, 1986, "Search May Yield Problem Boosters."
33. *Los Angeles Times*, February 4, 1986, "Salvage Crew Turn Attention to 17 Objects on Ocean Floor."
34. Hughes Aircraft Company Bulletin No. 3315, February 4, 1986, "Scholarship Fund."
35. *Los Angeles Times*, February 4, 1986, "Salvage Crew Turn Attention to 17 Objects on Ocean Floor."
36. *Countdown* 4, no. 4, April 1986.
37. *Ibid.*
38. *Ibid.*
39. *Countdown* 4, no. 12, December 1986.
40. *Countdown* 4, no. 4, April 1986.
41. *Ibid.*
42. *Countdown* 4, no. 12, December 1986.
43. *Countdown* 4, no. 4, April 1986.
44. Associated Press, February 9, 1986, "Navy Divers Hunt for Shuttle Rocket."
45. *Countdown* 4, no. 4, April 1986.
46. *US News & World Report*, February 10, 1986, "Frank Borman, Former Astronaut."
47. *US News & World Report*, February 10, 1986, "Poll: Forge Ahead."
48. *Countdown* 4, no. 4, April 1986.
49. *Countdown* 4, no. 12, December 1986.
50. *Aviation Week & Space Technology*, February 17, 1986, "NASA Assesses External Tanks Role in *Challenger* Accident."
51. *Countdown* 4, no. 12, December 1986.

52. *Countdown* 4, no. 4, April 1986.
53. Associated Press, February 14, 1986, "McAuliffe's Sub Would Accept Trip in Shuttle."
54. *Los Angeles Times*, February 14, 1986, "Head of NASA Defends Staff as Conscientious."
55. *Ibid*.
56. *Los Angeles Times*, February 14, 1986, "Photos Show Rocket Smoking Early in Flight."
57. *USA Today*, February 18, 1986, "Verification Next Step in Shuttle Hunt."
58. *Countdown* 4, no. 4, April 1986.
59. *USA Today*, February 18, 1986, "Shuttle Chief, Others, Likely to Leave Probe."
60. *Countdown* 4, no. 12, December 1986.
61. *Countdown* 4, no. 4, April 1986.
62. *Los Angeles Times*, February 18, 1986, "Rocket Experts to View Photos of Debris."
63. *Countdown* 4, no. 4, April 1986.
64. *Ibid*.
65. *Aviation Week & Space Technology*, February 24, 1986, "Officials Expect Right Booster to Yield Clues to Accident."
66. *Countdown* 4, no. 4, April 1986.
67. *Ibid*.
68. Associated Press, February 21, 1986, "Thiokol Opposed Launch."
69. *Countdown* 4, no. 4, April 1986.
70. *New York Times*, February 22, 1986, "NASA Mum About Any Human Remains."
71. *Countdown* 4, no. 4, April 1986.
72. Associated Press, February 24, 1986, "Piece of Shuttle's Fuel Tank Recovered."
73. *Ibid*.
74. *Countdown* 4, no. 4, April 1986.
75. *Ibid*.
76. *Ibid*.
77. *Ibid*.
78. *Ibid*.
79. *Ibid*.
80. *Countdown* 5, no. 1, January 1987.
81. *Ibid*.
82. *Countdown* 4, no. 5, May 1986.
83. *Ibid*.
84. *Ibid*.
85. *Ibid*.
86. *Ibid*.
87. Associated Press, July 15, 1986, "Flight Suit Marked the Spot."
88. *Ibid*.
89. David J. Shayler. *Disasters and Accidents in Manned Space Flight*. Praris Publishing, 2000.
90. *Countdown* 4, no. 5, May 1986.
91. *Ibid*.
92. *Ibid*.
93. *Ibid*.
94. *Daily Breeze*, March 10, 1986, "Remains of Shuttle Crew Recovered."
95. *Aviation Week & Space Technology*, March 17, 1986, "Additional Shuttle Crew Remains Recovered from Ocean Floor."
96. *Countdown* 4, no. 5, May 1986.
97. *Aviation Week & Space Technology*, March 17, 1986, "Additional Shuttle Crew Remains Recovered from Ocean Floor."
98. *Ibid*.
99. *Ibid*.
100. *Ibid*.
101. *Ibid*.
102. *Countdown* 4, no. 5, May 1986.
103. *Countdown* 5, no. 1, January 1987.
104. *Aviation Week & Space Technology*, March 17, 1986, "Additional Shuttle Crew Remains Recovered from Ocean Floor."
105. *Countdown* 4, no. 5, May 1986.
106. *Aviation Week & Space Technology*, March 17, 1986, "Additional Shuttle Crew Remains Recovered from Ocean Floor."
107. *Countdown* 4, no. 5, May 1986.
108. *Ibid*.
109. *Countdown* 5, no. 1, January 1987.
110. *Countdown* 4, no. 5, May 1986.
111. Associated Press, March 20, 1986, "Experts Examine Critical Booster Debris."
112. *Countdown* 4, no. 5, May 1986.
113. Associated Press, March 20, 1986, "NASA Phone Logs Studied for Launch Pressure."
114. *Countdown* 4, no. 5, May 1986.
115. *Ibid*.
116. *Ibid*.
117. *Ibid*.
118. *Ibid*.
119. Associated Press, March 28, 1986, "Shuttle Salvage Fleet Reports Little Progress."
120. *Countdown* 4, no. 5, May 1986.
121. *Ibid*.
122. Associated Press, April 2, 1986, "Shuttle Puzzle Pieces May Yield Cause."
123. *Countdown* 4, no. 6, June 1986.
124. *Ibid*.
125. *Ibid*.
126. Associated Press, April 7, 1986, "Shuttle Find."
127. *Countdown* 4, no. 6, June 1986.
128. *Ibid*.
129. *Ibid*.
130. Shayler.
131. *Countdown* 4, no. 6, June 1986.
132. *Ibid*.
133. *Countdown* 5, no. 1, January 1987.
134. *Countdown* 4, no. 6, June 1986.
135. Associated Press, April 20, 1986, "Search Ends for Remains of Shuttle Crew."
136. *Countdown* 4, no. 6, June 1986.
137. *Countdown* 4, no. 10, October 1986.
138. *Ibid*.
139. Shayler.
140. *Countdown* 4, no. 6, June 1986.
141. Shayler.
142. *Ibid*.
143. *Ibid*.
144. *Countdown* 4, no. 6, June 1986.
145. *Ibid*.

Chapter Nine

1. *Los Angeles Times*, February 4, 1986, "12 Members of Panel Experts in Their Fields."
2. *Aviation Week & Space Technology*, February 17, 1986, "Marshall Officials Review Data for Solid Booster Anomalies."
3. *USA Today*, February 14, 1986, "USA Still Sending Donations."
4. *Los Angeles Times*, February 12, 1986, "Shuttle Launch Was Warned About Cold."
5. *Ibid*.
6. *Ibid*.
7. Associated Press, February 13, 1986, "Shuttle Panel Shifts Investigation to Space Center."
8. *Countdown* 4, no. 12, December 1986.
9. *Ibid*.
10. *Ibid*.
11. Associated Press, February 16, 1986, "Panel: Decision to Launch Shuttle Possibly "Flawed."
12. *Ibid*.
13. *Time*, February 24, 1986, "Zeroing In on the O Rings."
14. *Ibid*.
15. *USA Today*, February 14, 1986, "USA Still Sending Donations."
16. Associated Press, February 14, 1986, "Shuttle Hired to Orbit 3 Intelsat Satellites."
17. *Countdown* 4, no. 4, April 1986.
18. *USA Today*, April 24, 1986, "NASA Accused of Waste Under Nominee's Leadership."
19. *Daily Breeze*, February 19, 1986, "Engineer: I Lost Shuttle Argument."
20. *Ibid*.
21. *Los Angeles Herald Examiner*, February 24, 1986, "Top NASA Officials Kept in Dark on Shuttle Seals."
22. *Countdown* 4, no. 12, December 1986.
23. Associated Press, February 21, 1986, "Thiokol Opposed Launch."
24. *Ibid*.
25. *Los Angeles Times*, February 23, 1980, "NASA, Thiokol in Contact Talks at Time of Blast."
26. *Ibid*.
27. *Ibid*.
28. Associated Press, February 26, 1986, "Rocket Engineers Say Bosses Ignored Their Shuttle Fears."

29. *Ibid.*
30. Associated Press, February 27, 1986, "Rockwell Questioned Shuttle Launch."
31. Associated Press, February 26, 1986, "Rocket Engineers Say Bosses Ignored Their Shuttle Fears."
32. *Time*, March 3, 1986, "The Questions Get Tougher."
33. *Los Angeles Times*, February 27, 1986, "NASA Officials Deny Firm Was Pressured to Launch."
34. *Ibid.*
35. Associated Press, February 28, 1986, "Marshall Team Defends NASA."
36. *Los Angeles Times*, February 28, 1986, "NASA Procedures Flawed, Panel Says."
37. *Los Angeles Times*, March 4, 1986, "NASA Aide's Testimony to Be Examined."
38. Associated Press, March 7, 1986, "Fletcher Returns to Lead NASA."
39. *New York Times*, April 23, 1986, "Chinks in NASA Armor Come to Light After *Challenger* Disaster."
40. *Daily Breeze*, March 9, 1986, "Launch Pace 'a Safety Hazard.'"
41. *Los Angeles Times*, March 14, 1986, "Rep. Nelson Plans Shuttle Safety Inquiry."
42. Associated Press, April 3, 1986, "Astronauts Have Their Day in Shuttle Probe."
43. *New York Times*, March 30, 1986, "NASA Confident of 1-Year Timetable."
44. Associated Press, April 3, 1986, "Astronauts Have Their Day in Shuttle Probe."
45. *Ibid.*
46. *Ibid.*
47. *Ibid.*
48. *Countdown* 5, no. 1, January 1987.
49. Associated Press, April 17, 1986, "Shuttle Hunk Confirms Rocket Leak Hypothesis."
50. *New York Times*, April 23, 1986, "Chinks in NASA Armor Come to Light After *Challenger* Disaster."
51. *USA Today*, April 24, 1986, "NASA Accused of Waste Under Nominee's Leadership."
52. *Ibid.*
53. *Ibid.*
54. *New York Times*, April 24, 1986, "NASA Cut *Challenger*'s Safety Tests, Audits Show."
55. *New York Times*, May 11, 1986, "Rocket Engineers Tell of Demotions."
56. *Ibid.*
57. *Aviation Week & Space Technology*, May 19, 1986, "NASA Investigates Claims That Morton Thiokol Demoted Engineers for Disclosing Launch Opposition."
58. Associated Press, May 12, 1986, "Shuttle Panel Chief Shocked by Transfers."
59. Associated Press, May 14, 1986, "NASA Probes Charges of Retribution."
60. *New York Times*, May 13, 1986, "NASA Knew of Rocket Seal Danger, Document Shows."
61. *Los Angeles Times*, May 20, 1986, "No Important Documents Missing, Shuttle Inquiry Members Indicate."
62. Associated Press, June 4, 1986, "McDonald Will Head Booster Redesigning."
63. Associated Press, June 5, 1986, "Shake-Up at NASA."
64. *Ibid.*

Chapter Ten

1. *Countdown* 4, no. 9, September 1986.
2. *Ibid.*
3. Associated Press, November 15, 1988, "Nelson Denies Cover-Up After Shuttle Blast."
4. Associated Press, September 7, 1988, "Space Lab Bolts Removed as Safety Precaution."
5. *Aviation Week & Space Technology*, June 16, 1986, "Criminal Negligence Charges Weighed in 51-L Crew Deaths."
6. *Ibid.*
7. *Wall Street Journal*, May 15, 1986.
8. Associated Press, June 18, 1986, "Thiokol Chairman Criticized."
9. Associated Press, June 18, 1986, "Senator Asks: Who Was Responsible for Shuttle Blast?"
10. Associated Press, April 4, 1987, "Morton Thiokol Subject of FBI Criminal Probe."
11. Associated Press, April 16, 1987, "Federal Judge Ordered the Release of Court Documents."
12. *Los Angeles Times*, January 29, 1987, "Shuttle Booster Builder Sued for $1 Billion."
13. *Aviation Week & Space Technology*, September 7, 1987, "NASA Panel Would Withhold Accident Witness Accounts."
14. Associated Press, January 25, 1987, "Some Blame NASA."
15. *Ibid.*
16. *Aviation Week & Space Technology*, July 28, 1986, "Aerojet Claims Single-Cast Boosters Can Be Ready for Next Shuttle Flight."
17. Associated Press, September 6, 1986, "5 Firms Vie to Design Booster."
18. Associated Press, November 19, 1986, "Panel Says NASA Paid Thiokol Award Despite Booster's Flaws."
19. *Countdown* 4, no. 8, August 1986.

Chapter Eleven

1. *Aviation Week & Space Technology*, February 10, 1986, "Booster Investigation Forces Examination of Procedures by NASA and Contractors."
2. Associated Press, July 3, 1986, "NASA: New Booster 'Extremely Resistant to Failure.'"
3. Associated Press, July 11, 1986, "Kennedy Space Center Chief Retires."
4. *Los Angeles Times*, August 13, 1986, "NASA Unveils Proposed $300-Million Redesign of Shuttle Rocket Joint."
5. *Ibid.*
6. *Daily Breeze*, September 20, 1986, "NASA to Release Transcript."
7. NASA, *Columbia* Accident Investigation Board Report, Volume 1, August 2003, Chapter 1.
8. *Aviation Week & Space Technology*, October 20, 1986, "Oversight Panel Endorses NASA Plan for Fixing SRBs."
9. *Aviation Week & Space Technology*, January 5, 1987, "Shuttle Escape System Given Tentative Approval."
10. Associated Press, January 10, 1987, "5 Named as New Crew for Shuttle."
11. *Los Angeles Times*, January 16, 1987, "Shuttle Captain Places Safety Over Time."
12. Associated Press, January 12, 1987, "NASA Starts a Standby Launch Team."
13. Associated Press, January 23, 1987, "NASA to Pay Thiokol $350 Million."
14. *Ibid.*
15. *Los Angeles Times*, January 27, 1987, "6th Rocket Test May Delay Next Shuttle Flight."
16. *Los Angeles Times*, January 29, 1987, "Shuttle Booster Builds Sued for $1 Billion."
17. Associated Press, January 15, 1987, "Time Short for Shuttle Changes."
18. Associated Press, February 25, 1987, "Morton Thiokol Trims Rocket Fees, Profits."
19. *Ibid.*
20. Associated Press, March 13, 1987, "Study Suggests Revision of Shuttle Strategy."
21. *USA Today*, March 20, 1987, "Astronaut Calls for More Tests of Booster."
22. *Aviation Week & Space Technology*, April 6, 1987, "NASA Will Compete New Shuttle Booster."

23. Associated Press, April 15, 1987, "Countdown Test Urged."
24. Associated Press, April 22, 1987, "New Tests to Delay Launch of Shuttle Several Weeks."
25. Associated Press, April 29, 1987, "Money Problems May Keep Shuttle Grounded into '88."
26. *Countdown* 10, no. 3, March 1992.
27. Associated Press, May 28, 1987, "Shuttle Booster Test Goes Off Without a Hitch."
28. Associated Press, June 4, 1987, "NASA Must Release *Challenger* Tapes."
29. Associated Press, July 1, 1987, "New Shuttle Design Has Escape Hatch."
30. Associated Press, July 17, 1987, "NASA Plans 3 More Tests of New Rocket."
31. Associated Press, August 27, 1987, "New Rocket's Motor Fails in Test Firing."
32. *Time*, August 31, 1987, "Getting NASA Back on Track."
33. *Daily Breeze*, September 30, 1987, "Shuttle Payload."
34. United Press International, September 30, 1987, "11 Experiments Slated for Shuttle to Include 2 Lost on *Challenger*."
35. NASA, September 15, 1987, "Crew Station Procedures Development Crew Egress/Escape System."
36. Associated Press, October 9, 1987, "The Third in a Series of Tests."
37. *Daily Breeze*, October 23, 1987, "NASA Plans 19 Shuttle Flights."
38. *Aviation Week & Space Technology*, October 26, 1987, "New Shuttle Launch Manifest Integrates Shuttle, Expendable Booster Missions."
39. *Ibid.*
40. *Los Angeles Times*, December 14, 1987, "Shuttle Emergency Escape System Being Tested."
41. *Aviation Week & Space Technology*, December 21, 1987, "Space Shuttle Program Managers Concerned About Staffing Levels for Three-Orbiter Processing."
42. Associated Press, December 24, 1987, "Shuttle Flights Returning."
43. *New York Times*, December 30, 1987, "Shuttle Rocket Test Discloses Flaw That Will Delay Scheduled Flight."
44. Associated Press, December 30, 1987, "Pressure to Replace Thiokol."
45. *Ibid.*
46. Associated Press, January 2, 1988, "Shuttle Booster Rocket Test Defended."
47. *New York Times*, January 6, 1988, "Second Flaw Seen in Shuttle Rocket."
48. *Aviation Week & Space Technology*, January 18, 1988, "NASA Selects Northrup Strip Alternate Shuttle Landing Site."
49. *Aviation Week & Space Technology*, April 11, 1988, "NASA Selects Telescoping Pole for Shuttle Crew Escape System."
50. Associated Press, April 21, 1988, "Booster Passes Exam."
51. *Aviation Week & Space Technology*, April 25, 1988, "Test Firing Indicates Booster Ring Design Can Tolerate Joint Flaws."
52. Associated Press, May 20, 1988, "Plant Loss May Cut Shuttle Flights."
53. *Aviation Week & Space Technology*, May 9, 1988, "Loss of Oxidizer Plant Will Not Hinder Near-Term Shuttle, Defense Programs."
54. *Time*, July 4, 1988, "Getting Ready to Try Again."
55. Associated Press, July 13, 1988, "NASA-Financed Study Estimates Shuttle Crash Chances at 1 in 70."
56. *Los Angeles Times*, August 11, 1988, "Oft Delayed Shuttle Test Successful."
57. Associated Press, August 19, 1988, "Shuttle Test Hailed as Success."
58. *Orange County Register*, September 4, 1988, "The Countdown Begins for *Discovery*."
59. Reuters, September 27, 1988, "Engineers Claim Shuttle Rockets Unsafe."
60. *Ibid.*
61. Associated Press, September 29, 1988, "Air Force Jumpers Responsible for Rescuing Crewmen."
62. *Countdown* 4, no. 9, September 1986.
63. *Countdown* 4, no. 10, October 1986.
64. *Aviation Week & Space Technology*, January 5, 1987, "Justice Dept. Settles Claims with Four *Challenger* Families."
65. *Los Angeles Times*, January 29, 1987, "Shuttle Booster Builders Sued for $1 Billion."
66. Associated Press, January 31, 1987, "Father Files Claim Over *Challenger*."
67. Associated Press, May 7, 1987, "Shuttle Pilot's Widow Sues Thiokol."
68. Associated Press, May 8, 1987, "The Widow."
69. *Daily Breeze*, February 18, 1988, "Settlement Reached in Astronaut's Death."
70. Associated Press, February 27, 1988, "Widow of *Challenger* Pilot Cannot Sue U.S. for His Death, Court Says."
71. Associated Press, March 7, 1988, "Thiokol, U.S. Bought Annuities to Settle."
72. Associated Press, August 23, 1988, "Last Lawsuit Settled in Rocket Blast."

Chapter Twelve

1. NASA, *Columbia* Accident Investigation Board, Report Volume 1, August 2003, Synopsis.
2. *Ibid.*
3. *Ibid.*, Chapter 7.
4. *Countdown* 9, no. 3, March 1991.
5. NASA, *Columbia* Accident Investigation Board, Report Volume 1, August 2003, Chapter 6.
6. *Ibid.*
7. *Ibid.*
8. *Countdown* 11, no. 2, February 1993.
9. NASA, *Columbia* Accident Investigation Board, Report Volume 1, August 2003, Chapter 6.
10. *Ibid.*
11. *Ibid.*
12. Associated Press, April 15, 1993, "Pliers Found Stuck on Shuttle Booster."
13. NASA, *Columbia* Accident Investigation Board, Report Volume 1, August 2003, Chapter 6.
14. *Ibid.*, Chapter 5.
15. Associated Press, July 23, 1995, "Shuttle Returns; Quick Launch Is Questioned."
16. NASA, *Columbia* Accident Investigation Board, Report Volume 1, August 2003, Chapter 5.
17. *Ibid.*
18. *Ibid.*
19. *Ibid.*, Chapter 3.
20. *Ibid.*
21. *Washington Post*, May 23, 1991, "Shuttle Dodged Engine Failure, Space Official Says."
22. *Aviation Week & Space Technology*, June 3, 1991, "Space Shuttle Quality Review Examined Discipline in Program."
23. NASA, *Columbia* Accident Investigation Board, Report Volume 1, August 2003, Chapter 5.
24. *Ibid.*
25. *New York Times*, April 6, 1986, "Plans to Develop Aerospace Plane Take Off."
26. NASA, *Columbia* Accident Investigation Board, Report Volume 1, August 2003, Chapter 5.
27. *Ibid.*
28. *Ibid.*
29. *Ibid.*
30. *Time*, July 15, 1996, "High-Tech Pie in the Sky."
31. *Washington Post*, March 2, 2001, "NASA Ends X-33 Project That Sought to Cut Spaceflight Costs."
32. NASA, *Columbia* Accident Investigation Board, Report Volume 1, August 2003, Chapter 5.

33. *Ibid.*
34. NASA, *Columbia* Accident Investigation Board, Report Volume 1, August 2003.
35. *Ibid.*

Chapter Thirteen

1. *Los Angeles Times*, January 29, 2003, "Crew Honors Those Who Died Aboard *Challenger*."
2. *Los Angeles Times*, February 2, 2003, "A Somber Bush Leads a Nation in Grieving."
3. *New York Times*, February 2, 2003, "Shuttle Destined for 1-in-50 Failure."
4. Reuters, January 15, 2003, "Security Extra Tight for Shuttle Launch."
5. Associated Press, May 17, 2002, "Israeli Astronaut Isn't Worried About NASA Security."
6. *Los Angeles Times*, February 22, 2003, "NASA Resisted Concerns, Engineer's E-mails Say."
7. *Los Angeles Times*, March 20, 2003, "Key Data Recorder on *Columbia* Found."
8. NASA, *Columbia* Accident Investigation Board Report, Volume 1, August 2003, Chapter 5.
9. *Los Angeles Times*, March 29, 2003, "Satellite Photos of Shuttles Planned."
10. NASA, *Columbia* Accident Investigation Board, Report Volume 1, August 2003, Chapter 5.
11. *Ibid.*
12. *Los Angeles Times*, March 18, 2003, "Amateur Videos Help NASA Trace *Columbia*'s Path."
13. NASA, *Columbia* Accident Investigation Board Report, Volume 1, August 2003, Chapter 6.
14. *Aviation Week & Space Technology*, March 10, 2003, "Rescue Effort."
15. *New York Times*, February 2, 2003, "16 Minutes from Landing Space Shuttle Disintegrates."
16. *Los Angeles Times*, February 3, 2003, "1,200 Clues Strewn Across Texas."
17. Associated Press, February 2, 2003, "Shuttle Crew's Remains Found."
18. *Fort Worth Star Telegram*, February 2, 2003, "Remains Scattered Over E. Texas."
19. *Los Angeles Times*, February 3, 2003, "1,200 Clues Strewn Across Texas."
20. *USA Today*, February 4, 2003, "Grim Search Interrupts Life in Small Texas community."
21. *Long Beach Press Telegram*, February 4, 2003, "Liftoff Fault Tied to Tragedy."
22. Associated Press, February 5, 2003, "Remains of Israeli Astronaut Found."
23. Associated Press, February 14, 2003, "NASA says Remains of All 7 Found."
24. NASA, *Columbia* Accident Investigation Board Report, Volume 1, August 2003, Chapter 6.
25. *Los Angeles Times*, February 7, 2003, "Shortage of Spare Parts Hinders NASA."
26. *Ibid.*
27. *Ibid.*
28. *Los Angeles Times*, February 6, 2003, "Tile Repair Kits for Astronauts Rejected."
29. *Long Beach Press Telegram*, February 5, 2003, "Insulation Needed Testing."
30. *Ibid.*
31. *Ibid.*
32. NASA, *Columbia* Accident Investigation Board Report, Volume 1, August 2003, Chapter 8.
33. *Los Angeles Times*, February 22, 2003, "NASA Resisted Concerns, Engineer's E-mails Say."
34. *Ibid.*
35. *Ibid.*
36. *Ibid.*
37. *Los Angeles Times*, March 29, 2003, "Satellite Photos of Shuttles Planned."
38. *New York Times*, April 1, 2003, "Shuttle Warning E-mail Unsent."
39. *Ibid.*
40. *Los Angeles Times*, February 22, 2003, "NASA Resisted Concerns, Engineer's E-mails Say."
41. *Los Angeles Times*, February 22, 2003, "NASA Worker Says E-mails Misinterpreted."
42. *Ibid.*
43. *Los Angeles Times*, February 27, 2003, "New Questions on Shuttle."
44. United Press International, January 18, 1990, "Navigational Problem Solved, Shuttle Crew Goes Back to Bed."
45. Associated Press, December 14, 1983, "Shuttle Landing Fire Revealed."
46. *Los Angeles Times*, December 9, 1983, "*Columbia* Lands Eight Hours Late."
47. Associated Press, January 3, 1986, "The Rocket Steering Problem."
48. Associated Press, August 6, 1986, "Fatigue Blamed for Fuel Error in Scrubbed Launch."
49. Associated Press, July 26, 1999, "NASA Suspects Shuttle Was Leaking Fuel."
50. *Long Beach Press Telegram*, July 21, 1999, "Shuttle Crew Will Try Again Thursday."
51. *Los Angeles Times*, February 9, 2003, "Cost Cutting Trumped Key NASA Safety Upgrades."
52. *Los Angeles Times*, February 8, 2003, "Scrutiny on NASA Decision Process."
53. *Los Angeles Times*, March 15, 2003, "NASA Is Planning to Launch Next Shuttle as Early as Fall."
54. *New York Times*, March 2, 2003, "Challenges Facing Shuttle Disaster Inquiry Include Culture at NASA."
55. *New York Times*, February 9, 2003, "Is There a Future in Space?"

Chapter Fourteen

1. *Los Angeles Times*, September 9, 2003, "Shuttle Safety Plans Detailed."
2. Reuters, September 17, 2003, "Next Shuttle Flight Won't Meet March Launch Date."
3. Associated Press, December 12, 2003, "Sensors on Shuttles Will Detect Blows."
4. Associated Press, September 11, 2004, "Fixing Up Shuttle Fleet Doubles to $2.2 Billion."
5. Associated Press, February 4, 2005, "Use of Space Station as Shuttle Shelter Is Questioned."
6. *Los Angeles Times*, February 11, 2005, "Shuttle's Return to Space Is on the Horizon."
7. Associated Press, April 1, 2005, "Space Shuttle Begins Crawl to Launch Pad."
8. *Los Angeles Times*, April 30, 2005, "Safety Concerns Delay NASA Shuttle Launch."
9. *Ibid.*
10. *Los Angeles Times*, May 21, 2005, "Shuttle Passes Test with Flying Colors."
11. Associated Press, June 15, 2005, "New NASA Boss Seeks to Hasten Next Spacecraft."
12. Cox News Service, July 1, 2005, "*Discovery* 'Go for Launch' in July."
13. *Chicago Tribune*, July 14, 2005, "Faulty Fuel Gauge Grounds Shuttle."
14. *Los Angeles Times*, July 25, 2005, "Launch Still on Despite Glitch."
15. *Ibid.*
16. *Los Angeles Times*, July 27, 2005, "*Discovery* Relaunches Shuttle Program."
17. *Los Angeles Times*, July 28, 2005, "Safety Worries Ground Shuttles."
18. *Los Angeles Times*, July 29, 2005, "Behind Chunks of Foam, a Failure to Confront Hazard."
19. *Los Angeles Times*, August 1, 2005, "Shuttle May Get Repair in Space."

20. *Los Angeles Times*, July 31, 2005, "Shuttle Sends Handymen for Spacewalk."
21. *Los Angeles Times*, August 5, 2005, "NASA Says *Discovery* Is Clear for Landing."
22. Associated Press, August 12, 2005, "Shuttle Foam Problem to Push Launches Back."
23. *Ibid.*
24. *Ibid.*
25. *Orlando Sentinel*, August 18, 2005, "NASA Seen as Improving But with a Ways to Go."
26. *Ibid.*
27. Reuters, February 18, 2006, "New Shuttle Crew Expects Foam Debris and Isn't Worried."
28. *Los Angeles Times*, June 1, 2006, "Fuel Tank Alterations Clear Way for Shuttle."
29. *Los Angeles Times*, June 18, 2006, "Shuttle Launch Is Set in Spite of Concerns."
30. Associated Press, July 4, 2006, "Shuttle OK'd for Today."
31. Associated Press, July 5, 2006, "July 4th Sees Rocket's Red Glare."
32. Associated Press, July 18, 2006, "An 'Awfully Good Day' for a Landing."

Chapter Fifteen

1. NASA, *Columbia* Accident Investigation Board Report, Volume 1, August 2003, Synopsis.
2. *Los Angeles Times*, September 6, 1988, "U.S. Space Program at Crossroads."

Bibliography

Government Sources

CHALLENGER REPORT

Report of the Presidential Commission on the Space Shuttle Challenger *Accident*, Volume 1, Washington, D.C.: Government Printing Office, 1986

COLUMBIA REPORT

Columbia *Accident Investigation Board*, Volume 1, Washington, D.C.: Government Printing Office, 2003

CONGRESSIONAL HEARINGS

Committee on Science, Space and Technology, U.S. House of Representatives: Assured Access to Space, 1986

Committee on Science, Space and Technology, U.S. House of Representatives: Investigation of the *Challenger* Accident, 1986

Committee on Science, Space and Technology, U.S. House of Representatives: NASA's Response to the Committee's Investigations of *Challenger* Accident, February 1987

NATIONAL AERONAUTICS AND SPACE ADMINISTRATION

NASA Activities, "Piggyback Aircraft Concepts Have Long History," September 1976

NASA Activities, "Rocket-Powered Aircraft Makes Final Flight," October 1975

NASA Activities, "Space Shuttle Solid Rocket Motor Fired"

NASA Activities, "Test Pilots Named for Space Shuttle," May 1976

NASA Crew Station Procedures Development Crew Egress/Escape System, September 15, 1981

NASA Crew Station Procedures Development Crew Egress/Escape System, September 15–18, 1987

NASA Facts, "*Challenger*" (STA-099, OV-99)

NASA Facts, "The Crew of the *Challenger* Shuttle Mission in 1986"

NASA Facts, "Dr. Joseph P. Kerwin Letter to Rear Admiral Richard H. Truly, July 28, 1986"

NASA Facts, "Enterprise" (OV-101)

NASA Facts, "51-L"

NASA Facts, "HL-10 Lifting Body," February 1994

NASA Facts, "Launch Complex 39-A and 39-B"

NASA Facts, "The Lifting Bodies"

NASA Facts, "Main Propulsion Test Article" (MPTA-098)

NASA Facts, "Shuttle Program Chronology"

NASA Facts, "61-C"

NASA Facts, "Space Shuttle," January 8, 1970

NASA Facts, "Transcript of the *Challenger* Crew Comments from the Operational Recorder April 17"

"NASA Flight Data File Crew Activity Plan, STS 51-L," December 23, 1985

NASA Historical Summary: Mission 51-L Mishap Investigation, May 1986

NASA live coverage of the Apollo 16 moonwalk, April 20, 1972

NASA live coverage of the launch of STS 51-L, January 28, 1986

NASA live coverage of the launch of STS-107, January 16, 2003

NASA live coverage of the landing of STS-107, February 1, 2003

NASA Press Kit, "America's Return to Space Flight, STS-26," September 1988

NASA Press Kit, "First Shuttle Orbiter Free Flight Test"

NASA Press Kit, "Mission 51-L," January 1986

NASA Press Kit, "Mission 107," December 16, 2002

NASA Press Kit, "STS-1 First Space Shuttle Mission," April 1981

NASA Press Kit, "Space Shuttle Orbiter Captive Manned Test Flight Series," June 1977

"NASA Public Affairs Contingency Operations Plan," March 30, 1984

NASA Teacher in Space Project, 1985

NASA's Actions to Implement the Rogers Commission

Recommendations After the Challenger *Accident*, July 14, 1986
National Space Transportation System Reference, June 1988
National Space Transportation System Reference, Volume 2, Operations, September 1988

President Ronald Reagan

President's speech on the *Challenger* Disaster, Oval Office, White House, January 28, 1986

News Services

Associated Press

August 2, 1973: "X-24B Makes Debut"
September 9, 1976: "Star Trek Fans Torpedo Name for Space Shuttle"
March 11, 1978: "Arrives at Houston"
May 2, 1979: "Inching Forward"
December 30, 1980: "A Giant Step"
December 14, 1983: "Shuttle Landing Fire Revealed"
February 3, 1984: "Perfect Shuttle Liftoff No. 10, Then Crew Gets Down to Work"
January 25, 1985: "Military Space Shuttle Operates Flawlessly as Satellite Launch Nears"
July 1, 1985: "10 Teacher Finalists for Berth on Shuttle"
July 19, 1985: "Teacher Picked for Space Trek"
January 3, 1986: "The Rocket Steering Problem"
January 26, 1986: "Space Shuttle Liftoff Delayed Until Monday"
January 28, 1986: "Words Were Routine Then Silence"
February 1, 1986: "Big Metal Object on Ocean Floor May Be Shuttle Cabin"
February 3, 1986: "The Peril of Solid Rockets"
February 3, 1986: "NASA Studying Spurt of Fire"
February 5, 1986: "Search May Yield Problem Boosters"
February 9, 1986: "Navy Divers Hunt for Shuttle Rocket"
February 10, 1986: "Theory of Shuttle Disaster Detailed"
February 12, 1986: "NASA: Booster Rockets Not Tested at Launch Pad"
February 13, 1986: "Shuttle Panel Shifts Investigation to Space Center"
February 14, 1986: "Shuttle Hired to Orbit 3 Intelsat Satellites"
February 14, 1886: "McAuliffe's Sub Would Accept Trip in Shuttle"
February 16, 1986: "Panel: Decision to Launch Shuttle Possibly 'Flawed'"
February 21, 1986: "Thiokol Opposed Launch"
February 24, 1986: "Piece of Shuttle's Fuel Tank Recovered"
February 26, 1986: "Rocket Engineers Say Bosses Ignored Their Shuttle Fears"
February 27, 1986: "Rockwell Questioned Shuttle Launch"
February 28, 1986: "Marshall Team Defends NASA"
March 4, 1986: "NASA May Go to Unmanned Space Missions"
March 7, 1986: "Fletcher Returns to Lead NASA"
March 12, 1986: "County Demands Astronaut's Remains"
March 20, 1986: "Experts Examine Critical Booster Debris"
March 20, 1986: "NASA Phone Logs Studied for Launch Pressure"
March 25, 1986: "Relatives of Crew 'Thank the World'"
March 26, 1986: "Remains of 6 of the 7 Astronauts Identified"
March 28, 1986: "Shuttle Salvage Fleet Reports Little Progress"
April 2, 1986: "Shuttle Puzzle Pieces May Yield Cause"
April 3, 1986: "Astronauts Have Their Day in Shuttle Probe"
April 7, 1986: "Shuttle Find"
April 15, 1986: "Crews Recover Shuttle Piece with Burned Joint"
April 17, 1986: "Shuttle Hunk Confirms Rocket Leak Hypothesis"
April 20, 1986: "Search Ends for Remains of Shuttle Crew"
May 12, 1986: "Shuttle Panel Chief Shocked by Transfers"
May 13, 1986: "Fletcher Sets Shuttle Date: July 1987"
May 14, 1986: "NASA Probes Charges of Retribution"
May 20, 1986: "NASA Shuttle Investigators Say They Have Copies of Destroyed Documents"
June 2, 1986: "NASA Accepted Report on 'Fail-safe' Boosters"
June 4, 1986: "McDonald Will Head Booster Redesigning"
June 6, 1986: "Shake-Up at NASA"
June 13, 1986: "Reagan Order: Implement NASA Report"
June 18, 1986: "Thiokol Chairman Criticized"
June 18, 1986: "Senators Ask: Who Was Responsible for Shuttle Blast?"
July 3, 1986: "NASA: New Booster 'Extremely Resistant to Failure'"
July 11, 1986: "Kennedy Space Center Chief Retires"
July 15, 1986: "Flight Suit Marked the Spot"
August 6, 1986: "Fatigue Blamed for Fuel Error in Scrubbed Launch"
September 6, 1986: "5 Firms Vie to Design Boosters"
October 14, 1986: "Shuttle Booster Redesign Approved"
November 19, 1986: "Panel Says NASA Paid Thiokol Award Despite Booster's Flaws"
December 10, 1986: "3rd Test of Redesign Booster Uneventful"
January 10, 1987: "5 Named as New Crew for Shuttle"
January 12, 1987: "NASA Starts a Standby Launch Team"
January 15, 1987: "Time Short for Shuttle Changes"
January 23, 1987: "NASA to Pay Thiokol $350 Million"
January 25, 1987: "Some Blame NASA"
January 31, 1987: "Father Files Claim over *Challenger*"
February 4, 1987: "Astronaut Doubts Adequacy of Shuttle Test"
February 24, 1987: "O-rings Withstand Pressure in 4th Test"
February 25, 1987: "Firm to Waive Profit on Shuttle Booster Repairs"

February 25, 1987: "Morton Thiokol Trims Rocket Fees, Profits"
March 13, 1987: "Study Suggests Revisions of Shuttle Strategy"
April 4, 1987: "Morton Thiokol Subject of FBI Criminal Probe"
April 15, 1987: "Countdown Test Urged"
April 16, 1987: "Federal Judge Ordered the Release of Court Documents"
April 22, 1987: "New Tests to Delay Launch of Shuttle Several Weeks"
April 29, 1987: "Money Problems May Keep Shuttle Grounded into '88"
May 7, 1987: "Shuttle Pilot's Widow Sues Thiokol"
May 8, 1987: "The Widow"
May 28, 1987: "Shuttle Booster Test Goes off without a Hitch"
June 4, 1987: "NASA Must Release *Challenger* Tapes"
July 1, 1987: "NASA Designing Shuttle Escape System"
July 1, 1987: "New Shuttle Design Has Escape Hatch"
July 17, 1987: "NASA Plans 3 More Tests of New Rocket"
August 27, 1987: "New Rockets Motor Fails in Test Firing"
August 31, 1987: "Shuttle Rocket Test Roaring Success"
October 9, 1987: "The Third in a Series of Tests"
December 24, 1987: "Shuttle Flights Returning"
December 30, 1987: "Pressure to Replace Thiokol"
December 31, 1987: "NASA: Shuttle Delay is Matter of Weeks"
January 2, 1988: "Shuttle Booster Rocket Test Defended"
January 12, 1988: "Shuttle Will Fly 'When It's Safe'"
March 3, 1988: "Shuttle Crew Escape System Passes Big Test"
February 27, 1988: "Widow of *Challenger* Pilot Cannot Sue U.S. for His Death, Court Says"
March 7, 1988: "Thiokol, U.S. Bought Annuities to Settle"
April 21, 1988: "Booster Passes Exam"
May 20, 1988: "Plant Loss May Cut Shuttle Flights"
July 13, 1988: "NASA-Financed Study Estimates Shuttle Crash Chances at 1 in 70"
July 15, 1988: "Flight Suit Marked the Spot"
August 19, 1988: "Shuttle Test Hailed as Success"
August 23, 1988: "Last Lawsuit Settled in Rocket Blast"
September 7, 1988: "Space Lab Bolts Removed as Safety Precaution"
September 22, 1988: "Outdated Ring Used in Shuttle Test"
September 29, 1988: "Air Force Jumpers Responsible for Rescuing Crewmen"
November 15, 1988: "Nelson Denies Cover-Up After Shuttle Blast"
April 15, 1993: "Pliers Found Stuck on Shuttle Booster"
July 23, 1995: "Shuttle Returns; Quick Launch Is Questioned"
July 26, 1999: "NASA Suspects Shuttle Was Leaking Fuel"
May 17, 2002: "Israeli Astronaut Isn't Worried About NASA Security"
January 17, 2003: "Shuttle *Columbia* Heads Skyward with Israeli Astronaut on Board"
February 2, 2003: "Shuttle Tragedy Is Merely Their Latest"
February 2, 2003: "Shuttle Crew's Remains Found"
February 2, 2003: "Early Concern over Landings Had Waned"
February 4, 2003: "Shuttle Nose Cone Found"
February 5, 2003: "Remains of Israeli Astronaut Found"
February 6, 2003: "Hunt for Shuttle Wreckage Turns Up Junk"
February 14, 2003: "NASA Says Remains of All 7 Found"
February 16, 2003: "Shuttle Inquiry, Search Proceed on New Fronts"
February 17, 2003: "Shuttle Overhaul Examined"
February 20, 2003: "*Columbia*'s Nose Landing Gear Found"
February 27, 2003: "NASA E-mails Reveal Concern"
March 2, 2003: "NASA Disputes Probe Request"
March 31, 2003: "Foam Theory Is Bolstered by Shuttle Data"
April 11, 2003: "Astronauts Recover *Columbia* Debris"
May 7, 2003: "Probe Concludes Tiles Led to Shuttle Disaster"
May 30, 2003: "Foam Test Bolsters Shuttle Theory"
June 15, 2003: "Sally Ride Hears Echoes in *Columbia* Disaster"
July 2, 2003: "NASA Erred in Photographing *Columbia* Launch, Panel Says"
July 8, 2003: "Findings Called 'Smoking Gun'"
August 27, 2003: "Report: Shuttle Imperfect, But Safe"
December 12, 2003: "Sensors on Shuttle Will Detect Blows"
September 11, 2004: "Fixing Up Shuttle Fleet Doubles to $2.2 Billion"
February 4, 2005: "Use of Space Station as Shuttle Shelter Is Questioned"
April 1, 2005: "Space Shuttle Begins Crawl to Launch Pad"
June 15, 2005: "New NASA Boss Seeks to Hasten Next Spacecraft"
August 12, 2005: "Shuttle Foam Problem to Push Launches Back"
September 9, 2005: "Hurricane Damage May Delay Launch of Shuttle"
October 5, 2005: "Shuttle Workers May Have Erred"
December 16, 2005: "New Fuel Tank May Delay Space Shuttle Again"
July 4, 2006: "Shuttle Launch OK'd for Today"
July 5, 2006: "July 4th Sees Rocket's Red Glare"
July 18, 2006: "An 'Awfully Good Day' for a Landing"

COX NEWS SERVICE

July 1, 2005: "*Discovery* 'Go for Launch' in July"

KNIGHT–RIDDER NEWS SERVICE

February 12, 1984: "*Challenger* Lands Like 'Dream,' Ends First Round Trip"

REUTERS

January 15, 2003: "Security Extra Tight for Shuttle Launch"

September 17, 2003: "Next Shuttle Flight Won't Meet March Launch Date"
January 31, 2004: "Shuttle *Columbia*'s Debris on View at NASA Facility"
August 14, 2004: "NASA Blames Application of Foam for Shuttle Disaster"
September 22, 2005: "NASA Evacuates Johnson Space Center in Houston"
February 18, 2006: "New Shuttle Crew Expects Foam Debris and Isn't Worried"

Reuters and Agence France-Presse

September 27, 1988: "Engineers Claim Shuttle Rockets Unsafe"

United Press International

January 6, 1972: "Shuttle Job Manna from Washington"
April 20, 1972: "House Votes on Shuttle"
January 24, 1986: "Space Shuttle Crew Arrives for Delayed Launch"
January 30, 1986: "Control Panel from Shuttle Recovered"
April 16, 1986: "Clue to Explosion"
May 1, 1986: "Wreckage Confirms Shuttle Joint Failure, Top Investigator Says"
May 28, 1987: "Shuttle Booster Fired in First of Six Ground Tests"
September 30, 1987: "Experiments Slated for Shuttle to Include 2 Lost on *Challenger*"
January 18, 1990: "Navigational Problem Solved, Shuttle Crew Goes Back to Bed"

Periodicals

Air & Space

April/May 1991: "The Space Shuttle Family Tree"
April/May 1991: "The Legacy of the Lifting Body"
Oct/Nov 1993: "At the Threshold of Space"
Dec 1993/Jan 1994: "Securing the High Ground"

Air Force Magazine

November 1985: "Slick Six"
July 1985: "High Space Heats Up"

Aviation Week & Space Technology

September 19, 1966: "Prime Lifting Body Undergoes Systems Test"
September 12, 1966: "Five Orbital Shots Planned in MOL Tests"
November 14, 1966: "Titan 3C Passes 6th Test, Furnishes MOL Support"
October 27, 1969: "Reusable Space Shuttle Effort Gains Momentum"
March 20, 1972: "Single Shuttle Contractor Planned"
June 19, 1972: "North American Rockwell Alters Shuttle"
July 31, 1972: "Shuttle Costs Remain $5 Billion"
September 4, 1972: "X-24B Lifting Body Nearing Completion"
January 21, 1974: "Shuttle Orbiter Nears Test Flight Decision"
January 28, 1980: "Shuttle Boosts NASA Budget to Record"
February 25, 1980: "Orbiter Protective Tiles Assume Structural Role"
April 28, 1980: "Shuttle Launch Delays Cost Impact Assessed"
June 30, 1980: "Vandenberg Shuttle Operations Delayed"
July 21, 1980: "NASA Presses to Hold Shuttle Schedule"
August 11, 1980: "NASA Tightens Shuttle Schedule Again"
November 16, 1981: "Second Flight Orbiter Work Advances"
October 22, 1984: "*Challenger* Readied for Defense Department Mission"
November 5, 1984: "Military to Withhold Shuttle Lift-off Time"
November 12, 1984: "Tile Problems Delay Mission 51-C Six Weeks"
December 3, 1984: "Garn, Teacher Orbiter Flights Expected in Next Realignment"
January 14, 1985: "NASA, Air Force Decide to Delay Initial Vandenberg Shuttle Launch"
February 18, 1985: "Space Shuttle Mounted on Vandenberg Launch Pad"
March 4, 1985: "Launch of Shuttle Mission 51-E Slips Further to March 7"
March 11, 1985: "TDRS Design Deficiencies Cancel Shuttle Mission 51-E"
January 20, 1986: "25th Shuttle Launch Will Inaugurate New Pad"
February 3, 1986: "Shuttle Destroyed, Killing Crew; Manned Space flights Halted"
February 3, 1986: "Search Continues for Shuttle Debris off Coast of Florida"
February 3, 1986: "Interim Team Formed to Review Accident Investigation Procedures"
February 10, 1986: "Ruptured Solid Rocket Motor Caused *Challenger* Accident"
February 10, 1986: "Search Team Forces Efforts on Retrieving Orbiter Wreckage"
February 10, 1986: "Booster Investigation Forces Examination of Procedures by ASA and Contractors"
February 17, 1986: "Presidential Accident Commission to Broaden Scope of Investigation"
February 17, 1986: "Marshall Officials Review Data for Solid Booster Anomalies"
February 17, 1986: "NASA Assesses External Tank's Role in *Challenger* Accident"
February 24, 1986: "Telemetry Details Rupture in Booster"
February 24, 1986: "Wind Tunnel Data Indicated Potential for Frozen Seals"
February 24, 1986: "Officials Expect Right Booster to Yield Clues to Accident"
March 17, 1986: "Additional Shuttle Crew Remains Recovered from Ocean Floor"
March 17, 1986: "Ground Controllers Cancelled Mission 61-C Thruster Tests Because of Explosion Danger"

March 31, 1986: "Accident Board Members Believe Booster Had Design Flaws"
April 7, 1986: "Shuttle Booster Section Raised from Ocean Floor"
April 14, 1986: "Booster Certification Process Raises Further Shuttle Design Flaw Concerns"
April 21, 1986: "White House Panel Moves Toward Key Shuttle Decisions"
April 28, 1986: "Navy Ends *Challenger* Crew Recovery Operations"
May 5, 1986: "USAF Prepares West Coast Site for Space Shuttle Processing"
May 19, 1986: "Commission Evidence Shows Marshall Problems Contributed to Shuttle Failure"
May 19, 1986: "NASA Investigates Claims that Morton Thiokol Demoted Engineers for Disclosing Launch Opposition"
June 16, 1986: "Rogers Commission Charges NASA with Ineffective Safety Program"
June 16, 1986: "Criminal Negligence Charges Weighed in 51-L Crew Deaths"
July 28, 1986: "Air Force Presses to Mothball Vandenberg Shuttle Complex"
July 28, 1986: "Aerojet Claims Single-Cast Booster Can Be Ready for Next Shuttle Flight"
August 4, 1986: "NASA Releases *Challenger* Transcript"
October 20, 1986: "Oversight Panel Endorses NASA Plan for Fixing SRBs"
November 3, 1986: "Solid Booster Test Duplicates *Challenger* Accident"
January 5, 1987: "Shuttle Escape System Given Tentative Approval"
January 5, 1987: "Justice Dept. Settles Claims with Four *Challenger* Families"
February 2, 1987: "*Challenger* Debris Stored in Missile Silos"
April 6, 1987: "NASA Will Compete New Shuttle Booster"
May 11, 1987: "Kennedy Moves Ahead with Plans for Wet Countdown Demonstration Test"
July 6, 1987: "NASA Nears Key Shuttle Tests, Changes Initial Payload Schedule"
August 17, 1987: "Problems Discovered in Shuttle Booster Test Hardware"
September 7, 1987: "NASA Panel Would Withhold Accident Witness Accounts"
September 7, 1987: "Thiokol Evaluates First Full-Scale Test of Redesigned Shuttle Booster"
September 14, 1987: "Disassembled SRM Field Joint Shows No Damage After Test"
October 26, 1987: "New Space Launch Manifest Integrates Shuttle, Expendable Booster Mission"
December 7, 1987: "NASA Modifying Shuttle Launch Pad for Safety, Efficiency"
December 21, 1987: "Space Shuttle Program Manager Concerned About Staffing Levels for Three-Orbiter Processing"
January 4, 1988: "Next Shuttle Launch Delayed Following SRM Nozzle Failure"
January 18, 1988: "Congress Cut Air Force Funding for Funding for Vandenberg Shuttle Complex"
January 18, 1988: "NASA Selects Northrop Strip Alternate Shuttle Landing Site"

February 1, 1988: "NASA to Smooth Shuttle Runway"
April 11, 1988: "NASA Selects Telescoping Pole for Shuttle Crew Escape System"
April 25, 1988: "Test Fixing Indicates Booster Ring Design Can Tolerate Joint Flaws"
May 2, 1988: "Vandenberg Shuttle Complex Will Go into Mothball Status"
May 9, 1988: "Loss of Oxidizer Plant Will Not Hinder Near-term Shuttle Defense Programs"
June 3, 1991: "Space Shuttle Quality Review Examined Discipline in Program"
February 10, 2003: "USAF Imagery Confirms *Columbia* Wing Damaged"
February 17, 2003: "Growing Evidence Points to *Columbia* Wing Breach"
February 17, 2003: "Numbers, Words Tell Different Roam Stories"
February 24, 2003: "Rough Wing + Debris = A Fatal Combination?"
March 3, 2003: "*Columbia* Revelations"
March 10, 2003: "Shuttle Probe Intensifies"
March 10, 2003: "Rescue Effort"
March 10, 2003: "Safety Process Questioned"
March 17, 2003: "Critical *Columbia* Tests"
March 24, 2003: "Water Damage Probed"
March 24, 2003: "Return to Flight"
April 21, 2003: "*Columbia* Probe Shifts"
April 28, 2003: "Echoes of *Challenger*"
May 5, 2003: "Assigning Risk"
May 12, 2003: "NASA's Eroding Safety"
May 26, 2003: "*Columbia* Accident Probe Widens"
June 2, 2003: "Foam Jars T-Seal"
June 16, 2003: "Accident Impact"
June 23, 2003: "Buckling Down"
June 30, 2003: "Final Draft"
July 7, 2003: "Recasting Shuttle"

CHICAGO TRIBUNE

July 14, 2005: "Faulty Fuel Gauge Grounds Shuttle"

COUNTDOWN
(MAIN STAGE PUBLICATIONS, INC.)

March 1985: Volume 3, No. 3
January 1986: Volume 4, No. 1
February 1986: Volume 4, No. 2
March 1986: Volume 4, No. 3
March 1986: Volume 4, No. 3
April 1986: Volume 4, No. 4
May 1986: Volume 4, No. 5
June 1986: Volume 4, No. 6
July 1986: Volume 4, No. 7
August 1986: Volume 4, No. 8
September 1986 Volume 4, No. 9
October 1986: Volume 4, No. 10
December 1986: Volume 4, No. 12
January 1987: Volume 5, No. 1
May 1988: Volume 6, No. 5
March 1991: Volume 9, No. 3
March 1992: Volume 10, No. 3
February 1993: Volume 11, No. 2
October 1993: Volume 11, No. 10

DAILY BREEZE
(LOS ANGELES, SOUTH BAY)

October 21, 1969: "Shuttle Bus Is Vital to Space Plans"
February 20, 1970: "Edwards Air Force Base"
December 30, 1970: "Space Shuttle Landing Sites Sought by Agency"
July 14, 1971: "Shuttle Contract Cheered"
October 6, 1971: "Pilots Pave Way for Astronauts Accomplishments"
January 3, 1972: "Air Force Eyes Military Space Shuttle"
January 8, 1972: "Space Shuttle Jobs May Number 25,000"
February 12, 1972: "NASA Official Urges Reusable Space Shuttle"
April 2, 1972: "NAR Gets Okay on Shuttle Pact"
April 14, 1972: "Two Bases Win Space Shuttles"
April 15, 1972: "Space Payload Seen as Big Area Boost"
April 20, 1972: "House Votes on Shuttle"
April 21, 1972: "Shuttle Vote Elates Bayons"
June 2, 1972: "Shuttle Plan Brightens Area Space Picture"
July 27, 1972: "Shuttle Contract News Helps Industry Morale"
June 27, 1973: "NASA Defends Shuttle Program"
July 30, 1973: "Space Plan Under Fire"
August 2, 1973: "X-24B Makes Debut"
February 5, 1974: "Plans to Build the Virtus"
June 7, 1974: "Regular Orbital Trips Due"
July 15, 1974: "Space Shuttle Will Meet Future Needs"
August 21, 1975: "Space Shuttle Test Works"
February 27, 1977: "Shuttle 3rd Piggyback Flight a Success"
August 31, 1984: "Shuttle Offer Has Area Teaches Riding High"
December 1, 1984: "Space Plumbers"
March 7, 1985: "Hermosa Astronaut Is Bumped from Flight"
June 24, 1985: "International Crew of the Shuttle *Discovery* Returns Home"
November 12, 1985: "*Challenger* May Be Delayed"
November 28, 1985: "Vandenberg Shuttle Launch Grounded Until Summer"
January 28, 1986: "*Challenger* Explodes"
January 29, 1986: "Shuttle Returns Unscathed from Its Orbit"
January 30, 1986: "Remnants of Shuttle Recovered"
February 1, 1986: "Major *Challenger* Clue Reported"
February 2, 1986: "Colleagues Bid Farewell to Shuttle 7"
February 15, 1986: "Shuttle Panel Examines Launch Pad, Wreckage"
February 16, 1986: "Panel: Decision to Launch Shuttle Possibly Flawed"
February 19, 1986: "Engineer: I Lost Shuttle Argument"
March 1, 1986: "Military Shuttle Delayed 1 Year"
March 9, 1986: "Launch Pace 'A Safety Hazard'"
March 10, 1986: "Remains of Shuttle Crew Recovered"
March 14, 1986: "Divers Recover *Challenger*'s Flight Recorders, Computers"
March 28, 1986: "Shuttle Salvage Fleet Reports Little Progress"
April 30, 1986: "Key Piece of Shuttle Found"
June 17, 1986: "Widow to NASA: Terrible Judgment"
June 18, 1986: "Senators Ask: Who Was Responsible for Shuttle Blast?"
June 19, 1986: "Air Force Chief: Most Problems at Vandenberg Resolved"
July 11, 1986: "Kennedy Space Center Chief Retires"
September 20, 1986: "NASA to Release Transcript"
October 4, 1986: "Booster — Joint Tests Continue"
October 4, 1986: "Vandenberg Space Center Put on Caretaker Status"
December 18, 1986: "Military Space Shuttle Mission Cloaked in Secrecy"
September 19, 1987: "Booster Test Successful"
September 30, 1987: "Shuttle Payload"
October 23, 1987: "NASA Plans 19 Shuttle Flights"
January 26, 1988: "New Problem for Shuttle"
February 18, 1988: "Settlement Reached in Astronaut's Death"

FORT WORTH STAR-TELEGRAM

February 2, 2003: "Remains Scattered over E. Texas"

THE HERALD (AUSTRALIA)

September 29, 1988: "Shuttle Is Ready to Fly"

HUGHES AIRCRAFT COMPANY BULLETIN

Number 3315, February 4, 1986

HUGHES NEWS

July 6, 1984: "Lucky Two Picked for Ride into Space"
January 31, 1986: "Shuttle Tragedy Hits Home"

LONG BEACH PRESS TELEGRAM

January 30, 1986: "Space Disaster: Questions, No Answers"
April 24, 1986: "NASA Cut *Challenger* Safety Tests, Audits Show"
July 29, 1986: "Uh, Oh ... the Last Words from Shuttle"
July 21, 1999: "Shuttle Crew Will Try Again Thursday"
February 4, 2003: "Liftoff Fault Tied to Tragedy"
February 5, 2003: "Insulation Needed Testing"

LOCKHEED STAR GAZER

August 6, 1987: "*Discovery* 'Power Up' Marks Major Milestones Achieved"

LOS ANGELES HERALD EXAMINER

August 26, 1971: "Supersonic Launch a Success"
January 29, 1986: "America Asks, 'Why?'"
January 29, 1986: "Text of Reagan's Speech"
February 3, 1986: "Shuttle Could Fly Again by June"
February 4, 1986: "Lockheed Shuttle Role Criticized"
February 18, 1986: "Analysts View Photos of Possible Shuttle Booster Parts in Atlantic"
February 20, 1986: "Searchers Find Part of Suspect Rocket Booster"
February 24, 1986: "Top NASA Officials Kept in Dark on Shuttle Seals"

July 16, 1986: "*Challenger* Pilots' Family Files $15 Million Claim"
July 29, 1986: "*Challenger* Crew Survived Blast"
September 29, 1988: "We're Back! Shuttle Rockets into Orbit after 1 1/2 Hour Delay"

LOS ANGELES TIMES

May 11, 1967: "Wingless Craft Crashes During NASA Glide Test; Pilot Injured"
October 24, 1968: "Rocket Failure Spoils Wingless Craft's Test"
March 15, 1972: "Space Shuttle May Get Solid Rocket Motors"
September 9, 1976: "Star Trek Fans Torpedo Name for Space Shuttle"
September 18, 1976: "Space Shuttle Put on Display; New Era Hailed"
February 16, 1977: "Space Shuttle Takes First Piggyback Ride"
February 19, 1977: "Space Shuttle Goes Aloft on 747 — All's A-OK"
June 29, 1977: "Shuttle Completes 2nd Manned Flight"
August 13, 1977: "Space Shuttle Sails Through Solo Flight"
November 27, 1977: "Hit by Cutbacks, NASA Lapses into Conservatism"
August 13, 1977: "Space Shuttle Sails Through Solo Flight"
September 24, 1977: "Space Shuttle 3rd Free-Fall Flight a Success"
January 3, 1979: "Space Shuttle Engine Explodes in Test"
January 21, 1979: "Space Shuttle Launching Set for Nov. 9"
March 20, 1979: "Bolts and Tape Ground Giant Space Shuttle"
March 21, 1979: "Heading East"
March 21, 1979: "Storms Force Shuttle to Land in El Paso"
February 21, 1981: "Officials Jubilant as Space Shuttle Engines Pass Test"
March 22, 1981: "The Investigation"
April 13, 1981: "Space Shuttle in Orbit, Functioning Smoothly"
April 13, 1981: "Shuttle Loses Some of Heat Shield"
November 12, 1981: "Shuttle Repairs Completed for Liftoff Today"
November 15, 1981: "*Columbia* Flies Home to Questions"
November 16, 1981: "Shuttle Is in Superb Shape, Slayton Says"
December 9, 1983: "*Columbia* Lands Eight Hours Late"
February 14, 1984: "Shuttle Mission Cancelled"
June 18, 1985: "Off to Perfect Start, *Discovery* Deploys Satellite"
January 29, 1986: "Shuttle Explodes; Crew Killed"
January 30, 1986: "NASA Opens Hunt for Clues to Space Disaster"
January 30, 1986: "Boosters Had to be Destroyed When One Headed for Towns"
January 31, 1986: "Pieces of Shuttle Pulled from Sea"
February 3, 1986: "Shuttle Carried No Booster Sensors"
February 4, 1986: "Reagan Names 12 for Shuttle Probe"
February 4, 1986: "12 Members of Panel Experts in Their Fields"
February 4, 1986: "Salvage Crew Turn Attention to 17 Objects on Ocean Floor"
February 5, 1986: "Debris May Be Shuttle Booster"
February 9, 1986: "Special Report: *Challenger*, Countdown to Disaster"
February 12, 1986: "Shuttle Launch Was Warned About Cold"
February 14, 1986: "Head of NASA Defends Staff as Conscientious"
February 14, 1986: "Photos Show Rocket Smoking Early in Flight"
February 17, 1986: "Parts Found by Sub May Solve Shuttle Puzzle"
February 18, 1986: "Rocket Experts to View Photos of Debris"
February 23, 1986: "NASA, Thiokol in Contract Talks at Time of Blast"
February 27, 1986: "NASA Officials Deny Firm Was Pressured to Launch"
February 28, 1986: "NASA Procedures Flawed, Panel Says"
March 4, 1986: "NASA Aid's Testimony to Be Examined"
March 11, 1986: "Winds Hamper Search for Crew of Lost Shuttle"
March 13, 1986: "Parts of Shuttle Deck, More Crew Remains Believed Found"
March 14, 1986: "Rep. Nelson Plans Shuttle Safety Inquiry"
April 9, 1986: "NASA to Give Preliminary Shuttle Report Thursday"
April 23, 1986: "Chinks in NASA Armor Come to Light after *Challenger* Disaster"
May 13, 1986: "NASA Knew of Rocket Seal Danger, Document Shows"
May 20, 1986: "No Important Documents Missing, Shuttle Inquiry Members Indicate"
May 22, 1986: "W. Coast Shuttle Launch in 1980s Termed Doubtful"
June 4, 1986: "Poor Management for Shuttle Charged"
June 9, 1986: "Astronauts Needn't Have Died"
June 10, 1986: "Panel Blames Shuttle Disaster on Poor Design Management"
June 18, 1986: "Senator Critical of Vandenberg as Shuttle Site"
July 29, 1986: "Vandenberg's Shuttle Launch Plans on Hold"
August 13, 1986: "NASA Unveils Proposed $300-Million Redesign of Shuttle Rocket Joint"
January 16, 1987: "Shuttle Captain Places Safety Over Time"
January 27, 1987: "6th Rocket Test May Delay Next Shuttle Flight"
January 29, 1987: "Shuttle Booster Builder Sued for $1 Billion"
February 18, 1987: "Truly Sees More Difficulty in Resuming Flights in Year"
August 31, 1987: "Shuttle Rocket Looks Flawless in Firing Test"
December 14, 1987: "Shuttle Emergency Escape System Being Tested"

December 30, 1987: "First Post-*Challenger* Shuttle Flight Delayed"
January 8, 1988: "Shuttle Rockets to Be Shipped to Florida Site"
April 8, 1988: "Shuttle to Have Device Enabling Crew to Escape"
May 5, 1988: "Blasts Destroy Rocket Fuel Plant; 1 Killed"
August 11, 1988: "Oft-Delayed Shuttle Test Successful"
September 6, 1988: "U.S. Space Program at Crossroads"
September 29, 1988: "Sheer Number of Modifications Produces New Debate on Safety"
January 29, 2003: "Crew Honors Those Who Died Aboard *Challenger*"
February 2, 2003: "'*Columbia* Is Lost'"
February 2, 2003: "A Somber Bush Leads a Nation in Grieving"
February 3, 2003: "1,200 Clues Strewn Across Texas"
February 3, 2003: "Shuttle Inquiry Points to Heat"
February 4, 2003: "Investigation Turns to Shuttle Heat Tiles"
February 4, 2003: "NASA Report Warned of Shuttle Safety"
February 5, 2003: "NASA Considering Space Hit"
February 6, 2003: "NASA Says Tank Foam Not to Blame"
February 6, 2003: "Tile Repair Kits for Astronauts Rejected"
February 7, 2003: "Shortage of Spare Parts Hinders NASA"
February 8, 2003: "Scrutiny on NASA Decision Process"
February 8, 2003: "Shuttle Wing Piece Could Be Key Clue"
February 9, 2003: "Cost Cutting Trumped Key NASA Safety Upgrades"
February 13, 2003: "NASA Engineer Warned of 'Catastrophic' Scenarios"
February 13, 2003: "*Columbia* Panel Autonomy Avowed"
February 14, 2003: "Breach in Shuttle Suspected"
February 17, 2003: "Glitches, Close Calls Haunted *Columbia*"
February 19, 2003: "Board to Search for Flaws at NASA"
February 22, 2003: "NASA Resisted Concerns, Engineers' E-Mails Say"
February 22, 2003: "NASA Worker Says E-Mails Misinterpreted"
February 26, 2003: "Shuttle Insulation Drew '79 Warning"
February 26, 2003: "Most of *Columbia* Will Never Be Found, Officials Concede"
February 27, 2003: "Tile Safety Raised"
February 27, 2003: "New Questions on Shuttle"
March 2, 2003: "Challenges Facing Shuttle Disaster Inquiry Include Culture at NASA"
March 4, 2003: "Adhesive Problem with *Columbia* Tank Is Cited"
March 7, 2003: "NASA Seen as Unable to Handle Safety Issues"
March 9, 2003: "Clues Point to Shuttle Wing Edge"
March 11, 2003: "NASA Engineer Was Confident of Safe Landing"
March 12, 2003: "Shuttle Investigators Look at Possibility of Weakened Wing"
March 15, 2003: "NASA Is Planning to Launch Next Shuttle as Early as Fall"
March 18, 2003: "Amateur Videos Help NASA Trace *Columbia*'s Path"
March 19, 2003: "Shuttle Inquiry Leans to Leading Edge Theory"
March 20, 2003: "Key Data Recorder on *Columbia* Found"
March 29, 2003: "Satellite Photos of Shuttles Planned"
March 30, 2003: "This Door NASA Will Keep Closed"
April 16, 2003: "Shuttle Disaster Linked to Unseen Flaw"
April 24, 2003: "Shuttles' First Engineers Exasperated"
May 21, 2003: "Investigators Want Big Changes at NASA"
June 22, 2003: "Foam Issue May Delay Resumption of Shuttle Flights"
July 3, 2003: "3 Shuttle Managers Reassigned"
July 15, 2003: "Too Many Astronauts with Too Little to Do, Report Finds"
August 27, 2003: "Foam Was to Blame, Says Shuttle Study"
September 9, 2003: "Shuttle Safety Plans Detailed"
September 24, 2003: "All 11 Members Resign from NASA Safety Panel"
February 11, 2005: "Shuttle's Return to Space Is on the Horizon"
April 1, 2005: "Space Shuttle Begins Crawl to Launch Pad"
April 23, 2005: "NASA Denies Claims It's Shading Shuttle Problems"
April 30, 2005: "Safety Concerns Delay NASA Shuttle Launch"
May 21, 2005: "Shuttle Passes Test with Flying Colors"
June 28, 2005: "Panel Cites Flaws, but Backs July Launch"
July 25, 2005: "Launch Still On Despite Glitch"
July 27, 2005: "*Discovery* Relaunches Shuttle Program"
July 28, 2005: "Safety Worries Ground Shuttles"
July 29, 2005: "Foam Bit Might Have Hit Right Wing"
July 29, 2005: "Behind Chunks of Foam, a Failure to Confront Hazard"
July 31, 2005: "Shuttle Sends Handymen for Spacewalk"
August 1, 2005: "Shuttle May Get Repair in Space"
August 5, 2005: "NASA Says *Discovery* Is Clear for Landing"
August 18, 2005: "NASA Seen as Improving but with a Ways to Go"
November 23, 2005: "NASA Is Still Puzzled by *Discovery*'s Foam Loss"
December 3, 2005: "NASA Chief Isolates Likely Cause of Shuttle Trouble"
June 1, 2006: "Fuel Tank Alterations Clear Way for Shuttle"
June 18, 2006: "Shuttle Launch Is Set in Spite of Concerns"

NATIONAL GEOGRAPHIC

March 1981: "When the Space Shuttle Finally Flies"

NEW YORK TIMES

November 10, 1984: "NASA Lobbying and a Trip in Space"

December 20, 1984: "Security, Freedom of Information Debated after Shuttle Secret Is Out"
December 27, 1984: "Shuttle Secrecy Announcement Tipped off Soviets"
January 28, 1985: "Silence Attends Shuttle Landing Ending 3-Day Trip"
June 28, 1985: "Air Force, Unhappy with Shuttle, Seeking Backup Rocket Program"
July 13, 1985: "T-Minus 3 Seconds: Shuttle Mission Aborted as Computer Spots Problem"
January 29, 1986: "Computers Hold Key to *Challenger* Failure"
February 6, 1986: "Rough Seas Hamper Look at Possible Rocket Debris"
February 9, 1986: "NASA Was Warned That Booster Seals Hazardous"
February 22, 1986: "NASA Mum about Any Human Remains"
March 24, 1986: "NASA Gives Priority to Military Missions"
March 30, 1986: "NASA Confident of 1-year Timetable"
April 6, 1986: "Plans to Develop Aerospace Plane Take Off"
April 23, 1986: "Chinks in NASA Armor Come to Light after *Challenger* Disaster"
April 24, 1986: "NASA Cut Back on Safety Spending for Shuttle Program Audit Shows"
April 24, 1986: "NASA Cut *Challenger*'s Safety Tests, Audits Show"
May 11, 1986: "Rocket Engineers Tell of Demotions"
May 13, 1986: "NASA Knew of Rocket Seal Danger, Document Shows"
May 20, 1986: "No Important Documents Missing, Shuttle Inquiry Members Indicate"
June 5, 1986: "Head of NASA's Heavily Criticized Rocket Division Resigns"
June 5, 1986: "NASA Urged to Open Bids on New Rockets"
June 14, 1986: "Video Shows NASA Knew of Booster Risks Last Year"
July 29, 1986: "*Challenger* Transcript"
January 21, 1987: "Escape System May Be Added to Next Shuttle"
January 27, 1987: "6th Rocket Test May Delay Next Shuttle Flight"
December 30, 1987: "Shuttle Rocket Test Discloses Flaw That Will Delay Scheduled Flight"
January 6, 1988: "Second Flaw Seen in Shuttle Rocket"
August 18, 1988: "Changes in Shuttle Rocket Get Biggest Test Today"
August 7, 1989: "Pentagon Pulling Out of NASA Program"
February 2, 2003: "Shuttle Destined for 1-in-50 Failure"
February 2, 2003: "16 Minutes from Landing Space Shuttle Disintegrates"
February 9, 2003: "Is There a Future in Space?"
March 2, 2003: "NASA Disputes Probe Request"
April 1, 2003: "Shuttle Warning E-mail Unsent"
February 16, 2006: "Hotz on the Trail of NASA"

Newsweek

February 7, 1977: "Launching the Shuttle"
February 10, 1986: "We Mourn Seven Heroes"
February 17, 1986: "What Happened"
May 5, 1986: "A Hurried Return for the Shuttle?"
January 11, 1988: "Trouble for the Shuttle"
February 10, 2003: "Out of the Blue"
February 10, 2003: "A Tragic Mission"
February 17, 2003: "Falling to Earth"

Orange County Register

September 4, 1988: "Anatomy of a Booster"
September 4, 1988: "The Countdown Begins for *Discovery*"
September 29, 1988: "Shuttle Escape System Ready"

Orlando Sentinel

August 18, 2005: "NASA Seen as Improving but with a Ways to Go"

Rockwell International News

July 1, 1977: "Successful Second Flight"

Science News

March 21, 1981: "*Columbia* Countdown"

Space Flight

April 1981: "Space Shuttle at Vandenberg"

Spaceflight News

May 1988: "X-Plane Veteran Still Going Strong"

Time

January 17, 1972: "A Boost for NASA"
January 12, 1981: "Aiming High in '81"
March 5, 1984: "New Pad for the Space Shuttle"
December 31, 1984: "Shrouding Space in Secrecy"
May 13, 1985: "Roger, Houston ... Er, Colorado"
February 10, 1986: "A Nation Mourns"
February 10, 1986: "'They slipped the surly bonds of earth to touch the face of God'"
February 24, 1986: "Zeroing In on the O-rings"
March 3, 1986: "The Questions Get Tougher"
March 10, 1986: "A Serious Deficiency"
March 24, 1986: "Painful Legacies of a Lost Mission"
June 9, 1986: "Fixing NASA"
June 23, 1986: "NASA Takes a Beating"
August 31, 1987: "Getting NASA Back on Track"
January 11, 1988: "Still Grounded"
February 1, 1988: "Putting Schedule over Safety"
February 29, 1988: "Can They Escape Next Time?"
July 4, 1988: "Getting Ready to Try Again"
July 15, 1996: "High-Tech Pie in the Sky"
February 10, 2003: "Seven Astronauts, One Fate"
February 10, 2003: "What Went Wrong?"
February 10, 2003: "*Columbia*'s Final Flight"
February 17, 2003: "Fragments of a Mystery"
March 24, 2003: "Those Last Few Seconds"

USA TODAY

February 14, 1986: "Investigators Eye Shuttles 'Odd' Smoke"
February 14, 1986: "USA Still Sending Donations"
February 18, 1986: "Verification Next Step in Shuttle Hunt"
February 18, 1986: "Shuttle Chief, Others, Likely to Leave Probe"
April 24, 1986: "NASA Accused of Waste under Nominee's Leadership"
June 10, 1986: "NASA 'to Regain Uur Honor'"
March 20, 1987: "Astronaut Calls for More Tests of Booster"
January 28, 1988: "NASA Aims for Aug. 4 Launch"
February 4, 2003: "Grim Search Interrupts Life in Small Texas Community"

U.S. NEWS AND WORLD REPORT

February 10, 1986: "Out of *Challenger*'s Ashes — Full Speed Ahead"
February 10, 1986: "Frank Borman, Former Astronaut"
February 10, 1986: "Poll: Forge Ahead"

WALL STREET JOURNAL

November 16, 1981: "NASA's Confidence in Shuttle Wavers, and Pilots Land Craft 3 Days Early"

WASHINGTON POST

June 17, 1986: "Shuttle Pilot's Widow Attacks NASA Decisions"
May 23, 1991: "Shuttle Dodged Engine Failure, Space Official Says"
March 2, 2001: "NASA Ends X-33 Project That Sought to Cut Spaceflight Costs"

WORLD SPACE FLIGHT NEWS

December 1984: "Secret Military Flight Set for Launch"

Books

Chant, Christopher. *Space Shuttle*. Hong Kong: McLaren Publishing, 1984.
Lewis, Richard. *The Voyages of Columbia*. New York: Columbia University Press, 1984.
McConnell, Malcolm. *Challenger: A Major Malfunction*. Garden City, NY: Doubleday, 1987.
McDougall, Walter A. *The Heavens and the Earth*. New York: Basic Books, 1985.
Shayler, David J. *Disasters and Accidents in Manned Spaceflight*. Chichester, United Kingdom: Praris Publishing, 2000.
Trento, Joseph J. *Prescription for Disaster*. New York: Crown Publishers, 1987.

Other

NORTHROP CORPORATION

Northrop Corporation History, M2-F2/F3, HL-10 Lifting Bodies, pages 228–236

Index

ABC news 106, 138
Abel, Richard 61
Abort-to-Orbit (ATO) 75
Acheson, David C. 155, 181
Acton, Loren 260
Adamson, James C. 264
Aerojet General Corporation 176
Aerojet Solid Propulsion Company 41, 203
Aerojet Strategic Propulsion Company 176
Aerospace Corporation 57
Agee, Steve 192
Agnew, Vice Pres. Spiro T. 17
Air Force Consolidated Space Operations Center 58
Aldrich, Arnold 94, 127, 140, 158
Aldridge, Edward C., Jr. 50, 51, 66, 67
Aldrin, Buzz 59
Allen, Andrew M. 265
Allen, Joseph P. 255, 258, 263
Alpha 80
Al-Saud, Sultan Salman 73, 162, 260
Altman, Scott 266
American West Airlines 191
Ames Research Center 6, 8
Anderson, Michael P. 212, 235, 266, 278
Anderson, William C. 227
Arlington National Cemetery 149
Armstrong, Neil 56, 59, 155, 163–164
Apollo-1 (204) 22, 27, 153, 183, 222
Apollo-8 247
Apollo-10 85
Apollo-11 17, 56
Apollo-13 27
Apollo-16 21, 59
Apollo-17 21, 22
Apollo-18 17
Apollo-19 17
Apollo applications 14, 17
Apollo program 16, 17, 22, 26, 27, 59, 85, 176, 182, 186, 199, 200, 207, 210, 238, 241, 250, 251

APUs 224, 228
Arabsat 1B 73
Ashby, Jeff 266
ASSET 13
Astro Ultraviolet Telescope 188
astronaut remains: *Challenger* 135, 136, 137 141, 142, 143, 144, 145, 146, 147, 148, 149; *Columbia* 221–222
ASTP 27–28, 85, 210
Atlantic Research Corporation 176
Atlantis (OV-104) 27, 138, 200, 201, 203, 205, 208, 221, 224, 238, 240
ATLAS-2 203
The Augustine Committee 200
Aussat 60

B1 Bomber 20
Bagian, James P. 143, 264
Bailey, Terry 143
Bailie, Carleton T. 122
Ball, Jim 86
Baker, Ellen S. 264
Baker, Michael A. 264
Barksdale Air Force Base 222
Barry, Maj. Gen. John L. 222
Bartholomew, Capt. Charles 147
Bartoe, John-David 260
Baudry, Patrick 260
Bean, Alan L. 232
Beggs, James M. 141
Benedict, Howard 123
Bieringer, Maj. Gerald F. 117
Bjornstad, John 33
"Black areas" 21
Blaha, John E. 265
Blue Gemini 56
Bluford, Guion S. 256, 261
Bobko, Karol J. 57, 255, 259, 261
Boisjoly, Roger 73, 78, 93, 94, 158, 160, 164, 165, 175, 183, 192, 195
Bolden, Charles 225, 262, 264
Borman, Frank 138
Bowersox, Kenneth D. 204, 264, 265
Brady, Charles 265
Brand, Vance D. 142, 255, 257, 263, 264

Brandenstein, Daniel 228, 256, 260, 264
Braun, Dr. Werner von 14, 39
Bridges, Roy 260
Briegleb, Gus 6
Brinton, Boyd 93
Brown, David M. 212, 235, 266, 278
Brown, George E, Jr. 59
Brown, Mark N. 264
Buchanan, Jack 95
Buchli, James F. 59, 259, 261
Buckey, Jay C. 266
Buran 176, 182, 193, 204, 248
Bush, Pres. George H.W. 124, 127, 128
Bush, Pres. George W. 210, 238

C-130 123, 132
Cabana, Robert "Bob" 222, 265
Camarda, Charles 238
Cape Canaveral 30, 62, 84, 117, 129, 130, 132, 133, 134, 136, 138, 139, 141, 142, 143, 146, 147, 148, 228
Cape Canaveral Air Force Station 117, 145, 183
Cape Kennedy 158
Carey, Duane 266
Carter, Pres. Jimmy 24, 27
Carter, Manley L. "Sonny," Jr. 98
Casablanca 82, 88
CBS news 137
Casper, John H. 265
Cenker, Robert 262, 264
Chaffee, Roger 22
Challenger (OV-099) 13, 20, 25, 26, 27, 39, 40, 42, 43, 47, 50, 55, 57, 60, 62, 65, 68, 69, 72, 73, 76, 77, 81, 82, 83, 84, 85, 86, 87, 88, 171, 172, 173, 176, 181, 185, 186, 187, 191, 193, 200, 201, 204, 205, 206, 207, 210, 222, 223, 228, 229, 231, 232, 236, 239, 241, 248; development 34–37; disaster costs 202; fire alarms 80; investigation 153–170; launch of 51-L 106–116; pre-launch count down 93–105;

search and recovery operations 129–149; search and rescue timeline 122–128; search area 148; sensor defects 75
Chandra X-Ray Telescope 229
Chang-Diaz, Franklin 262, 264, 265
Chawla, Dr. Kalpana 212, 235, 266, 278
Cheli, Maurizio 265
Chiao, Leroy 265
China, communist 22, 139, 181, 251
China Lake Naval Weapons Center 188
CIA 181
Clark, Dr. Laurel 212, 235, 266, 278
Clary, Ron 193
Cleave, Mary 261
CNN 48, 106, 123, 126, 131, 133
Coates, Keith 48
Coats, Mike 258
Cockrell, Kenneth 265
USS *Coelsch* 131
Cohen, Aaron 200
Cole, Forrest 33
Coleman, Cady 266
Coleman, Catherine G. 265
Collins, Eileen 237, 240, 266
Columbia (OV-102) 20, 21, 23, 26, 29, 30, 31, 32, 33, 34, 36, 39, 46, 84, 85, 86, 87, 88, 131, 137, 139, 160, 161, 162, 163, 165, 184, 186, 199, 200, 202, 203, 204, 205, 206, 207, 209, 210–211, 224, 231, 236, 237, 239, 240, 241, 248
Columbia Accident Investigation Board (CAIB) 20, 212, 222, 231, 232, 236, 237, 238, 240, 244
Columbus, Christopher 22
Congress 15, 17, 18, 19, 21, 22, 24, 27, 41, 43, 56, 85, 142, 161, 174, 175, 177, 179–180, 181, 182, 185, 186, 192, 205, 223, 230, 231, 250
Constellation program 252
Consumer Product Safety Commission 42
Convair 240 188
Convair 990 19
Cook, Richard C. 75, 157
Cooper, Gordon 210
Covert, Dr. Eugene E. 155
Covey, Richard D. 113, 122, 183, 189, 241, 260
Creighton, John 260
Crew Exploration Vehicle (CEV) 238
Crews, Albert H. 56
Crippen, Robert 34, 57, 140, 163, 177, 181, 200, 201, 254, 256, 257, 258, 263
Cronkite, Walter 252
Crossfield, Scott 55
Crouch, Roger 265

Dakar, Senegal 86, 88
USCG *Dallas* 130, 136
Dana, William 11
Daugherty, Robert L. 227
USCG *Dauntless* 131
Davis, B.K. 96
Davis, Christopher K. 225
Davis, Richard 193

Daytona Beach 134
Deal, Brig. Gen. Duane 222
DeLucas, Lawrence 264
Demonstration Motor 8 (DM-8) 186
Demonstration Motor 9 (DM-9) 189
Department of Defense 32, 39, 50, 51, 56, 60, 125, 126, 131, 159, 191, 201
Department of Energy 68
Department of Justice 232
Discovery (OV-103) 26, 27, 36, 59, 60, 62, 63, 64, 65, 69, 72, 73, 89, 98, 127, 138, 164, 179, 183, 187, 190, 191, 192, 200, 201, 203, 205, 207, 224, 230, 238, 239, 240, 241, 242, 243, 244; "Battlestar Discovery" 61; launch of STS-26 193
Dittemore, Ron 237
Doi, Takao 266
Dorsey, Edward 69, 175
Dover Air Force Base 222
Drake, Sir Francis 128
Dryden Flight Research Center 6, 8, 11, 19, 28
Duke, Charlie 21
Dunbar, Bonnie J. 261, 264
Durrance, Samuel 264

E-mails 226–227
Easter Island 59
Eastern Airlines 119
Ebay 223
Ebeling, Robert V. 79, 81, 93, 94
Edwards Air Force Base 6, 8, 12, 13, 14, 19, 27, 28, 57, 59, 73, 75, 80, 88, 190, 230, 241
Eggers, Dr. Alfred 6
Ehlers, John 225
Eisenhower, Pres. Dwight David 5
Ejection seats 20
ELINT 61
Endeavour (OV-105) 202, 224, 229
Engineering Test Motor 1A (ETM-1A) 185
England, Anthony 260
Engle, Joe 27, 28, 142, 254, 260, 263
Enterprise (OV-101) 12, 27, 28, 29, 32, 34, 36, 58, 66, 185
Espinosa, Ray 119
Eudy, Glenn 43
European Space Agency (ESA) 22, 139
Executive Order 154
Executive Order 12546 136
Explosive Ordnance Team 126
External Tank 23, 24, 28, 32, 33, 37, 58, 86, 112–117, 134, 135, 141, 143, 144, 148, 162, 208, 212, 225–226, 227, 238, 239, 241, 242, 243

Fabian, John M. 256, 260
Faget, Dr. Max 41
Falk, Brig. Gen. Rani 222
False Claims Act 175
Favier, Jean-Jacques 265
Fawsett, Patricia 195
FBI 167, 175, 181, 193
Federal Aviation Administration 137

Fettman, Martin 265
Feynman, Dr. Richard 64, 68, 153–154, 155, 156, 157, 158
First Lunar Outpost (FLO) 250
Fisher, Anna 258, 260
Fletcher, Dr. James 20, 39, 51, 67, 153, 161, 164, 177, 178, 180, 181, 185, 191, 204
Flyswatter 70
Ford, Pres. Gerald 27
Freedom of Information Act 175
Friendship-7 127
Frosch, Dr. Robert 20, 24, 32
Fuel cells 46
Fuller O'Brien Paint Company 42
Fullerton, Charles Gordon 27, 28, 57, 142
Fullerton, Gordon 255, 260, 263
Fulton, Fitzhugh, Jr. 28
Fuqua, Don 177
Furrer, Reinhard 261

Gaffney, F. Andrew 264
Galileo 188, 201
Gardner, Dale A. 256, 258
Gardner, Guy 264
Garn, Jake 69, 127, 128, 162, 259
Garneau, Marc 258
Garriott, Owen 256, 263
Gehman, Harold W., Jr. 222, 236, 240
Gemini program 41, 56, 176, 199, 210
General Accounting Office (GAO) 69, 164, 177, 178
General Dynamics 18
Gentry, Jerauld 7, 8, 11, 13
German, Charles D. 265
Gernhardt, Mike 265
Gerstennaies, Bill 243
Gibson, James 191
Gibson, Robert "Hoot" 88, 177, 184–185, 228, 257, 262, 264
Glenn, John H. 127, 128
Goldin, Dan 206, 227
Good Citizen Program 29–30
Gordon, Bart 232
Gordon, Henry C. 56
Grabe, Ronald 261
Graham, Dr. William 127, 128, 134, 135, 139, 143, 154, 157, 158
Greene, Harold 175
Greene, Jay 113, 119, 120, 121, 122, 131
Gregory, Frederick 259
Griffin, Michael D. 238, 239, 242, 244
Griggs, David 259
Grissom, Gus 22, 153
Grunsfeld, John 266
Guidoni, Umberto 265
Gutierrez, Sidney M. 264
G.W. Pierce 148

Haise, Fred 27, 29
Hale, N. Wayne, Jr. 215, 238, 239, 241, 242, 243
Halley's Comet 82, 87
Hallock, James 223
Halsell, James, Jr. 265

Ham, Linda 237
Hardy, George 43, 47, 48, 94, 95, 160
Harrington, James 127
Harris, Bernard A., Jr. 264
Harris, Hugh 89, 96, 98, 99–105, 107–108, 134, 136
Hart, Terry 257
Hartsfield, Henry W. 57, 142, 163, 200, 255, 258, 261, 263
Hassell, James D., Jr. 236
Hauck, Frederick H. 179, 183, 186, 189, 256, 258
Hawley, Steven A. 258, 262, 264, 266
Helium-3 250–251
Helm, Susan 265
"Help!" memo 80
Henize, Karl 260
Henricks, Terence 264, 265
Hercules Aerospace Company 176
Herrera, Nancy 232
Hess, Maj. Gen. Kenneth W. 222
Hieb, Richard 265
Hilmers, David C. 183, 189, 261
Hire, Kathryn P. 266
HL-10 8, 10, 11, 14, 28
HL-20 14
Hoag, Peter 11
Hoffman, Jeffery 259, 264, 265
Hollings, Ernest 140, 158
Horowitz, Scott 265
Hotz, Robert B. 155, 231
House Committee on Science and Astronautics 19
House Science and Technology Committee 45, 50
Houston, Cecil 94
Hubbard, Scott 222
Hubble Space Telescope 188, 201–202, 239
USS *Hubrey Fitch* 131
Hughes Aircraft Company 69, 106, 137
Hughes-Fulford, Millie 264
Hurricane Mesa, Utah 188
Husband, Evelyn 222
Husband, Rick 212, 222, 227, 232, 235, 266, 278

Ice Inspection Team 96, 98
USAF *Independence* 132, 138, 141
Inertial Upper Stage (IUS) 58, 135–136, 138
51-L Interim Mishap Review Board 129, 133, 134, 136, 140, 156
International Space Station 27, 80, 139, 204, 205, 206, 209, 212, 232, 236, 242, 252
Ivins, Marsha 264, 265

Jacksonville, Florida 136
Jacobs, Bob 222
Jarvis, Gregory B. 69, 88, 98, 99, 106, 137, 143, 146, 147, 148, 149, 172, 195, 262, 275
Javits, Jacob 22
Jernigan, Tamara E. 264, 265
Jet Propulsion Laboratory (JPL) 206

Johnson, Pres. Lyndon 18
Johnson, Norma H. 185
Johnson Space Flight Center 109, 120, 127, 140, 141, 189, 191, 200, 223, 226, 227
Jolly One 121, 124, 125
Jones, Col. Marvin 180
Jones, Thomas 265
Joos, Dan S. 201
Jordan, Frank 203
Jupiter 188

Kadenyuk, Leonid K. 266
Keegan, Sarah 137
Kelly, James 238
Kennedy, Pres. John F. 18, 41, 56, 248
Kennedy Launch Center 19, 23, 28, 29, 30, 36, 39, 61, 62, 68, 71, 82, 84, 85, 86, 87, 88, 89, 93, 94, 96–105, 106–116, 119, 127, 128, 129, 135, 136, 137, 140, 156, 159, 162, 181, 188, 190, 191, 192, 210, 224, 226, 229, 237, 241, 244
Kerwin, Dr. Joseph 121, 141, 171, 172, 181
KH-9 (Big Bird) 56
KH-11 17, 34, 159, 215
KH-13 57
Kilminster, Joseph 94, 95, 158, 161, 195
Kingsbury, James E. 162
Knight, William J. 56
Kohrs, Richard 128
Kostelnik, Michael 231
Kraft, Christopher Columbus 251
Kregel, Kevin 265, 266
Krikolev, Sergei 237
Kulcinski, Jerry 250
Kutyna, Gen. Donald J. 155, 160
Kyle, Lt. Joe 130

Lacrosse 188
Langley Research Center 8, 14, 226
Lawrence, Robert H. 57
Lawrence, Wendy 238
Lederer, Jerome 160
Lee, Jack 97
Leestma, David C. 258, 264
Lenneham, Richard 265
Lenoir, William B. 255, 263
Leslie, Fred W. 265
Leveson, Nancy 206
Liberty Station 249
Lichtenberg, Byron 256, 263
Lifting body 5, 6, 8, 9, 10, 11, 13, 14, 19, 28, 56
Lind, Don L. 259
Lindsey, Steven W. 242, 244, 266
Linnehan, Richard 266
Linteris, Greg 265
Locke, Charles S. 175, 209
Lockheed Aircraft Corporation 18, 34, 40, 88, 180, 225
Lockheed Space Operations 48
Logsdon, John 223
Lopez-Alegria, Michael 265
Lounge, John M. 183, 189, 260, 264
Lousma, Jack 255, 263
Love, Mike 13

Lovinggood, Judson 94, 97
Low, David 264
Lu, Edward 232
Lucas, William 93, 96, 127, 140, 160, 161, 167
Lucid, Shannon W. 260, 265
Lujan, Manuel, Jr. 189
Lunakod 247
Lund, Robert K. 75, 94, 195
Lunney, Glynn 47

M2-F1 6, 7
M2-F2 7, 8, 9, 10, 11, 28
M2-F3 10, 11, 12
MacLean, Steven 264
Maddox, Tom 222
Magellan 188, 201
Main Engine Cut Off (MECO) 87
Malenchenko, Yuri 232
Manke, John 11, 12, 13, 28
Manned Orbiting Laboratory (MOL) 34, 56, 57, 58, 65
Mara Smythii 250
Mars 14, 17, 186, 206, 239, 249, 250, 251–252
Marshall Space Flight Center 21, 39, 40, 42, 46, 48, 50, 72, 75, 78, 80, 82, 93, 111, 129, 140, 156, 158, 160, 163, 165, 167, 174, 175, 176, 181, 183, 226, 248
Martin Marietta 18, 23
Mason, Jerald 94, 195
Massimino, Mike 266
Mattingly, Thomas K. 59, 65–7, 255, 259, 263
McAllister, Mike 143
McArthur, William S., Jr. 265
McAuliffe, Sharon Christa 80, 88, 89, 97, 98, 105, 106, 134, 143–144, 149, 195, 262, 274
McBride, Jon A. 145, 176, 258
McCandless, Bruce 49, 257
McCarran International Airport 191
McCool, William C. 212, 227, 235, 266, 278
McCurdy, Howard 228, 236, 244
McDonald, Allen J. 81, 94, 95, 140, 141, 158, 159, 160, 164–165, 167, 183, 188, 189, 192, 201
McDonnell Douglas 18, 19, 56
McLucus, John J. 58
McMurty, Thomas 28
McNair, Cheryl 195
McNair, Ronald E. 98, 106, 143, 145, 149, 178, 195, 257, 262, 269
McNamara, Robert 56
Meade, Carl J. 190, 264
Merbold, Ulf 256, 263
Mercury program 9, 28, 41, 48, 56, 176, 210, 232
Messerschmid, Ernst 261
Metcalf, Maria 186
Miller, John 43
MIR 203, 249
Mitchell, Royce 200
Mizell, Jim 62, 137
Mondale, Walter 22, 24, 27
Moon 5, 14, 17, 24, 39, 41, 176, 230, 232, 239, 248, 249, 250

Moore, Jesse W. 73, 82, 87, 93, 126, 127, 133, 136, 140, 156, 157, 158
Morgan, Barbara 80, 139
Moron, Spain 86
Morton Thiokol Chemical Corporation 25, 40, 41, 42, 47, 48, 50, 64, 65, 72, 73, 78, 80, 81, 93–134, 135, 136, 153, 156, 157, 158, 159, 161, 162, 164, 165, 174, 181, 183, 186, 188, 192, 195, 199, 200; decision to launch 51-L 97; incentive pay 176; temperature tests 73; Thiokol Corporation 201
Moser, Tom 148
MPTA-098 24–25, 33
Mukai, Chiaki 265
Mullane, Mike 258
Mulloy, Lawrence 47, 50, 72, 72, 80, 93, 94, 95, 96, 156, 164, 165, 194, 195
Murthi, Aroon 192–193
Musgrave, Story 255, 260, 265
Myers, Dale 19

Nagel, Steven 260, 261, 264
NASA *Freedom Star* 120, 129, 132, 134, 136, 138
NASA *Liberty Star* 120, 129, 132
NASA Select TV 99
National Air and Space Museum 12, 14, 29
National Military Command Center (NMCC) 124
National Reconnaissance Office 215
National Research Council 182, 206
National Transportation Safety Board (NTSB) 129, 137, 146
NAVSTAR 188
NBC news 122, 137
Nelson, Bill 88, 161, 162, 172, 179
Nelson, George "Pinky" 183, 188, 190, 193, 257, 262, 264
Nelson, William 262, 264
Nesbit, Steve 109, 118, 120
New York Air National Guard 193
Newman, John 266
Nicholson, Lt. Col. Bob 62, 65, 124, 126, 131 133
Nicollier, Claude 265
Nixon, Pres. Richard M. 17, 18, 19, 40, 153
NOAX ("Goo") 240, 244
Noguchi, Soichi 238, 240
North American Aviation 55
North American Rockwell 18, 19, 20, 21, 24, 25, 27, 30, 34, 35, 36, 84, 141, 186, 191
Northrop Aircraft Corporation 8, 9, 10
NR-1 Submarine 140, 141

Oberth, Herman 80
O'Brien, John 128
Ockels, Wubbo 261
O'Connor, Bryan D. 175, 242, 243, 261, 264
O'Connor, Col. Edward A. 140, 163
Office of Technology Assessment 200, 202, 204

O'Keefe, Sean 215, 221, 227, 231, 237, 242
Olympic Torch 131
Onizuka, Ellison S. 59, 65, 99, 103, 104, 144, 145, 149, 156, 157, 158, 159, 161, 162, 165, 167, 172, 175, 176, 182, 183, 186, 189–190, 195, 200, 201, 202, 226, 259, 262, 269
USS *Opportune* 148
Orbital Sciences Corporation 207
O-ring Erosion/Potential Failure Criticality 73
O-rings 19, 37, 42, 43, 46, 48, 49, 61, 62, 63, 75, 81, 82, 93–97, 98, 142, 203, 204; launch failure 107–116; temperature tests 72; tests 44; Titan 41, 42, 50
Orion Crew Exploration Vehicle (OCEV) 252
Orr, Verne 159
Osheroff, Douglas 223
Overmyer, Robert F. 57, 128, 143, 255, 259, 263

Pacific Engineering and Production Company 191
Pailes, William 261
Pan American Airlines 48
Para-rescue Squadron, 1730th 123, 193
Parise, Ronald 264
Parker, Robert 256, 263, 264
Parker Hannifin Corporation 42
Patrick Air Force Base 123, 133, 193
Pawelczyk, James A. 266
Payton, Gary E. 59, 61, 65, 259
Pentagon 17, 24, 32, 57, 58, 65, 125
Perry, Brian 119
Personal Egress Air Packs (PEAPs) 121, 172
Peterson, Bruce 8, 10
Peterson, Donald H. 57, 255
Petrone, Dr. Rocco 141, 160
Petty, John 227
Philbin, Lt. John T. 132, 142
Pickard, Lowe & Garrick, Inc. 192
Pike, John 192, 229
Poindexter, John 124
USCG *Point Roberts* 129, 131
Ponce de Leon, Florida 136
Powers, Ben 94
Precourt, Charles J. 264
USS *Preserver* 138, 139, 141, 143, 144, 145, 147, 148
Presidential Commission on the Space Shuttle Challenger Accident 136, 139, 146, 153–154, 155, 176, 179, 199, 230
Presidential Space Task Group 17, 18
PRIME 13
Proxmire, William 22

Qualification Motor 6 (QM-6) 191
Qualification Motor 7 (QM-7) 191
Qualification Motor 8 (QM-8) 200
Quayle, Vice Pres. Dan 202

Ramon, Col. Ilan 212, 222, 235, 266, 278
Rand, Maj. Ron 139

Raob, Rocky 189
Ray, Leon 43
Readdy, William F. 226, 236
Reagan, Pres. Ronald Wilson 67, 106, 127–128, 131, 134, 136, 142, 145, 153, 161, 180, 181, 182, 227
Redman, Charles 134, 157, 158
Reeves, Dr. Ronald 172
Regan, Donald 153
Reinarty, Stanley 94, 95
Remotely Operating Vehicle (ROV) 126
Rescue parachutes 122–123, 133
Research and Application Module (RAM) 21
Resnik, Judith 98, 99, 103, 104, 105, 106, 149, 172, 195, 258, 262, 268
Return-to-Flight Task Group 238, 241
Return-to-Launch-Site (RTLS) 86
Richards, Richard N. 264
Ride, Dr. Sally K. 155, 186, 223, 256, 258
Ride Report 186
Riegle, Donald 175
Robinson, Stephen 238, 240
Rocha, Alan R. 227
Rockwell International 160, 177
Roe, Ralph 229
Rogers, Russell L. 56
Rogers, William P. 135, 136, 153, 155, 157, 158, 160, 161, 163, 165, 175
Rominger, Kent V. 265, 265
Roosevelt, Pres. Teddy 247
Ross, Jerry 261, 264
Rothenberg, Joseph 223
Rummel, Robert W. 155
Russell, Brian 81–82
Russell, Thomas 159
Russia 232, 236, 251

Sabotage 29
Sacco, Albert, Jr. 265
Salyut 16
A.O. Sammons Company 174
USS *Sampson* 131
Sandia National Laboratories 68
Saturn 1B 14, 85
Saturn V 14, 17, 18, 24, 39, 85, 176, 230
Savannah, Georgia 134, 136
Scheuer, James 47
Schlegel, Hans 264
Schnityer, Susan 175
Scobee, Francis R. "Dick" 13, 89, 98, 99, 102, 103, 104, 105, 106, 110, 112, 113, 121, 149, 172, 193, 195, 257, 262, 267
Scolese, Chris 242, 243
RPV *Scorpio* 136
Scott, Winston E. 266
Scully-Power, Paul 258
Sea Link II 139, 140 141
Seal Team Task Force 81
Searfoss, Richard A. 265, 266
Seddon, M. Rhea 259, 264, 265
Sensenbrenner, James 45
Shapley, Willis H. 22
Shaw, Brewster 204, 256, 261, 263, 264

Shelby, Michael T. 232
Shepard, William M. 264
Shriver, Loren J. 59, 259
Shuttle bailout 190
Shuttle-C 248–249
Shuttle failure studies 68
Shuttle main engines 24, 25, 33, 75, 87, 141, 147
Sieck, Robert 85, 128, 158
Sierra Study 68
Signal Intelligence Satellite (SIGINT) 61
Silverira, Milton 128, 179, 180
USS *Simpson* 131
Simpson, Commander James G. 130, 136
The Six Million Dollar Man 10
Skylab program 14, 17, 21, 26, 56, 85, 121, 141, 210
Skylab-2 85
Slayton, Donald K. 27–28, 68
SLC-6 ("Slick Six") 29, 59, 65–67, 159
Smith, Jane 195
Smith, Michael J. 83, 88, 97, 98, 99, 100, 101, 102, 103, 104, 105, 106, 110, 113, 114, 121, 149, 172, 194, 195, 262, 268
Smith, Richard G. 82, 127, 135, 140, 162, 181
Socobab, Milton 69
Socorro, New Mexico 57
Solid Rocket Motor Joint Leakage Study 43
Solid Rockets Boosters (SRBs) 28, 32, 39, 40, 48, 50, 64, 69, 70, 73, 75, 80, 81, 93, 129, 130, 131, 133, 134, 137, 140, 142, 144, 147, 148, 156, 158, 159, 161, 162, 163, 164, 174, 175–176, 177, 180, 186, 191, 229; hit by crane 85; launch failure on 51-L 107–116; proposals 41; recovery 87, 145; redesigned 199; temperatures 98
Sorely, Donald 8
Soviet Union 5, 13, 16, 61, 62, 69, 131, 159, 176, 181, 182, 193, 200, 249
Soyuz 85, 207, 232
Space Flight Safety Panel 175
Space Plane 5
Space Shuttle Approval 19–21
Space Shuttle Independent Assessment Team 223
Space Shuttle Mishap Interagency Investigation Board 222
Space Shuttle Transportation System (SSTS) 19, 21
Space Shuttle Verification/Certification Committee 43, 44, 46
Space Station 17, 206, 208, 238, 225
Space Station *Freedom* 80, 249
"Space Truck" 18
Spacelab 22, 23, 57, 72, 80, 127, 142, 174, 203, 204
Spacelab-2 57, 172
Spacelab-3 57
Spacelab-D1 57, 80
Spartan-Halley 82, 148
Speaks, Larry 124

Spring, Sherwood 261
Sputnik-1 237
Stafford, Tom 241, 242
Star Trek 27
Star Wars 61–62
START 13
Stein, Ed 239
Stena Workhorse 142, 145
Stevenson, Charles 96
Stewart, Robert 257, 261
Still, Susan 265
STS-1 27, 34, 208, 254, 263
STS-2 46, 47, 50, 59, 254, 263
STS-3 46, 255, 263
STS-4 59, 255, 263
STS-5 255, 263
STS-6 36–37, 57, 255
STS-7 208, 256
STS-8 256
STS-9 87, 109, 228, 256, 263
STS-26 107, 179, 183, 185, 186, 200
STS-27 200, 208
STS-28 264
STS-30 201
STS-32 202, 208, 228, 264
STS-35 208, 264
STS-40 264
STS-41 164
STS-41B 48, 106, 257
STS-41C 50, 106, 257
STS-41-D 58, 138
STS-41G 258
STS-42 208
STS-45 202, 208
STS-50 202, 208, 264
STS-51A 61, 258, 259
STS-51B 65, 72, 73, 81
STS-51C 51, 55–65, 93, 96, 106, 159, 259
STS-51D 50, 65, 69, 127, 259
STS-51E 69
STS-51G 65, 72, 260
STS-51F 65, 72, 73, 260
STS-51I 76, 113, 260
STS-51J 261
STS-51L 60, 72, 82, 84, 146, 178, 181, 262; launch 106–116; pre-launch countdown 93–15; search and recovery operations 129–149; search and rescue 122–128
STS-52 202, 209, 264
STS-55 202, 264
STS-56 203, 209
STS-58 265
STS-61A 80, 261
STS-61B 80, 81, 138, 261
STS-61C 85–88, 131, 161, 179, 183, 262, 264
STS-61F 164
STS-62 209, 265
STS-65 265
STS-70 203
STS-71 203
STS-73 204, 240, 265
STS-75 265
STS-78 265
STS-80 265
STS-83 265
STS-87 205, 209, 266
STS-90 266

STS-93 229, 266
STS-94 265
STS-102 205
STS-103 205
STS-107 88, 208, 209, 212–221, 266
STS-109 266
STS-112 208, 209
STS-114 209, 230, 237–240
STS-118 139
STS-300 238
Sullivan, Kathryn 258
Sutherland, Gary 69
Sutter, Joseph F. 155, 205
USCG *Sweetgum* 130, 131, 134
Syncom IV-3 76

TDRS 57, 60, 106, 138, 139, 145, 148, 162, 188, 192, 203
Tetrault, Roger E. 222
Thagard, Norman 256, 259
Thermal Protection System (TPS) 23, 26, 29–30, 34, 36, 46, 60, 80, 208, 224–225
Thirsk, Robert 265
Thomas, Andrew 238
Thomas, Donald 265
Thomas, Elmer 142
Thomas, Gene 89, 158
Thomas, James 82
Thomas, John 181
Thompson, Arnold 94
Thompson, J.R. 183
Thompson, Mickey 6
Thompson, Milton 6, 7, 8, 10, 56
Thornton, Kathryn C. 265
Thornton, William 256, 259
Thuot, Pierre J. 265
Tognini, Michel 266
Transoceanic Abort Landing (TAL) 86, 88
Trinh, Eugene H. 264
Truly, Richard H. 27, 28, 47, 57, 143, 148, 158, 172, 177, 178, 180, 188, 193, 204, 205, 254, 256, 263
Tureotte, Rear Admiral Stephen 222

Ulysses 164
USS *Underwood* 131
United Space Alliance 223, 226
United Space Boosters, Inc. 50
United States Air Force 24, 50, 55–65, 68, 81, 120, 126, 130, 131, 133, 135, 136, 139, 212
United States Air Force Academy 144
United States Air Force Strategic Command 215
United States Coast Guard (USCG) 125, 126, 130, 134–149
United States Navy 125, 126, 134, 141, 212
U.S. Space Foundation 155, 157
United Technologies Corporation 40, 50, 176
Utsman, Thomas 156

Van den Berg, Lodewijk 259
Vanderburg, Vince 85
Vandenberg Air Force Base 13, 29, 39, 56, 58, 59, 65–67, 159, 188
Van Hoften, James 257, 260

Veach, C. Lacy 264
Vela, Rodolfo Neri 261
Venture Star 207
Venus 131, 135, 188, 201
Virtus 18
Voss, Janice 265

Walker, Dr. Arthur B.C., Jr. 155
Walker, Charles D. 258, 259, 261
Walker, David M. 258
Walker, Robert 175, 180
Wallace, Steven B. 223
Walter, Ulrich 264
Walz, Carl 265
Wang, Taylor 259
Watterson, John Brett 67
Wear, Lawrence 81, 93
Webb, James 22
Weeks, L. Michael 47

Weitz, Paul 36, 163, 200, 255
Welch, Amber 222
Wetherbee, James 228, 264
Wheelon, Albert D. 155
White, Bob 221
White, Douglas 215
White, Ed 22
White Sands, New Mexico 190
Widnall, Sheila 223
Wiggams, Calvin 94, 195
USS *William Sims* 131, 134
Williams, Dafydd Rhys 266
Williams, Donald 259
Williams, Larry 203
Williams, Walt 127
Winterhalter, David 190
Wolf, David A. 265
Wood, James W. 56
Wyrick, Roberta 100

X-1 8
X-15 6, 10, 28, 55, 56
X-20 Dyna Soar 55–56
X-24A 13, 207
X-24B 13, 28
X-30 National Aerospace Plane 207
X-33 207, 249
X-34 207
X-38 Crew Return Vehicle 207
X-Planes 5

Yahoo 223
Yeager, Charles E. 8, 155
Yeltsin, Boris 204
Young, John 6, 21, 34, 89, 97, 143, 161, 163, 178, 183, 192, 199, 200, 228, 254, 256, 263

www.ingramcontent.com/pod-product-compliance
Ingram Content Group UK Ltd.
Pitfield, Milton Keynes, MK11 3LW, UK
UKHW050543150426
5217IPUK00026B/2056